OXIDATIVE ENZYMES IN FOODS

OXIDATIVE ENZYMES IN FOODS

Edited by

D. S. ROBINSON

Procter Department of Food Science, University of Leeds, Leeds LS2 9JT

and

N. A. M. ESKIN

Department of Foods & Nutrition Faculty of Human Ecology, University of Manitoba, Winnipeg, Canada R3T 2N2

ELSEVIER APPLIED SCIENCE
LONDON and NEW YORK

ELSEVIER SCIENCE PUBLISHERS LTD
Crown House, Linton Road, Barking, Essex IG11 8JU, England

Sole Distributor in the USA and Canada
ELSEVIER SCIENCE PUBLISHING CO., INC.
655 Avenue of the Americas, New York, NY 10010, USA

WITH 26 TABLES AND 43 ILLUSTRATIONS

British Library Cataloguing in Publication Data

Oxidative enzymes in foods.
1. Food. Additives
I. Robinson, David S. (David Stuart) *1935–* II. Eskin, N. A. M.
664.06

ISBN 1-85166-613-3

Library of Congress Cataloging-in-Publication Data

Oxidative enzymes in foods/edited by D. S. Robinson and N. A. M. Eskin.
p. cm.
Includes bibliographical references and index.
ISBN 1-85166-613-3
1. Oxygenases. 2. Food—Composition. I. Robinson, D. S.
II. Eskin, N. A. M. (Neason Akivah Michael)
QP603.085085 1991
664—dc20 91–13478
CIP

Printed in Northern Ireland by The Universities Press (Belfast) Ltd.

Preface

Oxidative enzymes were among the first enzymes identified in plant food sources. Although relatively small in number, compared to hydrolases, increased attention has been focussed on their effect on the colour, flavour and nutrient content of foods. Considerable advances in our understanding of the biochemistry of oxidative enzymes over the past decade pointed to the need for a concise yet comprehensive text on the major oxidative enzymes affecting food quality and stability. The increased demand for fresh food products by consumers has resulted in a greater use of storage and transportation. Under these conditions, oxidative enzymes present in foods can quickly reduce their quality. This is particularly evident when the cellular tissue is disrupted as, for example, during the production of fruit juices, chopped vegetables and comminuted meat products.

Each chapter in this book provides a critical discussion of the isolation and characterisation of the major oxidative enzymes. These include peroxidases, superoxide dismutases, lipoxygenases, polyphenol oxidases, amino acid oxidases, carbohydrate oxidases and lactoperoxidases. The roles of peroxidases in plant foods and lactoperoxidases in milk are reviewed in separate chapters. The mode of action of peroxidases as free radical generators is elucidated while the biochemical action of lactoperoxidase in milk as an antibacterial agent in the presence of thiocyanate and halide anions is also covered. The possible suppression of lipid autooxidation by superoxide dismutases and their biochemical role is reviewed, although the latter still remains controversial. Each of the authors present a comprehensive, yet concise coverage of the latest developments in our understanding of these enzymes, including their potential application in biotechnology and genetic engineering.

We hope that the overall organisation of this text will prove useful not only to food scientists but also to students, teachers and researchers in many fields in which oxidative enzymes are important. We are particularly indebted to the staff at Elsevier Science Publishers for their constant encouragement and invaluable assistance in the publication of this text. We are most appreciative of the support given by our wives and families in allowing us the time to complete what we hope will be a useful addition to the food science library.

It is with deep regret that we record the death of one of our authors, James Burt, on June 30th, 1991. His chapter 'Amine Oxidases and

Amino Acid Oxidases', is a fitting memorial to his work as a biochemist principally concerned with muscle tissue in marine species.

DAVID S. ROBINSON
N. A. MICHAEL ESKIN

Contents

Contributors

S. A. Barker
School of Chemistry, The University of Birmingham, Edgbaston, Birmingham B15 2TT, UK

C. G. Biliaderis
Department of Food Science, University of Manitoba, Winnipeg, Manitoba, Canada R3T 2N2

J. R. Burt
Ministry of Agriculture, Fisheries & Food, Torry Research Station, PO Box 31, 135 Abbey Road, Aberdeen AB9 8DG, UK

J. K. Donnelly
Procter Department of Food Science, The University of Leeds, Leeds LS2 9JT, UK

N. A. M. Eskin
Department of Foods & Nutrition, The University of Manitoba, Faculty of Human Ecology, Winnipeg, Manitoba, Canada R3T 2N2

D. N. Kamau
Department of Biochemistry, The University of Alabama at Birmingham, UAB Station, Birmingham, Alabama 35294, USA

K. M. Pruitt
Department of Biochemistry, The University of Alabama at Birmingham, UAB Station, Birmingham, Alabama 35294, USA

D. S. Robinson
Procter Department of Food Science, The University of Leeds, Leeds LS2 9JT, UK

J. R. Whitaker
Department of Food Science & Technology, University of California, Davis, California 95616, USA

J. Zawistowski
Department of Food Science and Agri-Food Biotechnology Laboratory, Manitoba Research Council, University of Manitoba, Winnipeg, Manitoba, Canada R3T 2N2

1

Peroxidases and Catalases in Foods

David S. Robinson

Procter Department of Food Science, The University of Leeds, Leeds, UK

INTRODUCTION

Peroxidases and catalases are widely distributed in the plant and animal kingdoms and have been found in all higher plants that have been investigated. Both enzymes are found mainly as haemoproteins and use hydrogen peroxide as the oxidising substrate, although other more unusual peroxidases have been shown recently to contain either metal ions, such as selenium and vanadium, or a flavin prosthetic group (Table 1). Nevertheless, it is the peroxidases in plants that are of major interest to the food scientist. Among all enzymes their

Table 1. Classification of Peroxidases

Prosthetic group	*Common name*	*Source*	*Potential substrates*[a]
Haem	Peroxidases	Plants/animals	Test substrates including guaiacol and *o*-dianisidine
	Ascorbic acid peroxidase	Plants	Ascorbic acid
	Cytochrome *c* peroxidase	Plants	Cytochrome *c*
	Lactoperoxidase	Animals	Thiocyanate
	Lignin peroxidase	Plant cell walls	Lignin precursors including cinnamic and ferulic acids
	Myelo-peroxidase	Animals	Test substrates and thiols
	Eosinophil peroxidase	Animals	Test substrates and thiols
	Thyroid peroxidase	Animals	Phenols including tyrosine
Selenium	Glutathione peroxidase	Plants	Reduced glutathione
Vanadium	Bromo-peroxidase	*Ascophyllum nodosum*	Halogens
Manganese	Mn-haem peroxidase	*Phanerochaete chrysosporium* (white rot fungus)	Lignin
Flavin	Flavo-peroxidases	Micro-organisms	—

[a] Although some peroxidases may have been given substrate related names, the peroxidatic action may not necessarily be restricted to the named substrate.

precise mode of action in living and harvested foods has proved to be the most elusive. Different types of plant peroxidases have been recognised since the beginning of this century and an ever-increasing number of functions have been ascribed to them. Peroxidase (EC 1.11.1.7) is a member of a large group of enzymes called the oxidoreductases. It is now fully accepted that peroxidases in plant tissues exist in groups of isoenzymes which can be separated and detected on gels after electrophoresis. There is now evidence which indicates that these isoenzymes are coded for by separate structural genes located on different chromosomes. The question of how many isoenzymes catalyse the oxidation of various substrates at different rates and their related physiological functions remains unresolved. There have been numerous reports in the plant physiology literature of the specific involvement of peroxidases in lignin biosynthesis, oxidation of indoleacetic acid, ripening of fruits and protection against fungal and bacterial attack: peroxidases have been claimed to influence a large array of physiological functions. In harvested plant foods peroxidases are believed to be responsible for loss of colour, flavour and texture as well as nutritional attributes. During the last 30 years peroxidase activity has been repeatedly related to the quality of processed foods, because of the enzyme's well-known thermostability. There is a need for detailed studies at the molecular level to determine both the precise reasons for the occurrence of such a wide range of isoperoxidases within a given plant species and the relative mode of action of the isoenzymes on various natural substrates. To this end substantial progress is now being made through the use of cultured cells which enables larger and hence more useful amounts of isoperoxidases to be isolated.

PEROXIDASE ACTIVITY AND FOOD QUALITY

During the ripening of fruits and particularly during the climacteric it has been found that peroxidase activity increases alongside the activity of other enzymes, such as polygalacturonases and cellulases which are normally associated with the ripening process. Indeed, it has been suggested that peroxidase activity could be a possible parameter in the ripening and senescence of Golden Delicious apples (Gorin & Heidema, 1976). Also, after harvesting, the peroxidase activity of apricots has been reported to increase four- to fivefold whereas for comparison, the level of polyphenol oxidase activity increased by only 10% (Vamos-Vigyazo *et al.*, 1985). However, the exact physiological functions of peroxidases in the ripening process remain obscure.

Physiological stress, wounding, fungi and viral infections bring about changes in peroxidase isoenzymes. During these complex physiological changes the chemical reactions catalysed by peroxidases have not been fully elucidated. Peroxidases promote a large number of reactions and therefore have a versatility not surpassed by any other enzyme; peroxidase does not exist as a single enzyme in fruits and vegetables, like many other plant enzymes, peroxidase activity is found in the form of a number of discrete isoenzymes, the isoperoxidases, which are now readily detected by isoelectric focusing; there are many naturally occurring phenolic compounds in all plant tissues, each of which can be oxidised by peroxidase in the presence of small amounts of hydroperoxide. Due to the diversity of compounds which are susceptible to oxidations catalysed by peroxidases, the range of products formed is very extensive. Consequently, food scientists have found it difficult to correlate perceived post-harvest changes, like loss of flavour and texture, with specific chemical reactions. Furthermore, peroxidases can exist in different oxidation states and catalyse at least three different types of oxidation. Also all plant tissues contain substrate-specific peroxidases, namely ascorbic acid peroxidase and glutathione peroxidase. Although the role of these enzymes is primarily the removal of hydrogen peroxide, the presence of ascorbic acid peroxidase in foods is likely to lead to additional loss of vitamin C.

The relationship of peroxidase activity to off-flavours and off-colours in raw and unblanched vegetables is still largely empirical: in post-harvest plant foods, peroxidases are believed to be responsible for deterioration in flavour, colour, texture and nutritional qualities of raw and processed foods (Burnette, 1977; Haard, 1977). A relationship between peroxidase activity and off-flavour development in peas is well established (Wagenknecht & Lee, 1958; Pinsent, 1962). Active enzyme can spoil fruits and vegetables at temperatures as low as −18°C and at low moisture levels, the development of off-flavours is often associated with the oxidising effect of peroxidases on indigenous lipids and the phenolic constituents of foods (see Sessa & Anderson, 1981). For orange juice, it has been suggested that the significant negative correlation between peroxidase activity and the flavour scores of high- and low-yield orange juice indicates that it might be possible to use peroxidase activity as an index of potential adverse flavour (Bruemmer *et al.*, 1976). An increased peroxidase activity and content of pulp in the juice due to processing conditions was accompanied by lower juice quality. For string beans, the qualitative peroxidase test was found to be the best index of adequate blanching (Bedford & Joslyn, 1939). For scalded asparagus, Joslyn & Bedford (1940), using gum guaiacol as substrate, found that flavour retention closely related to the inactivation of peroxidase activity. Scalding of asparagus in water for 4 min at 92°C or 3 min at 100°C was sufficient for retention of flavour. Wagenknecht & Lee (1958) found that off-flavours developed when horse-radish peroxidase was added to pea slurries. Zoueil & Esselen (1959) also showed that the addition of horse-radish peroxidase to sterilised green bean purée caused an increase in the amount of acetaldehyde in stored samples. Similarly, earlier work by Arighi *et al.* (1936) and Joslyn (1949) showed the accumulation of volatile aldehydes in the raw and unblanched peas. Joslyn & David (1952) claimed that peas containing more than 10 ppm of volatile aldehydes were objectionable. However Halpin & Lee (1987) have stated that for blanched green peas stored at −23°C peroxidase activity did not correspond with quality. The use of peroxidase activity as an indicator of blanching treatment is continuously being appraised in view of the higher temperatures required to inactivate peroxidases. It is frequently suggested that such temperatures are not needed for the blanching of vegetables: the activity of other less stable enzymes, such as polyphenol oxidase, lipoxygenase and catalase, as blanching indicators is tempting and more attractive. Nevertheless, Halpin *et al.* (1989) have restated that these enzymes are not more appropriate as blanching indicators and peroxidase will continue to serve this requirement. It is important to record that peroxidase catalyses both the peroxidatic and oxidative reactions through the involvement of free radicals, which by their very nature can initiate chain reactions and thus allow a wide range of substrates to be oxidised.

Products of the action of peroxidases in both horseradish and milk were first observed approximately 180 years ago: it was observed that 'extract of guaiacum' turned blue in the presence of what we now know to be peroxidase. Fundamental investigations began in 1920 with the work of Onslow (1920) who observed that the activity of oxidising enzymes varied with the ripeness of fruit and Willimott & Wokes (1926) found that peroxidase was present in both the flavedo and albedo of citrus fruits. During the 1930s, Keilin and co-workers observed the importance of the haem component (Keilin & Mann, 1937). With advances in spectral techniques and protein purification methods Chance (1951) showed the existence of the peroxidase–hydrogen peroxide complexes, compound I and compound II. These are important intermediates formed between peroxidase and the oxidant in peroxidase-catalysed reactions (Chance, 1952). Peroxidases from horseradish have been most extensively studied because of its abundance and consequently its availability. The elucidated mode of action of horseradish peroxidase together with knowledge of its molecular structure have served to illustrate the probable structure and role of many other peroxidases universally present in fruits and vegetables. However, because of the widely differing nature of plant tissues, the occurrence of isoenzymes and the potential diverse action of these on different substrates, it does not suffice to apply data obtained from studies on horseradish peroxidase to other plant foods. To date the precise roles and physiological functions of peroxidases in living plants and post-harvest products are not fully understood.

For various peanut cultivars Cherry & Ory (1973) suggested that conditions in the mature peanut may be conducive to the synthesis of storage isoperoxidases which may vary with cultivars grown in different geographical regions. Gardner *et al.* (1969) showed that peroxidase activity in maize varied with respect to the anatomical part. Davis (1942) reported that the greatest peroxidase activity was in the inner seed coat in citrus fruits. Peroxidase is now also suspected to have a role in the degradation of chlorophyll, senescence in plants, the ripening of fruits, lignification and also, surprisingly, the degradation of lignin by white rot fungi.

OCCURRENCE AND DISTRIBUTION IN PLANT CELLS

Catalases

Large amounts of catalase are located within peroxisomes for eukaryotes. The term 'peroxisome', while not indicating the presence of catalase, emphasises the role of the organelle in peroxidatic reactions.

Catalase in peroxisomes can be demonstrated by electron microscopy and staining with 3,3″-diaminobenzidine. The reason for an apparent large excess of catalase in most peroxisomes is unknown, but certainly ensures that hydrogen peroxide is not allowed to accumulate within the organelle: the need to quickly remove hydrogen peroxide is achieved by the compartmentalisation of the flavin oxidase and catalase. The excess of catalase is so great that the non-enzymatic oxidation of glyoxylate to carbon dioxide and formic acid is avoided. The general hypothesis is that all reactions involving glyoxylate biosynthesis are held in microbodies to ensure that formic acid is not formed in large amounts in cells. Glyoxysomes also contain catalase which again is responsible for the removal of hydrogen peroxide formed by the action of a flavin–enzyme, fatty acyl oxidase.

Peroxisomes do not contain peroxidases, although mitochondria contain both peroxidases and catalases. Peroxisomes in animal cells have been reported to contain xanthine dehydrogenase, urate oxidase, a fatty acid beta-oxidation pathway and allantoinase in birds. It has been suggested that the microbodies, peroxisomes and glyoxysomes present in plant tissues are formed by budding from the endoplasmic reticulum. However it is not known how the specific enzymes are selected for packaging into the microbodies.

Plant peroxisomes, together with the glyoxysomes, can be separated after careful equilibrium sucrose density centrifugation at a buoyant density of 1·24–1·26 g/cm^3. These microbodies have only a single outer membrane, unlike the double membrane of mitochondria and chloroplasts, and therefore are facile and fragile. Thus, after homogenisation of tissues considerable amounts of catalase are present in the soluble fraction with small quantities adsorbed to the other subcellular fractions. However, there may be a significant cytoplasmic pool of catalase prior to homogenisation.

Three isoenzymes of catalase have been found in maize encoded by three unlinked structural genes (Redinbaugh *et al.*, 1990). Electrophoretic analysis has shown that CAT-1 occurs mainly in the milky endosperm of the developing kernel, whereas CAT-2 is mainly present in the scutellum after germination, which also has the highest enzymic activity, and CAT-3 is present in the coleoptile and leaf shoots.

Peroxidases

Isoperoxidases seem to be located in almost all parts and organelles of the cell, although they vary in number and activity with physiological changes: peroxidase activity has been detected in vacuoles, tonoplasts, plasmalemmae, mitochondria, microsomes and cell walls. Within the last few years the tissues, especially of lupin (Ros Barcelo *et al.*, 1987), petunia (Hendriks *et al.*, 1985), tobacco (Pang *et al.*, 1989) and suspension cultures of peanut cells (van Huystee, 1989) have been used to study the transport and distribution of isoperoxidases in plant cells. It is accepted that peroxidases are mainly located in cell walls, cytoplasm and vacuoles, although there is little unanimity on the location of individual isoperoxidases in plant cells.

Acidic and Cationic Isoperoxidases

Much of the difficulty in understanding the effects of peroxidase activity in plant foods is due to the presence of a large number of isoperoxidases: early studies on horseradish peroxidases demonstrated the existence of a complex system of isoenzymes (Shannon, 1968). Fractionation of intracellular membranes by isopynic sucrose density gradient centrifugation (Thomas *et al.*, 1981) has indicated that lupin anionic isoperoxidases are associated principally with membranes of the endoplasmic reticulum and to a lesser extent with the Golgi apparatus and plasmalemma-derived membranes. These locations are in accord with the biosynthesis of peroxidases on polysomes attached to the rough endoplasmic reticulum and with the transport pathway: endoplasmic reticulum → Golgi → plasmalemma. Following biosynthesis and transport of peroxidases, it has been claimed (Gaspar *et al.*, 1986) that the final locations of peroxidases are likely to be cell walls, intracellular free spaces and vacuoles. The transport of peroxidases towards their actual site of action may be considered as a control mechanism of cellular peroxidase activity. Tomato plants are thought to contain up to 13 isoperoxidases. Although it has been claimed that only one anionic isoperoxidase is present in the tomato fruit, it is likely that this is dependent on the variety and the stage of ripening. Cytochemical studies (Thomas *et al.*, 1981) have indicated that a tomato fruit peroxidase may be localised on the inner membrane surface of the tonoplast and the cell walls, while other researchers have concluded that part of the tomato cellular peroxidase is bound to the cell wall (see Thomas *et al.*, 1981): peroxidase was not found in the cytoplasm of tomato fruit cells but only on the cell walls and the inner-membrane surface of the tonoplasts. For tomatoes, it has been assumed that soluble peroxidases obtained after extraction are derived from the tonoplast owing to the weak attachment of the enzymes to the membranes (Thomas *et al.*, 1981). The subcellular location of membrane-bound acidic isoperoxidases in etiolated lupin hypocotyls has been reported (Ros Barcelo *et al.*, 1988*a*).

Optically clear solutions containing peroxidase activity are generally obtained after relatively low-speed centrifugation (e.g. 800 *g*). Such a solution is normally designated the soluble or cytoplasmic peroxidase fraction, although the solution may still contain suspended intracellular membranes with bound peroxidases. However, for a complete extraction of peroxidase activity a high ionic strength is needed to solubilise ionically bound isoperoxidases. The 800 *g* pellet consists mainly of cell walls, possibly some large vacuoles, a few nuclei and partially broken cells. After extensive washing of the pellet the remaining ionically bound peroxidases can be extracted from the insoluble pellet using 1 M salt solutions (McLellan & Robinson, 1981; Moulding *et al.*, 1987; Ros Barcelo *et al.*, 1988*b*). These ionically bound isoperoxidases are often of the cationic type and rarely contain a substantial quantity of anionic isoenzymes (Thomas *et al.*, 1981; McLellan & Robinson, 1983; Moulding *et al.*, 1987). This simple fractionation procedure has proved valuable for the rapid separation of mainly the cationic peroxidases which seem to be ionically bound to crude cell wall material. A full complement of isoperoxidases, including both the anionic and cationic types, is present in the so-called soluble fraction, which would be expected to contain suspended intracellular endoplasmic reticulum, Golgi membranes and microsomes. However, it still remains possible that the cationic peroxidases are not primarily located in the cell walls, since binding may arise during homogenisation and in that sense their apparent location would be an artefact of cell breakage (Schloss *et al.*, 1987; Goldberg *et al.*, 1989). Thus, in homogenates the binding of isoperoxidases to cell walls and other membranes may be influenced by the ionic strength and the pH value of the extracting buffer, the isoelectric points of isoperoxidases and charged groups present in cell walls. For acidic isoperoxidases Ros Barcelo *et al.* (1988*b*) have shown that low pH (4·0–5·0) favours binding to cell walls, while a neutral pH favours release. At neutral pH both the cell walls and the acidic isoperoxidases are negatively charged and therefore ionic interactions are prevented. However, under these conditions the cationic isoperoxidases would be positively charged and therefore would be expected to bind to cell wall fragments during extraction. In the living plant the pH of cell walls may be modified from pH 4·5 to 7·5 during growth and this will clearly affect the binding of acidic isoperoxidases. The binding equilibrium as influenced by pH has been suggested to act as a control of both peroxidase activity and hydrogen peroxide production by peroxidases from NADH (Pedreno *et al.*, 1989). Thus protons may modulate the binding of acidic isoperoxidases to cell walls and therefore the level of enzyme in the intercellular space as well as the pH-dependent activity of the enzyme. This is a neat way in which both binding and enzymic activity may be controlled: significant changes in the Ca^{2+}/H^{+} ratio in cell walls would induce modifications of the cell wall pH and hence peroxidase activity (Ros Barcelo *et al.*, 1988*b*). The exact localisation of bound peroxidases within whole cells has not been sufficiently investigated, although Naveh *et al.* (1981) claimed to have developed a chemiluminescent method for the quantitative analysis of peroxidases.

Using an immunogold-labelling technique Hu *et al.* (1989) have claimed that the anionic isoenzyme of peanut cell cultures is predominantly located in the plasmalemma, but the significance of such a location is still open to question.

Through the use of cell cultures for peanuts it has been shown that the yield of extracellular isoperoxidases of the cationic type, normally associated with cell walls, can be increased (Chibbar & van Huystee, 1986). In suspended cultures both the lack of adhesion between cells and the dilution of the secreted enzymes increases the yield of cationic isoperoxidases: the medium may be considered as an extracellular space. Chibbar & van Huystee (1986) consider that transport is to the growth medium as well as cell walls in suspended cell cultures which in total represent the extracellular space in plant tissues. During the growth of suspension culture cells of *Capsicum annum* a sequential release of first basic isoperoxidases followed by acidic isoperoxidases has been reported (Cuenca *et al.*, 1989).

The occurrence of isoenzymes is generally evaluated after gel electrophoresis by zymogram patterns; the cationic isoperoxidases have isoelectric points in the pH range of 8·0–10·0, while the anionic isoperoxidases have isoelectric points in the range of pH 3·5–6·0 (McLellan & Robinson, 1984, 1987*a, b*; Moulding *et al.*, 1987, 1988). Ros Barcelo *et al.* (1987) have also shown that isomers may arise from interaction with phenolic compounds which induce changes in the isoelectric points and thus zymogram patterns: such interactions can be removed by gel filtration in the presence of $CaCl_2$. For petunias Hendriks *et al.* (1985) has shown that there are specific genes for isoperoxidases located on different chromosomes. However, for a given group of isoperoxidases, for example the acidic isoenzymes, the small differences in isoelectric points may reflect both small mutations and post-transcriptional modifications. Different electrophoretic patterns, found for different anatomical parts of plants and more mature tissues, may also reflect organ specificity and genetic variation and minor post-synthesis modification. Peroxidase activity is also expected to be controlled by genetic regulation of both apoenzyme synthesis and haem biosynthesis (van Huystee, 1989).

Plant peroxidases are glycoproteins and as for other proteins of this type, the site of glycosylation is presumed to be the Golgi apparatus. Because of the presence of covalently linked carbohydrate, lectin bound supports—based on their affinity for the carbohydrate moieties—are likely to be useful for recovery and isolation of peroxidases from tissue homogenates. Moreover, such types of interactions might account for the binding of the glycosylated isoperoxidases to intracellular membranes and organelles. Interestingly van Huystee (1989) has suggested that the peroxidases which constitute the ionic group of isoenzymes, which are more difficult to extract, are simply unglycosylated and bound to the microsomes of the Golgi apparatus. However, this conjecture rests on the unproven absence of other contaminating proteins in the cationic isoperoxidase preparation (see p. 20), and the relative antigenic activity of the cationic isoperoxidase.

Calcium ions influence both peroxidase activity and the release of the isoenzymes. For horseradish peroxidase, calcium is an integral part of the molecule and affects the conformational structure of the protein close to the haem group (Welinder, 1985). Calcium intensifies the release of enzyme protein from culture cells (Bakardjieva *et al.*, 1987) and increases peroxidase activity (Penel, 1986). There are grounds for believing that the calcium ions are a specific and necessary factor for the secretory function of plant cells and that they play a significant role in the control of enzymic activity. Ros Barcelo *et al.* (1988*b*) has suggested that the Ca^{2+}/H^{+} ratio controls both the binding of acidic isoperoxidases to cell walls and the enzymic activity. Calcium ions are also important for amylase activity and are believed to play a part in the secretion of this enzyme (Carbonell & Jones, 1984). The calcium-induced increase in peroxidase activity has been attributed to, especially, an increase in the activity of the cationic isoperoxidases initiated by stimuli-induced release of potassium (Gaspar *et al.*, 1985). The release of peroxidase activity by such a mechanism could occur within minutes of any physical damage or stress and if this was the case for peroxidases from all plant sources would thus affect the quality of any plant food product.

Given the complex-bound and facile states in which isoperoxidases exist in cells, it is to be expected that such enzymes will have a very variably catalytic action towards a wide range of compounds. Further, it is not surprising that food storage and the different methods of processing affect the action of these enzymes. Only a fuller understanding of both the specificity of the various isoperoxidases in a wide range of fruit and vegetables and the chemistry of the degradation intermediates, which often occur as free radicals, will facilitate the production of quality-controlled products.

THERMOSTABILITY OF ISOPEROXIDASES

Catalase is not considered to be a thermostable enzyme as it is normally inactivated at temperatures in excess of 50°C. The most thermostable enzymes present in fruits and vegetables are lipoxygenases and peroxidases. Consequently research has concentrated on the thermal properties as well as the occurrence and structure of the latter two enzymes. Both enzymes generate free radicals and therefore are capable, at least in theory, of causing the oxidation of a wide range of naturally occurring substrates. The thermal properties of peroxidases have been investigated during the last few decades because the action of the enzymes is widely suspected of causing unfavourable colour and flavour changes during the storage of frozen foods. For a long time peroxidases attracted attention because the level of residual enzymic activity after blanching has been used as an indicator of its effectiveness. High correlations between off-flavour development in frozen vegetables during storage and the residual peroxidase activity (Nebesky *et al.*, 1951; Guyer & Holmquist, 1954; Labbee & Esselen, 1954) has made peroxidases useful as an index for the degree of blanching. The thermostability of peroxidase activity has been investigated in turnips (Servant *et al.*, 1986), cabbages (McLellan & Robinson, 1981, 1983, 1987*b*,), apples (Moulding *et al.*, 1987, 1988), mangoes, pears (Moulding *et al.*, 1989), grapes (Robinson *et al.*, 1990) and citrus products (McLellan & Robinson, 1984). The flavour of mango is very delicate and is easily affected by thermal processing. Mango is marketed in the form of juice, pulp or slices canned in syrup, but the thermal processes followed at present for these canned products are mainly empirical (Nanjundaswamy *et al.*, 1973).

During the last 30 years both the food manufacturing and retail sectors have developed to serve the dietary needs of large centres of population. This has given rise to increasing use of blanching and freezing of a wide range of vegetable crops that include peas, beans, cauliflowers, root crops and *Brassica* species. Well-established blanching techniques generally use immersion in boiling water for fixed times followed by rapid cooling and blast freezing. Other methods of processing might include microwave heating which has not been sufficiently exploited for the blanching of plant foods. This subject has been reviewed by Thomas (1987). Gamma irradiation might be used which can also be expected to affect enzyme activity. Irradiation of pure solutions of horseradish peroxidase using ^{60}Co gamma rays caused a loss of enzymic activity which was highest in the presence of oxygen-saturated water and Metodiewa *et al.* (1987) suggested that hydroxy and peroxy radicals were responsible for the inactivation. It is reasonable to suppose that such inactivation is the result of secondary attack on peroxidase by free radicals produced from water during the irradiation process. The species involved are likely to include the hydroxy, peroxy and superoxide radicals. On the other hand, for irradiated citrus fruits the accumulation of phenolic compounds was thought to be due to enhanced activity of phenylalanine ammonia lyase which is a key enzyme for the biosynthesis of phenolics (Riov *et al.*, 1972). Increases of peroxidase activity 1 or 2 weeks after irradiation of citrus fruits has been reported (Monselise *et al.*, 1969). For climacteric fruits, the major objective of radiation is to extend the post-harvest storage life by slowing down the rates of first ripening and subsequently senescence, where the advantage resides in the inhibition of biosynthesis of a number of enzymes responsible for the ripening processes.

Effective blanching of some vegetables can be difficult, not only because of the presence of thermostable enzymes, but also because of either their large size, shape or density. For instance, for Brussels sprouts, due to their compact structure, deactivation of enzymes throughout the sprout may be difficult to achieve (McLellan & Robinson, 1987*a*). Similarly, corn on the cob is one of the most difficult vegetables to blanch effectively because of its large size and the structural characteristics of its different anatomical parts (Garrote *et al.*, 1987). Long blanch times are necessary to inactivate the enzymes at all points within the cob. Lee & Hammes (1979) determined significant correlations between residual peroxidase activity in the outer cob and kernels blanched at 100°C and the off-flavour development detected by a panel after 9 months storage at −18°C.

For heat inactivation studies, peroxidases are normally first extracted from the experimental material into aqueous solutions in which the peroxidase activity can then be easily measured spectrophotometrically. Low-speed centrifugation can be used to crudely separate the isoperoxidases into soluble and bound types. Generally, for *Brassica* species the soluble peroxidase activity has been shown to be more heat stable than the peroxidase activity associated with the cationic isoenzymes extracted from the ionically bound fractions (McLellan & Robinson, 1981, 1987*a*,*b*). Likewise, for apples (Moulding *et al.*, 1988), pears (Moulding *et al.*, 1989) and grapes (Robinson *et al.*, 1990), the extracted ionically-bound isoperoxidases, containing mainly cationic isoenzymes, were more heat labile. Attempts have been made by Chang *et al.* (1988) using electrophoretograms to identify the more thermostable horseradish isoperoxidases. However, although it was claimed for the mixture of horseradish isoperoxidases that the most heat stable isoenzymes had pI values of between 7·5 and 9·0, their proposal was based only on a non-quantitative visual assessment of enzymic activity, as revealed by staining, after separation of the isoperoxidases by electrophoresis.

Although small quantities of a thermostable isoperoxidase have been isolated from cauliflower (Lee & Pennesi, 1984), the heat inactivation plots for the isolated isoperoxidases were still non-linear. Of the three isoenzymes isolated from cauliflower the major isoenzyme, C, which accounted for 94% of the total peroxidase activity, has been shown to be very heat labile (Lee *et al.*, 1984): 98% of its activity was lost in 15 min at 50°C. The more heat-stable cauliflower isoenzyme, A, had a $t_{1/2}$ value of 30 min at 50°C compared to the more heat labile isoenzyme C, for which $t_{1/2}$ at 50°C was 3 min. Although the thermostable isoenzyme A accounted for only a small proportion of the total enzymic activity, because of its greater heat resistance it could still adversely affect the quality of blanched cauliflower.

For the *Brassica* species the purified anionic isoenzymes were more thermostable than the purified cationic group. The heat inactivation plots for the more thermostable isoperoxidases were clearly more non-linear than for the less thermostable cationic isoperoxidases (McLellan & Robinson, 1987*a,b*). Generally it seems that the higher the thermal stability of isoperoxidases the more non-linear is the heat inactivation plot. Although vegetable extracts do contain peroxidase isoenzymes with different heat stabilities, the presence of a mixture of isoperoxidases with different thermostabilities does not explain the non-linear character of heat inactivation plots for peroxidase in extracts of certainly *Brassica* species and probably other fruits and vegetables. For purified apple isoperoxidases, the anionic isoenzymes A1 and A2 were more thermostable than the cationic types and in particular isoperoxidase A2 retained at least 70% of its enzymic activity after heating at 70°C for 2 min (Moulding *et al.*, 1987). Such studies have confirmed the existence of distinct isoperoxidases with different thermal characteristics in a range of fruits and vegetables, nevertheless the distribution of even the identified isoperoxidases may well vary between varieties, the anatomical parts of plants and possibly samples.

Mechanisms for Heat Inactivation and Regeneration

It is almost self-evident that the rates of inactivation and reactivation of peroxidase activity should depend on many factors, such as temperature and time of blanching, pH value and salt concentration. Bruemmer *et al.* (1976) suggested that the loss of peroxidase activity during the pasteurisation of orange juice is pH-dependent. Therefore insight into the mechanism of heat inactivation can in principle be obtained by studying the effects of experimental variables like pH, ionic strength, and the concentrations of different polyols and anions such as SCN^- on the kinetics of inactivation. Lu & Whitaker (1974) showed that the rate of inactivation of horseradish peroxidase decreased some eight-fold as the pH was increased from 4 to 7 and then two-fold as the pH was increased to 10. As added hematin reduced the rate of inactivation of peroxidase activity, it was claimed by Lu & Whitaker (1974) that the marked effect of pH on the rate of heat inactivation of peroxidase was associated with the release of hemin and that the apoprotein, devoid of haem, was more heat labile than the haemoprotein complex. When studying the effect of ionic strength on the rate of inactivation Lu & Whitaker (1974) observed that in the the presence of NaCl there was a considerable decrease in the thermostability of horseradish peroxidase, with an eight-fold increase in the initial rate of inactivation of peroxidase in the presence of 0·6 M NaCl. Such a decrease in thermostability might be due to NaCl-induced changes in the conformation of the peroxidase enzyme and possibly the dissociation of thermostable aggregated molecules.

The changes in peroxidase activity for haricot seed have been reported (Kermasha *et al.*, 1988) to depend on many factors such as temperature, pH, time of blanching and salt concentration. Many investigations on the thermal inactivation of peroxidase have been made on the assumption that the mechanism of inactivation of the enzyme follows first-order reaction kinetics. However, the residual activity plots against heating time generally show non-linearity indicating a deviation from first-order kinetics. Several reasons and mechanisms have been suggested for the non-linear loss of peroxidase activity with heating time, but so far none of the theories has been validated.

The most obvious explanation for the deviation from linearity is the presence of a mixture of thermostable and heat-labile isoperoxidases (Yamamoto *et al.*, 1962; Lu & Whitaker, 1974; Ling & Lund, 1978; Naveh *et al.*, 1982; Yoon & Park, 1982; Lee & Pennesi, 1984). While it seems very likely that some of the deviation is probably accounted for by the presence of isoperoxidases with different susceptibilities to heat, this has not yet been clearly established as the main cause of the deviation from linearity of heat inactivation plots. Furthermore, it has not been shown conclusively that purified individual thermostable isoperoxidases, as opposed to heat-labile isoenzymes, exhibit linear first-order thermal inactivation kinetics. To date, it has proved difficult to completely purify and isolate sufficient quantities of the individual thermostable isoperoxidases. Nevertheless Naveh *et al.* (1982) for corn ears peroxidases has formulated a kinetic model to take account of a mixture of several isoperoxidases, with each assumed to be inactivated in a first-order manner but with different rate constants. An equation was given for calculating the total residual activity in the form of an integrated Arhenius equation. It has been claimed (Naveh *et al.*, 1982) that there is a good fit between the model described by the equation and corrected kinetic data which indicates that the model is sufficiently accurate for practical blanching calculations. The results suggest that the inactivation of peroxidase activity can be described by assuming two enzyme fractions with vastly different thermal stabilities. Other tenable explanations have been proposed for the deviation from linearity and include the formation of aggregates of the thermostable isoperoxidases which may possess an increased heat stability (Winter, 1971), or alternatively, a series-type enzyme inactivation kinetics induced by heat (Henley & Sadana, 1985). For other thermostable proteins it has been suggested that aggregation may be a stabilising mechanism (Merkler *et al.*, 1988).

An alternative explanation for thermostability may support the concept that heat-induced denaturation for such enzymes is a more reversible process. For peroxidases it has been known for a long time that these enzymes are capable of regenerating their enzymic activity when either held at room temperature in aqueous solution or held in the frozen state as in whole foods. Significant regeneration of enzymic activity after heat denaturation is unusual for enzymes generally. The ability of peroxidases to regenerate after heat denaturation varies not only between different plant species but also between the isoenzymes occurring within a single variety. Restoration of peroxidase activity is generally observed over a period of a few hours after heat treatment of either test enzyme solutions or whole vegetables.

At the molecular level, the mechanism of regeneration of peroxidases is not understood. However, in general terms it is considered that regeneration must require a reversal of denaturation with concomitant changes in the conformational structure of the protein moiety. As it has been shown that added haematin increases the amount of restored enzymic activity (Lu & Whitaker, 1974), it has been generally accepted that the haem moeity can recombine with the apoprotein to form active enzyme. Lu & Whitaker (1974) observed maximum restoration of enzymic activity at pH 9.

Lu & Whitaker (1974) showed that the percentage of original activity restored depended on the extent of inactivation and the process did not follow either first- or second-order kinetics but was more complex. Both the rates of inactivation and reactivation of peroxidase were found to be pH-dependent. Other components may also be required in some cases for restoration of peroxidase activity, as McLellan & Robinson (1987*a*) found that purified Brussels sprout anionic isoperoxidases, unlike peroxidase activity in crude extracts from Brussels sprouts and cabbage, did not regenerate following heat inactivation. On the other hand, for a purified apple isoperoxidase A2, up to 80% of the original activity was restored when the heat-treated isoenzyme was held at 30°C (Moulding *et al.*, 1988). Cationic isoperoxidases in crude extracts from *Brassica* species and apples (McLellan & Robinson, 1981, 1987*a*; Moulding *et al.*, 1988) were incapable of regeneration which implies that the structure of the protein primarily determines this property. Naveh *et al.* (1982) showed an interesting linear relationship between residual activity immediately after heat treatment and subsequent regenerated peroxidase activity (Fig. 1). The large slope of approximately 3·0 for the graph implies that more drastic heat processing would not only increase inactivation but also markedly decrease the extent of regeneration. Also, the proportionality between increasing thermostability and the enzyme's ability to regenerate supports the general observation that the more thermostable enzymes tend to exhibit a greater ability to regenerate. Tamura & Morita (1975), while studying the spectral changes on heating of concentrated solutions of peroxidases, observed decreases in absorbance at 404 nm (the maximum of the Soret band of peroxidase) and at 505 nm. Cooling for 20 h at room temperature after heat treatment led to a recovery in the absorbance at 404 nm coinciding

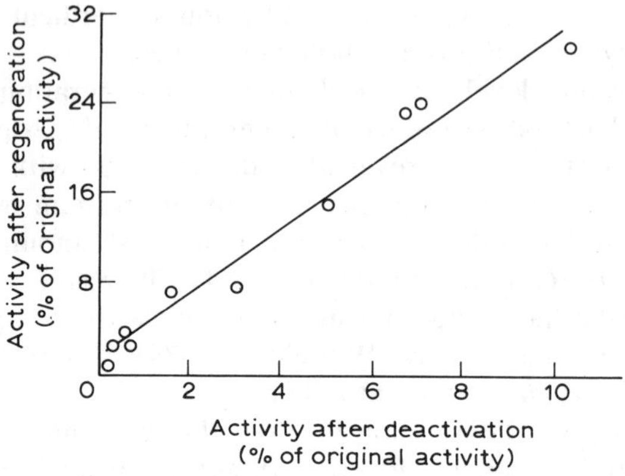

Fig. 1. Effect of extent of thermal deactivation on the reaction of peroxidase. Reprinted from Naveh *et al.* (1982).

with a recovery in enzymic activity. Inactivation of the enzyme was marked at both pH 5 and 9, whereas at pH 7, heat inactivation at 60°C was slight. Tamura & Morita (1975) showed that the degrees of regeneration were considerably lower at these pH values than at pH 7 after prolonged heat treatments. The work of Tamura & Morita (1975) and the earlier investigations of Lu & Whitaker (1974) all support the idea that heat inactivation of peroxidases involves the dissociation of the haem moiety from the apoprotein, as well as the separate denaturation of the peroxidase protein. Indeed, Tamura & Morita (1975) demonstrated that peroxidase activity can be restored by the addition of prohaemin to previously heat-treated apoperoxidase. Therefore, as suggested by Lu & Whitaker (1974) and Tamura & Morita (1975), it seems likely that the inactivation of peroxidases during blanching involves the release of the haem moiety, possibly its degradation chemically, and the separate denaturation of the liberated apoprotein. The reversibility of these processes will determine the extent of regeneration.

In recent years there has been increased interest in the structural features, domains of proteins and small molecules, and additives generally that may be involved in determining thermostability; the wider aspects of this subject have been reviewed by O'Fagain *et al.* (1988). Since, generally, polyols and sugars have been found to protect the native conformation of proteins against thermal denaturation, the effects of polyhydric alcohols and sugars have recently been extended to studies on the denaturation of peroxidases. Chang *et al.* (1988) showed, by the use of differential scanning calorimetry, that in the presence of sucrose the denaturation temperature of the protein increased. However, measurements of residual enzymic activity showed that in the presence of 10% sucrose the thermal stability of the enzymic activity was reduced. For a range of sugars tested, fructose was the most effective in reducing the thermostability of the enzymatic activity and it was suggested that this was due to the interaction of fructose, a reducing sugar, with the protein's amino acids. While attempting to explain these apparently contradictory results, Chang *et al.* (1988) suggested that the destabilising effect of low concentrations of sucrose might be due to an influence on the conformation of the enzyme's active site, which might then result in a change in thermostability, while the overall native structure of the enzyme remains unaltered below its denaturation temperature. Generally the effect of such substances on thermostability indicates that this property of peroxidases and possibly other enzymes in whole foods is likely to be affected by other constituents that may interact at the molecular level.

MOLECULAR STRUCTURE OF CATALASES AND PEROXIDASE

Catalases and peroxidases in plants and animals are mainly haem-containing enzymes where the prosthetic group is protophorphyrin IX. The haem moiety is involved in the active site of peroxidases. The spectra of all haem-containing proteins are similar and show absorption peaks at 640, 583, 548 and 498 nm (Fig. 2). Purity is often assessed by the ratio of absorption at 403/280 nm (RZ value). However, purity cannot be judged entirely on this criterion due to the dependence of absorption at 280 nm on aromatic amino acids which would be expected to vary between haem proteins. The concentration of the enzymes can be estimated using known extinction coefficients, of approximately $\varepsilon = 100\,\text{mM}\ \text{cm}^{-1}$ for concentrated solutions (Yamazaki *et al.*, 1981). However, such measurements may have little bearing on both the relative enzymic activity of the isoenzymes and their affinity towards a wide range of substrates. At pH 2·4, and low concentrations of acetone, total detachment of the haem occurs with a decrease in the solubility of the apoprotein. Thus, acetone powder preparations of peroxidases are of doubtful value for investigating enzymatic activity. Plant peroxidases have very high specific activities

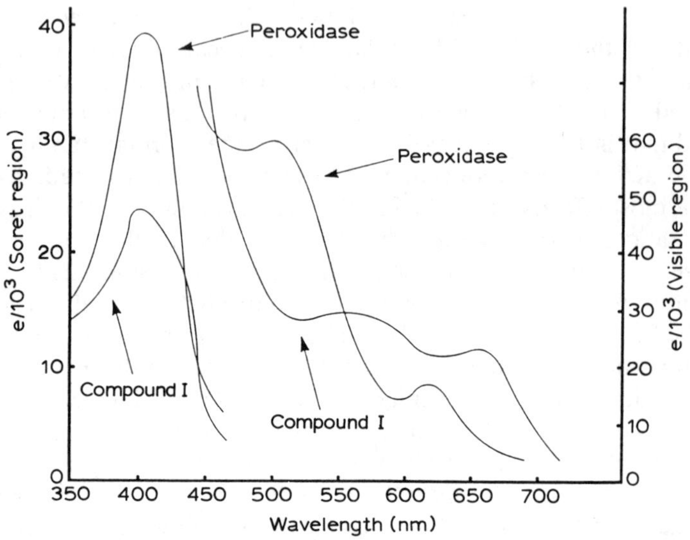

Fig. 2. The Soret and visible spectra of peroxidase and compound I. From Robinson (1987).

and thus while it has been easy to detect a range of isoperoxidases, it may be especially difficult to obtain sufficient enzyme to characterise the protein moiety. For this reason suspension cultures are being used to prepare larger amounts of pure isoperoxidases.

Peroxidases

Some of the practical problems for the isolation of plant isoperoxidases have been mentioned by Aibara *et al.* (1982). These authors quite rightly point out that purification procedures are time-consuming and frequently require expensive carrier ampholytes. Furthermore, claims for purity of proteins either generally, or even more so for enzymes, rest very heavily on the adequacy and sensitivity of the methods available for detection of protein at the time in question. During the last few years more sensitive staining methods using silver have been developed to detect low levels of protein on zymograms. This method is particularly suitable for peroxidases, where either the isoenzymes or contaminating proteins do not stain with Coomassie Blue. We have shown (Khan & Robinson, 1991), for mango isoperoxidases that contaminating proteins and the enzyme protein can only be observed after silver staining. Lagrimini *et al.* (1987) have also used silver staining to establish that a mixture of anionic tobacco isoperoxidases was free of contaminating proteins. The development of this technique and its proven use for detecting small amounts of protein now raises doubts about the purity of horseradish and turnip peroxidases previously used for amino acid sequencing (Welinder, 1985): van Huystee (1989) has remarked about the divergence of claimed amino acid sequences between peroxidases from different sources. Clearly this divergence could be explained partly as being due to undetected contaminating proteins leading to sequencing errors.

The secondary structure of yeast cytochrome *c* peroxidase has been determined by X-ray crystallography (Finzel *et al.*, 1984) and has provided a reference point for comparing and matching amino acid sequences for active site regions in horseradish and turnip peroxidases. For cytochrome *c* peroxidase it has been shown that the arginine and histidine side chains in the distal pocket are configured in such a way as to be able to serve as catalytic sites for reaction with hydrogen peroxide. NMR studies have also been reported on the electronic structure and molecular structure of specifically the haem cavity in horseradish peroxidase (Thanabal *et al.*, 1987). These investigations are less dependent on the absence of contaminating proteins in the preparations used, as the results relate directly to the labelling of the haem moiety. Using this technique Thanabal *et al.* (1987) have identified on the proximal side of the haem His 170, Leu 237 and Tyr 185 and on the distal side Arg 46, His 42 and Phe 41 as catalytically important residues in a cyano complex of horseradish peroxidase (Fig. 3). The inability to obtain single crystals of both horseradish and turnip isoperoxidases for X-ray diffraction has led to attempts to compare and correlate, for the remainder of the protein molecule, the amino-acid sequences with the identified secondary structures in cytochrome *c* peroxidase (Welinder, 1985). Although apparent homogeneity has been claimed for the isoperoxidases on criteria less demanding than would be used today, with this reservation, the attempts made by Welinder (1985) to identify important sequences and similar structural elements at the molecular level provide a qualified insight into the molecular structure of specific regions of the molecule. The gross size of plant peroxidases is similar to that of yeast cytochrome *c* peroxidase with approximately 300 amino-acid residues

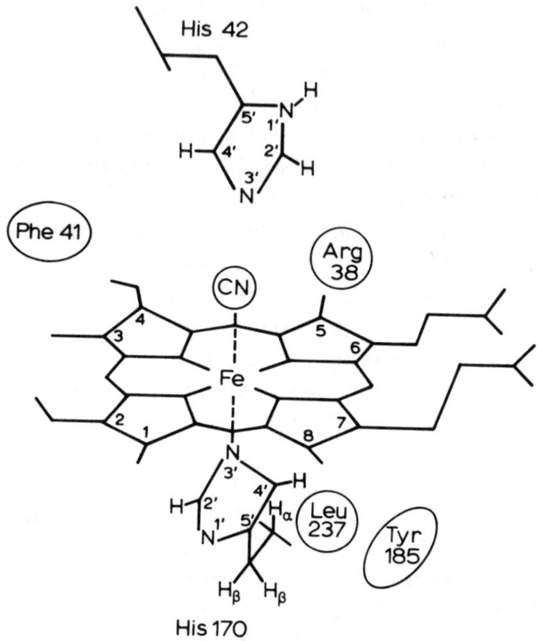

Fig. 3. Proposed haem-pocket of cyano horseradish peroxidase. From Thanabal *et al.* (1987).

per mole. Isoperoxidases do not differ significantly in size, with molecular weights generally in the range 40 000–50 000. The amino acid analyses reported for horseradish peroxidase are reproduced here in Table 2 from Welinder *et al.* (1972). Of particular importance is the claimed presence of four cystine residues, and three residues of histidine. Although the number of cystine residues is greater than that found for cytochrome *c* peroxidase, Lagrimini *et al.* (1987) have recently found eight cystine residues in tobacco anionic isoperoxidases. Structural studies for the haem-containing proteins, myglobin and haemoglobin, have shown that histidine residues are invariably ligated to the fifth and sixth coordinates of the haem iron. The distal histidine residue is No. 52 in cytochrome *c* peroxidase and No. 42 in both horse-radish C and turnip peroxidase TP7. The proximal histidine residue is No. 170 in horse-radish peroxidase, No. 175 in cytochrome *c* peroxidase, but No. 159 in turnip peroxidase TP7. The oxidation–reduction potential of the haem proteins is believed to be affected by

Table 2. Amino-Acid Analyses of Horse-radish Peroxidases

	Residues per mole	
	Mann HRP (lot U1929)	*Isoenzyme C*[a]
Lysine	5·8	6
Histidine	2·9	3
Arginine	19·7	21
Aspartic acid	46·5	50
Threonine	25·3[b]	23
Serine	25·7[b]	22
Glutamic acid	19·9	19
Proline	17·1	16
Glycine	17·0	15
Alanine	23·0	23
Half-cystine	7·9 (7·8)[c,d]	4
Valine	17·6[e]	17
Methionine	4·2 (3·3)[c]	3
Isoleucine	12·3[e]	13
Leucine	35·0	36
Tyrosine	5·9	5
Phenylalanine	20·2	23
Glucosamine	9·7[f]	

[a] The amino-acid analyses taken from the literature have been recalculated assuming a molecular weight of 32 000–33 000 for the protein moiety of HRP or nearly 300 amino-acid residues per mole of HRP.
[b] Extrapolation to zero time hydrolysis.
[c] Cysteic acid and methionine sulfone are average values obtained from four hydrolysates of oxidised HRP.
[d] Corrected against blanks of non-oxidised HRP amino-acid analyses.
[e] Seventy-two hours hydrolysis value.
[f] This value is obtained from 20-h-hydrolysed samples and is uncorrected. Seventy-two hours hydrolysis gave complete destruction of GlcN.
From Table 1, Welinder *et al.*, 1972.

the basicity of the axial ligand. Strong hydrogen bonding to the N delta proton of the ligated histidine in horseradish peroxidases leads to a pentacoordinated structure that tends to repel the bonded pair of electrons from a sixth ligand such as water. The E_m values for a range of peroxidases are given in Table 3.

Table 3. E_{m7} Values for Lignin Peroxidases, Mn-Dependent Peroxidases and Several Haem Proteins

Haem protein	E_{m7} *(mV)*
HRP	−278
Turnip peroxidase (P_1)	−223
CCP	−194
Lignin peroxidase H1	−142
Chloroperoxidase	−140
Lignin peroxidase H8	−137
Lignin peroxidase H2	−135
Lignin peroxidase H10	−127
Turnip peroxidase (P_7)	−100
Mn-dependent peroxidase H4	−93
Mn-dependent peroxidase H3	−88
Myeloperoxidase	+24
Metmyoglobin	+46

From Millis *et al.* (1989).

Amino acid sequences around the histidine residues of plant peroxidases point to similar helical structures for the haem-linked peptide chains to those identified by X-ray crystallography for cytochrome *c*. Welinder (1985) also found a sequence similarity for the suspected active site regions for plant and yeast peroxidases and cytochrome *b*-562. The distal histidine-ligated sequence for these three peroxidases included arginine 48, an aromatic residue 51 and histidine 52: in plants it is claimed that the aromatic residues are tyrosines. Attempts to align the yeast cytochrome *c* peroxidase, horseradish and turnip peroxidases using the proximal and distal histidines as identity points indicated less than 20% identity between the enzymes. The N-terminal and C-terminal sequences showed little similarity. This lack of correlation could in part be due to contamination of the enzyme preparations by extraneous proteins for the reasons given above related to the absence of a rigorous assessment of purity. The validity of these sequences can only be confirmed or otherwise when more information becomes available on the molecular structure of plant peroxidases perhaps derived from the parent nucleotide sequences. To date nucleotide and deduced amino-acid sequences have been reported for tobacco anionic isoperoxidases (Lagrimini *et al.*, 1987). Interestingly, cross-reactivity has been found between antibodies to the tobacco isoperoxidases and horseradish and turnip peroxidases, thus indicating the presence of common amino-acid sequences for peroxidases from unrelated sources. Nucleotide sequences have also been reported for the gene of porcine thyroid peroxidase (Magnusson *et al.*, 1987).

Plant peroxidases also differ from yeast cytochrome *c* peroxidases, in so far as they contain four disulphide bonds, bind two atoms of calcium per mole and are glycoproteins. Welinder (1985) has claimed that there are eight sites for attachment of carbohydrate in horse-radish peroxidases, while the turnip TP7 possesses only one site. Lagrimini *et al.* (1987) have detected four potential asparagine-linked glycosylation sites in tobacco anionic isoperoxidases. As for other glycoproteins the oligosaccharide groups vary in size from approximately 10 to 20 residues with mannose and glucosamine as the predominant sugars. Hu & van Huystee (1989*a*) have recently shown that deglycosylation of peroxidases from peanut oil cell cultures reduces enzymic activity and therefore stabilisation of the peroxidase molecules could be linked directly to the covalent bonding of the oligosaccharide moieties. Immunological studies with the cultured peanut peroxidases have indicated that the carbohydrate moieties of the cationic and anionic isoperoxidases are closely related (Hu & van Huystee, 1989*a,b*). The plant peroxidases also differ from cytochrome *c* peroxidase as they catalyse the oxidation of small aromatic molecules, whereas yeast cytochrome *c* peroxidase catalyses the oxidation of a macromolecule. For plant peroxidases the active sites and the binding sites of the enzymes have not been identified and therefore at present there is no obvious explanation at the molecular level for the different reactivities of the various isoenzymes towards different substrates.

The greatest progress on structural studies of peroxidases is likely to be made with cultured cells as exemplified by the work with peanut cell cultures, where large amounts of enzyme accumulate in the growth medium. For these cultures it seems that 2% of the total amount of protein synthesised could be peroxidases and that a single cationic isoenzyme is responsible for 75% of the peroxidatic activity (Lambien *et al.*, 1985). Suspension cell cultures are especially useful because of the higher yields of peroxidases and hence easier purification.

Catalases

The catalase molecule is usually composed of four identical subunits with a total molecular weight in the region of 240 000 containing 4 mol of protohaem. Although the overall structure of catalases has been known for a long time and crystals can be obtained for crystallographic work, the nature of the ligands attached to the haem does not seem to have been fully investigated.

Catalase from the pericarp of tomato fruit has been claimed to have been obtained in a purified form as determined by silver staining and shown to have a molecular weight of 225 000 with subunits of 55 000 (Inamine & Baker, 1989), which seems similar to the catalases found in cucumber, pumpkin and cotton, although larger subunits have been reported for catalases present in sweet potato and maize enzymes. Recently green catalases containing more novel haem components including an enzyme from spinach (Hirasawa *et al.*, 1987) and a catalase from *Neurospora crassa* (Jacob & Orme-Johnson, 1979) with a ferric dihydroporphyrin prosthetic group have been isolated.

Catalactic Reaction

The rate of dismutation of hydrogen peroxide into oxygen and water is approximately 10^{-7} per mol at pH 7 and 30°C.

$$2H_2O_2 \rightleftharpoons 2H_2O + O_2$$

Sources of hydrogen peroxide include fatty acid breakdown catalysed by flavin-linked enzymes and the oxidation of glycolate catalysed by glycolate oxidase:

$$\begin{array}{l}COOH\\|\\CH_2OH\end{array} + O_2 \rightleftharpoons \begin{array}{l}COOH\\|\\CHO\end{array} + H_2O_2$$

During the storage and processing of foods the compartmentalisation of catalases within cells is likely to break down, as peroxisomes are particularly fragile. Such uncoupling in heat-processed foods may result in the direct chemical oxidation by hydrogen peroxide of oxo-acids, like glyoxalate and α-ketoglutarate to formic and succinic acids, respectively, particularly if flavin-linked enzymes such as glycolate oxidase are not denatured.

$$\begin{array}{l}COOH\\|\\CHO\end{array} + H_2O_2 \rightleftharpoons CO_2 + H_2O + HCOOH$$

Catalases are used in the food industry for the removal of H_2O_2, the most well-known example being the oxidation of glucose by glucose oxidase and the coupled removal by catalases of the hydrogen peroxide. Catalases are not involved in the oxidation of other substrates and therefore do not have any peroxidatic role.

Peroxidatic Reaction

Peroxidases can promote a large number of separate reactions and therefore have a degree of versatility not exhibited by any other enzyme. Peroxidases are able to catalyse peroxidatic reactions, oxidatic reactions and hydroxylations. Peroxidases utilise both an oxidising substrate and a reducing substrate. The oxidising substrate is usually a peroxide or a peroxy acid, ROOH. Typically, peroxidases can use hydrogen peroxide, methyl-hydrogen peroxide and ethyl-hydrogen peroxide (Kermasha & Metche, 1988) as oxidising agents in the peroxidatic reaction. The distinctive absorption spectra due to the haem prosthetic group exhibits transient changes (Fig. 2) on mixing with H_2O_2. The Soret band and visible spectra change markedly on the binding of peroxide and the consequent formation of compound I, which has a ferryl-type structure carrying an oxygen atom acquired from the peroxide. Compound I absorbs at 583 and 548 nm. Peroxidases cause the heterolytic cleavage of the O—O bond. The iron-oxo compound (Fig. 4) is in the formal oxidation state of +5, the H_2O_2 having effectively removed two electrons from the iron atom of the enzyme. The second stage of the peroxidatic reaction involves the reduction of compound I by a reducing donor substrate. In Table 4 some typical donor test substrates are listed. Hydrogen abstraction readily occurs from phenolic substrates because of the resonance stabilisation of the resulting phenoxy radical.

Table 4. Action of Peroxidase–H_2O_2

Substrate	*Product*
Pyrogallol	Purpurogallin
Guaiacol	Tetraguaiacoquinone
Hydroquinone	Quinhydrone
Benzidine	*p*-Quinone di-imide
o-Phenylene diamine	Phenazine
Leucomalachite green	Malachite green
Catechol	*o*-Quinone
p-Cresol	Milky precipitate
o-Cresol	Green solution
m-Cresol	Flesh-colored solution
Tyrosine	Yellow solution
Adrenaline	Reddish solution
Bilirubin	Biliverdin

From Burnette (1977).

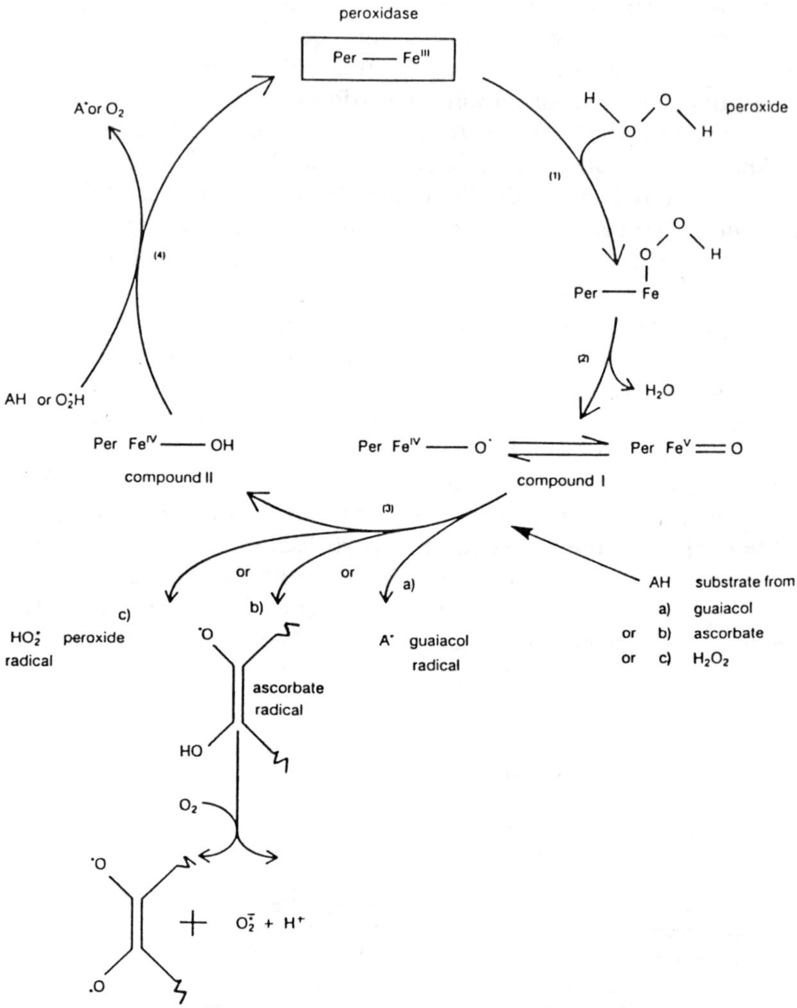

Fig. 4. Cyclic scheme for the formation of compound I and subsequent catalysed oxidation of electron-donating substrates. From Robinson (1987).

Generally, peroxidases are defined as enzymes that catalyse the following overall peroxidatic reaction:

$$ROOH + AH_2 \rightleftharpoons H_2O + ROH + A^{\cdot}$$

The first step (Fig. 4) of the initial oxidation of the donor substrate (AH_2) produces a free radical ($A^{\cdot}$). Hydrogen peroxide is generally the oxidising substrate (ROOH) and the rate constant is typically $10^7\,M^{-1}\,s^{-1}$ (Dunford, 1982). Compound I is reduced by the stepwise addition of two single electrons to form native peroxidase:

$$\underset{\text{compound I}}{\text{Per—Fe}^{V}=O} + e^- + H^+ \rightarrow \underset{\text{compound II}}{\text{Per–Fe}^{IV}\text{—OH}}$$

$$\underset{\text{compound II}}{\text{Per–Fe}^{IV}\text{—OH}} + e^- + H^+ \rightarrow \underset{\text{native enzyme}}{\text{Per–Fe}^{III}\text{—}H_2O}$$

Compound II absorbs at 498 and 640 nm. Reduction of compound II results in the reformation of the native peroxidase in oxidation state +3 (Fig. 4). Both reductions require a single electron and a proton. A broad range of reducing substrates usually convert compound I back to the native enzyme via the intermediate formation of compound II. The overall reaction involves a cycle of oxidation and reformation of the enzyme. The separate cationic free radicals ($A^{\cdot}$) derived from the donor substrate then react non-enzymically with each other to form dimers or higher polymers. At high concentrations of donor substrates, the liberated free radicals interact with each other to form polymers of the A—A type as in the case of the oxidation of guaiacol or *o*-dianisidine.

From the natural substrates, including the phenolic compounds, present in fruits and vegetables, a wide range of products may be formed through reactions involving free radicals. The natural substrates in orange juice alone include ascorbic, caffeic, genistic and coumaric acids. Peroxidatic oxidation of ascorbic acid to form dehydroascorbic acid, which is relatively unstable, ultimately results in loss of nutritional value. However, because the concentration of hydrogen peroxide is generally low in whole citrus fruits and other plant foods, even though the concentration of ascorbate is relatively high, the rates of oxidation by this route are likely to be low. Nevertheless, hydrogen peroxide is frequently used as a sterilising and cleaning agent and thus may be present in greater quantities in processed juices, The specific effects of peroxidase action are difficult to identify, becoming apparent only during prolonged storage of fresh and processed foods and may be confused with polyphenol oxidase-catalysed oxidations. Mono- and dihydroxyphenols and complex compounds like chlorogenic acid are potential substrates for peroxidases. Others include the flavonoids for which it has been claimed that a large number of products are formed through the action of peroxidases (Schreier & Miller, 1985) and that flavonoids act as radical scavengers (Darimont *et al.*, 1989). Other reducing agents for compound I include halides, the bisulphite ion and secondary aldehydes (Dunford, 1986). Specific ascorbic acid peroxidases (Gerbling *et al.*, 1984; Castillo *et al.*, 1986; Goldberg *et al.*, 1989) have been described which also seem to be active towards the common test substrates for peroxidases. Although ascorbate peroxidase may assist in detoxification by rapid removal of hydrogen peroxide, roles for this other enzyme are not known.

OXIDATIC REACTION

Oxidatic reactions catalysed by peroxidases accompanied by the uptake of oxygen, occur with dihydroxyfumarate (DHF), ascorbic acid, thiols and hydroquinones. The peroxidase–oxidase reactions are characterised by peroxidase-catalysed oxidations in the absence of added hydrogen peroxide. Investigations have been undertaken mainly with dihydroxyfumarate (Halliwell & de Rycker, 1978). Based on the powerful inhibition of the reaction by superoxide dismutase (Halliwell, 1977), as opposed to hydroxyl radical scavengers, an oxidative scheme involving superoxide radical and the continual oxidation of DHF has been widely accepted:

$$\text{Peroxidase} + H_2O_2 \rightarrow \text{compound I}$$

$$\text{compound I} + \text{DHF} \rightarrow \text{DHF}^{\cdot} + \text{compound II}$$

$$\text{compound II} + \text{DHF} \rightarrow \text{DHF}^{\cdot} + \text{native peroxidase}$$

and

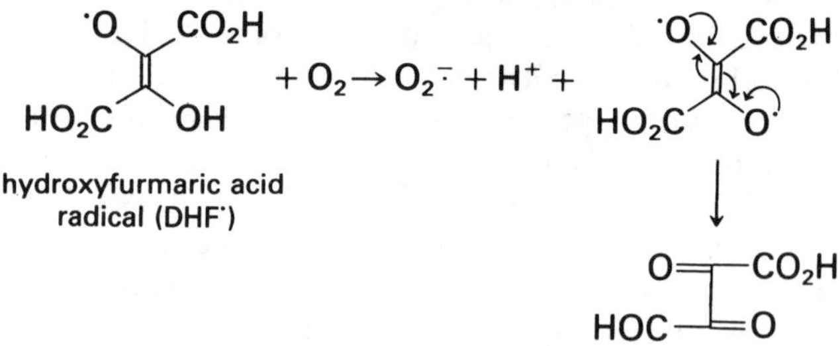

hydroxyfurmaric acid radical (DHF·)

diketosuccinic acid

One mole of O_2 is taken up per mole of DHF oxidised. The initial amount of H_2O_2 required for the formation of compound I is believed to be generated by the non-catalytic autoxidation of DHF:

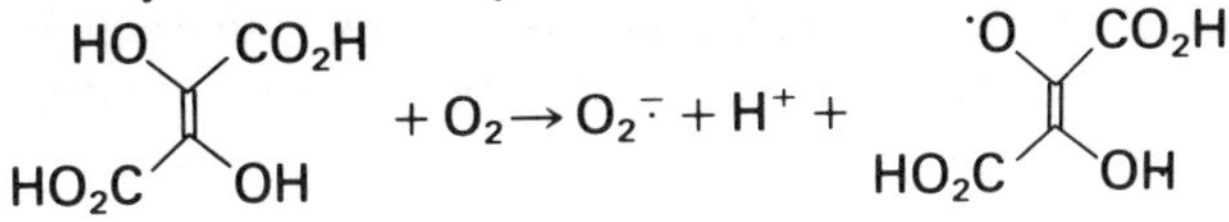

dihydroxyfurmaric acid

and

$$DHF + O_2^{\cdot -} + 2H^+ \rightarrow DHF^{\cdot} + H_2O_2$$

It is suggested that oxidase reactions occur when radicals such as $DHF^{\cdot}$ can reduce oxygen directly. Peroxidase also reacts with $O_2^{\cdot -}$ to form an oxygenated ferroperoxidase (compound III):

$$\text{Per–Fe}^{III} + O_2^{\cdot -} \rightleftharpoons \underset{\text{compound III}}{\text{Per—Fe}^{II}\text{—}O_2}$$

which can break down to form the native peroxidase and $O_2^{\cdot -}$ and subsequently H_2O_2.

In plants, a similar scheme is believed to operate for the oxidation by compounds I and II of at least NADH, pyridoxal compounds, indole-3-acetic acid (Srivastava & van Huystee, 1977) and possibly ascorbic acid. In NADH oxidation, as for oxidation of DHF, peroxidases act according to its peroxidatic cycle to produce NADH radicals.

$$\text{compound I} + \text{NADH} \rightleftharpoons \text{NADH}^{\cdot} + \text{compound II}$$

$$\text{compound II} + \text{NADH} \rightleftharpoons \text{NADH}^{\cdot} + \text{native peroxidase}$$

Both $NADH^{\cdot}$ and the semidehydroascorbic radical can reduce oxygen directly. Svensson (1988) has shown that although myloperoxidase and eosinophil peroxidases can oxidise the aminothiols, cysteamine and cysteine esters, these compounds are not oxidised by horseradish peroxidase and lactoperoxidase.

Mason *et al.* (1957) first reported that horseradish peroxidase in the presence of reducing equivalents of dihydroxyfumarate and oxygen could catalyse the hydroxylation of a wide range of aromatic compounds including tyrosine, phenylalanine and other phenolic compounds. *p*-Coumaric acid is hydroxylated to caffeic acid in the presence of DHF and O_2 by peroxidase. Through the use of scavengers for both the superoxide and hydroxyl radicals, Dordick *et al.* (1986) showed that both radicals are involved and suggested that compound III, the oxyperoxidase ($Fe^{II}—O_2$), is a key intermediate. Compound III is said to be formed from $O_2^{\cdot -}$, which itself is generated from the reduction of molecular oxygen by dihydroxyfumarate. The hydroxyl radical ($OH^{\cdot}$) is believed to be the hydroxylating species.

OXIDATION OF INDOLE-3-ACETIC ACID

Peroxidases are believed to play an important role in fruit ripening. It is generally accepted that the degradation of indole-3-acetic acid (IAA) by peroxidases controls the *in vivo* concentration of this phytohormone in plants, although other enzymes may also catalyse the degradation of IAA (Reinecke & Bandurski, 1988). Inactivation of the plant growth regulator occurs mainly by an oxygen-consuming oxidation catalysed by peroxidases. Diverse mechanisms have been proposed for the peroxidase-catalysed oxidation of the plant growth hormone (IAA). The complete sequence of reactions through an oxidase pathway, including the mechanism of oxygen activation, has been difficult to elucidate due to both the reported involvement of small amounts of H_2O_2 and phenolic compounds and because peroxidases in the absence of H_2O_2 catalyse oxidatically the degradation of IAA. The assay for IAA activity is normally performed in the presence of $MnCl_2$ and 2,4-dichlorophenol, which act as co-factors. Small amounts of added H_2O_2 have been claimed to abolish a lag period. Indole-3-methanol (IM) is the only rapidly appearing product of the peroxidase catalysed oxidation.

A reaction scheme proposed by Grambow (1986) accounts for the formation of indole-3-methanol (IM) and 3-hydroxymethyloxindole. After abstraction of an electron from IAA by peroxidase, the radical $IAA^{\cdot}$ and ferroperoxidase ($Per{-}Fe^{II}$) is believed to be formed, which then decarboxylates to form an indole-3-methyl radical ($I^{\cdot}$).

Ferroperoxidase has a high affinity for oxygen and forms compound III. The indole-3-methyl radical is then peroxidated by oxygenated peroxidase (compound III) and reduced peroxidatically to indole-3-methanol.

$$IOO^{\cdot} + Per{-}Fe^{III} \rightarrow IM + \underset{\text{compound III}}{Per{-}Fe^{IV}{=}O}$$

Both indole-3-methanol and 3-methyleneoxindole are now thought to be the main products (Grambow, 1986). The formation of 3-hydroxymethyloxindole may occur via a non-enzymatic rearrangement of the hydroperoxide. The reaction scheme includes the proposal that phenolic compounds, if present in sufficient quantity, regenerate native peroxidase by the reduction of compounds I and II peroxidatically. The results indicate that 3-hydroxymethyloxindole cannot be formed from indole-3-methanol and that the routes are controlled by the availability of phenolic compounds to regenerate native peroxidase. Activation of oxygen to form compound III may also produce a source of superoxide radical which may then also be available for oxygenase reactions (Smith *et al.*, 1982).

Although the mechanism for the peroxidase-catalysed degradation of IAA was proposed for horseradish peroxidase, there are many

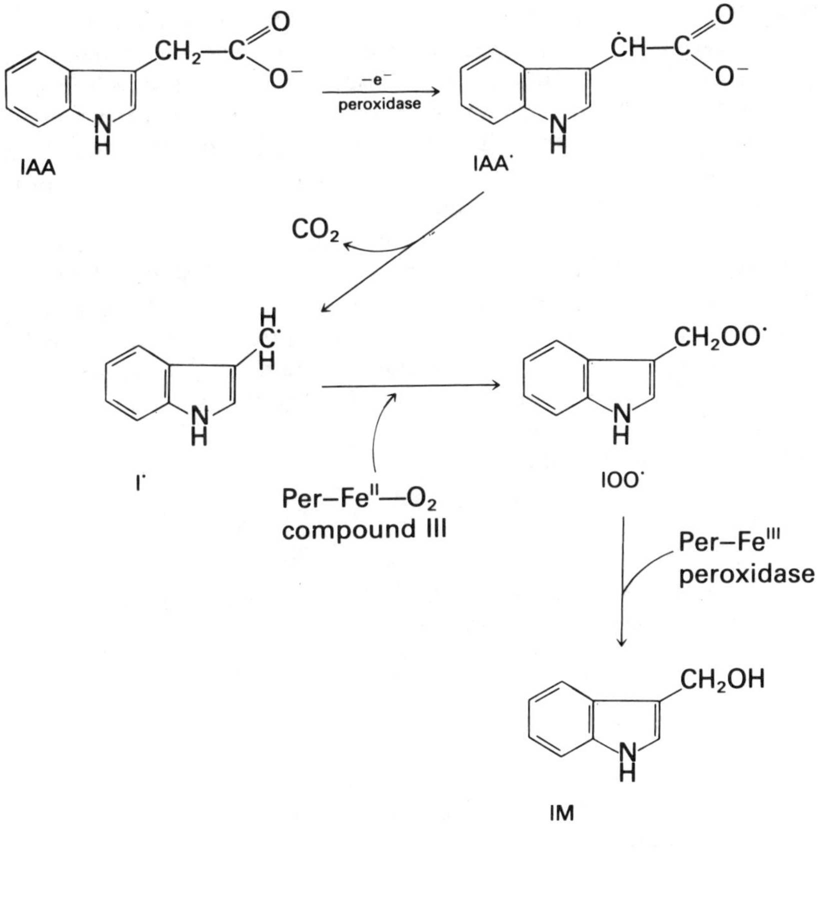

reports on the oxidation of IAA by peroxidases from other sources. Since numerous isoperoxidases typically exist, often with different properties and substrate specificities, it is difficult to assign precise physiological roles to each isoenzyme. Thomas *et al.* (1981) showed that IAA oxidase activity increases prior to the respiration climacteric. Since IAA is considered to be a naturally occurring retardant to fruit ripening, it is conceivable that IAA must be inactivated before fruit ripening can occur. Thomas *et al.* (1981) also reported an increase in IAA oxidase activity of tomato peroxidase during ripening. It has been suggested that cell wall-bound peroxidase may function in oxidising IAA as the phytohormone enters the cell. The IAA oxidase:peroxidase ratios in the bound fractions were found on average to be about 10 times greater than in the soluble fraction, thus indicating that the bound peroxidase has a greater capacity for oxidising IAA than the peroxidase found in the soluble fraction. Also, the isoenzyme patterns of the ionically bound fractions showed the appearance of two cationic peroxidases during fruit development, indicating that totally different isoenzyme forms are found in the cell wall. Brooks (1986) and Gaspar *et al.* (1982) indicated that basic isoperoxidases frequently have a higher capacity for IAA degradation than acidic isoperoxidases, although this may not be consistent.

The propensity of peroxidases to react with a wide range of compounds is also illustrated by the peroxidase-catalysed oxidation of malonaldehyde (MA). As for the oxidation of indole-3-acetic acid, $MnCl_2$ can act as a co-factor for the oxidation of malonaldehyde, although it can be replaced by hydrogen peroxide. MacDonald & Dunford (1989) have suggested that Mn^{II} reduces malonaldehyde peroxy radicals to produce an organic peroxide which then oxidises peroxidase to compound I.

$$\text{Per–Fe}^{III} + \text{ROOH} \rightarrow \underset{\text{compound I}}{\text{Per–Fe}^{V}{=}\text{O}} + \text{ROH}$$

It is suggested that malonaldehyde radicals are generated by reaction of malonaldehyde with compound I and II.

$$\text{Per–Fe}^{IV}{=}\text{O} + \text{MA} \rightarrow \text{Per–Fe}^{IV}\text{–OH} + \text{MA}^{\cdot}$$

$$\text{Per–Fe}^{IV}\text{—OH} + \text{MA} \rightarrow \text{Per–Fe}^{III} + \text{MA}^{\cdot}$$

It is proposed that the malonaldehyde radicals react directly with oxygen to form peroxy radicals and then through the involvement of Mn^{II} form malonaldehyde hydroperoxide.

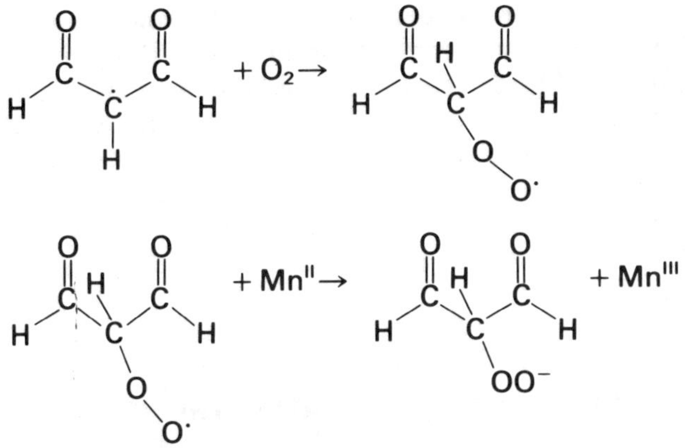

From MacDonald & Dunford (1989).

The mechanism proposed by MacDonald & Dunford (1989) is similar to that described above for the oxidation of dihydroxyfumarate to DHF by compounds I and II with overall consumption of oxygen. Also it still remains possible that $MnCl_2$ plays a similar role in the oxidation of indole-3-acetic acid. Malonaldehyde is a common product of lipid oxidation in foods and the above reactions indicate how further oxidation of malonaldehyde might be mediated by peroxidase. Likewise other aldehydes, such as those present as flavour substances in foods, might also be vulnerable to oxidatic reactions catalysed by peroxidase. It is also known that peroxidases catalyse the oxidation of NADH, isobutraldehyde, and other linear aldehydes.

OXIDATION OF CHLOROPHYLL

Peroxidases may play important roles in the colour changes associated with the ripening of fruits and the senescence of fruits and vegetables. Although two other enzymes, namely chlorophyllase and chlorophyll oxidase, are known to catalyse the degradation of chlorophyll, there is increasing circumstantial evidence that peroxidase plays a role in leaf senescence and the associated yellowing of fruits (Matile *et al.*, 1987; Abeles *et al.*, 1988). For instance, it has been claimed (Brennan & Frenkel, 1977; Lauriere, 1983) that peroxidase activity and levels of hydrogen peroxide increase during the senescence and ripening of fruits. Abeles *et al.* (1988) reported studies that implicate the biosynthesis of a 33-KDa peroxidase during ethene-induced senescence in cucumber cotyledons. Nevertheless, although Matile *et al.* (1987) have possibly detected some non-green, pink-pigmented degradation products, the biochemistry of chlorophyll catabolism is virtually unknown. Gaspar *et al.* (1986) suggested that peroxidase has a role as an oxidase in the conversion of 1-aminocyclopropane-1-carboxylic acid to ethene and supporting evidence is quoted in Paul (1987). However, the evidence is not conclusive (Boyer & De Jaegher, 1986).

ISOPEROXIDASES AND LIGNIN BIOSYNTHESIS

There have been numerous reports (Chibbar & van Huystee, 1984) in the literature with respect to the general involvement of peroxidases in lignin biosynthesis. Lignin is an abundant heterogeneous polymer which forms a three-dimensional network and is generally regarded as being responsible for toughness in vegetables and stringiness in beans. Lignins are strengthening materials that occur with cellulose and other polysaccharides in the cell walls of higher plants. The two reactions involved in the lignification process are oxidative polymerisation of coniferyl alcohol and the generation of hydrogen peroxide at the expense of NADH oxidation. Peroxidase first catalyses the oxidation of coniferyl alcohol to form phenoxy radicals which then polymerise nonenzymically to form lignin, a strong, resistant polymer with a high proportion of ether bonds and methoxy groups. Coupling can occur with other phenols and other cell constituents, including hemicelluloses, to create the three-dimensional lignin network. The formation of extensin and the cross-linking of diferulate to hemicelluloses or pectin also requires peroxidase. Tracer experiments have shown that lignin is synthesised from phenylalanine or tyrosine via cinnamic acids and the corresponding hydroxycinnamyl alcohols. These compounds are directly dehydrogenated by peroxidase with hydrogen peroxide to produce phenoxy free radicals. Although the precise nature of the reductant of oxygen is not known it seems to be NADH generated by membrane bound malate dehydrogenase (Gross *et al.*, 1977). This complex reaction may involve the superoxide radical which is formed by the reduction of oxygen by $NADH^{\cdot}$ as intermediate. The formation of $NADH^{\cdot}$ is thought to be initiated through the oxidation of NADH by phenoxy radicals formed in the peroxidase-catalysed reaction in the presence of manganese. This scheme can possibly explain how hydrogen peroxide may be synthesised directly at the proper site without a requirement for transportation of toxic hydroperoxide from the cytoplasm to the cell wall. It is claimed that lignification is always accompanied by an increased activity of most anionic isoperoxidases: a view has emerged in the last few years that lignin biosynthesis is catalysed by anionic peroxidases and the oxidation of indole-3-acetic acid by the cationic peroxidases.

Strengthening of cell walls by lignification during the growth and maturation of plants can largely account for the increased hardness and toughness in green beans, broad beans and other seed crops. Seed coat impermeability to water is important in the longevity of minute weed seeds in soil. During the maturation of seeds, peroxidase is instrumental in the conversion of soluble phenolics to insoluble lignin polymers. Egley *et al.* (1983) showed that peroxidase is instrumental in the development of water-impermeable seed coats in prickly sida seeds. Highest peroxidase activity of coat extracts was well correlated with the development stages for maximum conversion of soluble phenolics to soluble lignin polymers. Although seed extracts oxidised dihydroxyphenylalanine, this activity was eliminated by catalase, indicating that the oxidation of phenolics in the coat is catalysed by peroxidase. Kieffer pears, which contain a large number of sclerides, have been shown to contain more peroxidase than Bartlett fruit. Histochemical analysis with *o*-dianisidine and hydrogen peroxide showed that peroxidase activity was concentrated in the parenchyma cells, surrounding grit cells and on the cell walls throughout the pulp. The lower calcium levels in fruits possessing excessive scleroid development were consistent with the thesis that calcium inhibits peroxidase localisation and accordingly controls peroxidase-catalysed lignin deposition in pear fruit. Asparagus undergoes lignification and hence increases in toughness when stored at ambient temperatures. It has been suggested that a free radical mechanism is involved (Smith & Stanley, 1989) and trace amounts of peroxidase could well be responsible for initiating such free radical reactions.

Many tropical fruits are susceptible to chill injury and peroxidase may be involved in these commercially undesirable changes. Zauberman *et al.* (1988) showed that peroxidase and catalase activities in mango peel increased during the development of chill injury to much higher levels than in non-chilled fruits. Peroxidase activity was also enhanced in avocado fruit, another fruit known to undergo chill injury, when stored at 5°C. However, such empirical data do not provide an explanation for the mechanism of chill injury.

INDUSTRIAL APPLICATIONS

Research to date has shown that electrophoretograms can be used for surveying and comparing the occurrence of isoperoxidases between cultivars. The application of up-to-date purification techniques coupled to the specific assays of peroxidatic and oxidatic activity now being used should allow the definitive identification of unwanted isoperoxidases in our indigenous crops. Such comprehensive knowledge of the occurrence and enzymic properties of isoperoxidases is a prerequisite to any improvement that may be achieved through genetic selection of cultivars for improved storage and processing. Further in the future, but the foreseeable future, it will be possible by genetic engineering to modify the action and the amounts of peroxidases so as to control their action in post-harvest foods. One example would be the control of lignification by suppressing either the enzymic activity or biosynthesis of lignin peroxidase, for instance in asparagus. It is perhaps conceivable that the extent of ripening may also be influenced by the control of the oxidatic action of peroxidases on indoleacetic acid when the specific isoperoxidases for this phytohormone have been characterised. Likewise it may also be possible to identify the isoperoxidases responsible for the peroxidatic oxidation of flavonoids, other phenolics and susceptible flavour substances and thus decrease the deleterious action of the enzymes.

Elucidation of the mode of action and the availability of purified isoperoxidases increases the likelihood of their use for manufacturing purposes. The catalytic action of peroxidases is unusual in so far as these enzymes generate two well-defined chemically distinct oxidant enzyme–substrate complexes from hydrogen peroxide. Moreover the substrates that are converted to free radical intermediates may polymerise or react with other substances to form further products. For enzymes bound to solid supports thermostability is frequently a sought-after attribute and the natural occurrence in plants of such isoperoxidases augers well for their use as manufacturing agents. Just based on our knowledge of the action of peroxidases on phenolics, dihydroxy compounds like dihydroxyfumarate and heterocyclics like indoleacetic acid, a likely use of peroxidases is to provide sources of both simple and complex oxidised aromatic intermediates in the form of hydroperoxides or hydroxy compounds. The degradation of chlorophyll, catalysed by peroxidases, also points towards a possible use for the degreening of, as yet, unattractive leaf protein and other proteins of photosynthetic origin. Some more unusual peroxidases are already known that catalyse the oxidation of halogens to form hypochlorous and hypobromous acids and therefore there is known potential for their use as oxidising agents. Prior oxidation of aromatics in waste effluents by peroxidases may also be very beneficial in reducing the oxygen demands that such products place on lakes and rivers. Although peroxidases are ubiquitous in the plant kingdom the most likely sources of peroxidases for industrial use are plant tissue cultures and micro-organisms.

Lignin depolymerisation is brought about by peroxidases of white-rot basidomycetes such as *Phanerochaecte chrysosporium.* Lignin peroxidases and Mn-dependent peroxidases form the two predominant classes of isoenzymes produced by white-rot fungi (Bonnarme & Jeffries, 1990) together with a glycolate: hydrogen peroxide-generating system. The lignin peroxidases catalyse not only the oxidation of lignin and lignin model compounds but also a large number of other organic compounds, which cannot be oxidised by other peroxidases through one electron oxidation of aromatic rings to form aryl cation radicals. The Mn-dependent enzymes catalyse the oxidation of a number of phenolic compounds as well as amines. Both these types of peroxidases use hydrogen peroxide to attain the higher oxidation state of compound I, but the Mn-dependent enzymes uniquely use Mn^{II} as the reducing agent thus yielding the highly reactive oxidising Mn^{III} species (Millis *et al.*, 1989; Wariishi *et al.*, 1989). The fungal lignin peroxidases and Mn-dependent lignin peroxidases are distinct in their ability to oxidise high redox potential substrates and in their low pH optima for activity which suggests a haem environment that is more electron deficient than the one present in the plant peroxidases. The redox potentials (Table 3) of the Fe^{2+}/Fe^{3+} couple of lignin peroxidases is approximately 130 mV higher than that of horse-radish peroxidases (Millis *et al.*, 1989). Molecular cloning and sequencing of lignin peroxidase genes of *Phanerochaete chrysosporium* has been reported (Schalch *et al.*, 1989). Clearly such enzymes are likely to have a major part to play, not only for the possible conversion of degraded lignins to more valuable substances and as providers of chemical intermediates, but also in the utilisation of agricultural products such as waste plant residues from sugar cane and sugar beet and other leafy crops.

BIBLIOGRAPHY

Abeles, F. B., Dunn, L. J., Morgens, P., Callahan, A., Dinterman, R. E. & Schmidt, J. (1988). Induction of 33-kD and 60-kD peroxidases during ethylene-induced senescence of cucumber cotyledons. *Plant Physiol.*, **87,** 609–15.

Aibara, S., Yamashita, H., Mori, E., Kato, M. & Morita, Y. (1982). Isolation and characterisation of five neutral isoenzymes of horse-radish peroxidase. *J. Biochem.*, **92,** 531–9.

Arighi, A. L., Joslyn, M. A. & Marsh, G. L. (1936). Enzyme activity in frozen vegetables. *Ind. Eng. Chem.*, **28,** 595–8.

Bakardjieva, N. T., Izvorska, N. D. & Hristova, N. (1987). Influence of Ca^{2+} on the activity and release of peroxidase from tobacco callus tissues. *Comptes rendus l'Academie bulgare des Sciences,* **40,** 85–8.

Bedford, C. L. & Joslyn, M. A. (1939). Enzyme activity in frozen vegetables: String beans. *Ind. Eng. Chem.*, **31,** 751–8.

Bonnarme, P. & Jeffries, T. W. (1990). Mn^{II} regulation of lignin peroxidases and manganese-dependent peroxidases from lignin-degrading white rot fungi. *Appl. Envir. Microbiol.*, **56,** 210–17.

Boyer, N. & Jaegher, G. de (1986). Direct or indirect role of peroxidases in ethylene biosynthesis? In *Molecular and Physiological Aspects of Plant Peroxidases,* ed. H. Greppin, C. Penel & Th. Gaspar. University of Geneva, Switzerland, pp. 47–60.

Brennan, T. & Frenkel, C. (1977). Involvement of hydrogen peroxide in the regulation of senescence in pear. *Plant Physiol.,* **59,** 411–16.

Brooks, J. L. (1986). Oxidase reactions of tomato anionic peroxidase. *Plant Physiol.*, **80,** 130–3.

Bruemmer, J. H., Roe, B. & Bowen, E. R. (1976). Peroxidase reactions and orange juice quality. *J. Food Sci.*, **41,** 186–9.

Burnette, F. S. (1977). Peroxidase and its relationship to food flavour and quality: A review. *J. Food Sci.*, **42,** 1–6.

Carbonell, J. & Jones, R. L. (1984). A comparison of the effects of Ca^{2+} and gibberellic acid on enzyme synthesis and secretion in barley aleurone. *Physiol. Plant.*, **63,** 345–50.

Castillo, F. J., Cotton, G., Kevers, C., Greppin, H., Gaspar, Th. & Driessche, Th. V. (1986). Changes in ascorbic acid content and ascorbate peroxidase activity during the development of *Acetabularia mediterranea. Differentiation,* **33,** 17–23.

Chance, B. (1951). Enzyme substrate compounds. *Adv. Enzymol.*, **12,** 153–90.

Chance, B. (1952). The transition from primary to secondary peroxidase-peroxide complex. *Archs Biochem. Biophys.*, **37,** 235–7.

Chang, B. S., Park, K. H. & Lund, D. B. (1988). Thermal inactivation kinetics of horseradish peroxidase. *J. Food Sci.*, **53,** 920–23.

Cherry, J. P. & Ory, P. L. (1973). Characterization of isoperoxidase activity in *Arachis hypogaea* cultivars. *Phytochem.*, **12,** 1581–3.

Chibbar, R. N. & Huystee, R. B. van (1984). Characterization of peroxidase in plant cells. *Plant Physiol.*, **75,** 956–8.

Chibbar, R. N. & Huystee, R. B. van (1986). Immunochemical localisation of peroxidase in cultured peanut cells. *J. Plant Physiol.*, **123,** 477–86.

Chibbar, R. N., Cella, R. & Huystee, R. B. van (1984). The heme moiety in peanut peroxidase. *Can. J. Biochem. Cell Biol.*, **62,** 1046–9.

Cuenca, J., Garcia-Florenciano, E., Ros Barcelo, A. & Munoz, R. (1989). Sequential release of both basic and acidic isoperoxidases to the media of suspension cultured cells of *Capsicum annuum*. *Plant Cell Reports,* **8,** 471–4.

Darimont, E., Hofinger, M., Guillot, B. & Gaspar, Th. (1989). Flavonoids de *Chrysenthellum indicum*: substrats peroxydasiques. *J. Pharm. Belg.*, **44,** 325–8.

Davis, W. B. (1942). The distribution and preparation of citrus peroxidase. *Am. J. Botany,* **29,** 252–4.

Dordick, J. S., Klibanov, A. M. & Marletta, M. A. (1986). Horseradish peroxidase catalyzed hydroxylations: Mechanistic studies. *Biochemistry,* **25,** 2946–51.

Dunford, H. B. (1982). *Advances in Inorganic Biochemistry, Vol. 4,* ed. G. L. Eichhorn & L. G. Marzilli. Elsevier, New York/Amsterdam.

Dunford, H. B. (1986). Peroxidases: historical background. In *Molecular and Physiological Aspects of Plant Peroxidases,* ed. H. Greppin, C. Penel & Th. Gaspar. University of Geneva, Switzerland, pp. 15–23.

Egley, G. H., Paul, R. N. Jr, Vaughn, K. C. & Duke, S. O. (1983). Role of peroxidase in the development of water-impermeable seed coats in *Sidas spinosa* L. *Planta,* **157,** 224–32.

Finzel, B. C., Poulos, T. L. & Kraut, J. (1984). Crystal structure of yeast cytochrome *c* peroxidase refined at 1.7 Å. *J. Biol. Chem.*, **259,** 13027–36.

Gardner, H. W., Inglett, G. E. & Anderson, R. A. (1969). Inactivation of peroxidase as a function of corn processing. *Cereal Chem.*, **46,** 626–34.

Garrote, R. L., Luna, J. A., Silva, E. R. & Bertone, R. A. (1987). Prediction of residual peroxidase activity in the blanching-cooling of corn-on-the-cob and its relation to off-flavour development in frozen storage. *J. Food Sci.*, **52,** 232–5.

Gaspar, Th., Penel, C., Thorpe, T. & Greppin, H. (1982). Peroxidases 1970–1980. *A Survey of their Biochemical and Physiological Roles in Higher Plants*. University of Geneva—Centre Bot. Publ., Geneva, Switzerland.

Gaspar, Th., Penel, C., Castillo, F. J. & Greppin, H. (1985). A two-step control of basic and acidic peroxidases and its significance for growth and development. *Physiol. Plant.*, **64,** 418–23.

Gaspar, Th., Penel, C. & Greppin, H. (1986). Peroxidases: structures and catalytic reactions, biosynthesis, transport and location, physiological roles. *Bull. Liaison-Group Polyphenols,* **13,** 159–76.

Gerbling, K. P., Kelly, G. J., Fisher, K. H. & Latzlko, E. (1984). Partial purification and properties of soluble ascorbate peroxidases from pea leaves. *J. Plant Physiol.*, **115**, 59–67.
Goldberg, R., Kevers, C. & Gaspar, T. (1989). Guaiacol and ascorbate peroxidase compartmentation and gradient along the growing Mung bean hypocotyl. *Biochem. Physiol. Pflanzen*, **184**, 155–61.
Gorin, N. & Heidema, F. T. (1976). Peroxidase activity in Golden Delicious apples as a possible parameter of ripening and senescence. *J. Agric. Food Chem.*, **24**, 200–1.
Grambow, H. J. (1986). Pathway and mechanism of the peroxidase-catalysed degradation of indole-3-acetic acid. In *Molecular and Physiological Aspects of Plant Peroxidases*, ed. H. Greppin, C. Penel & Th. Gaspar. University of Geneva, Switzerland, pp. 31–41.
Gross, C. G., Jansen, C. & Elstner, E. F. (1977). Involvement of malate, monophenols and the superoxide radical in H_2O_2 formation by isolated cell walls from horseradish. *Planta*, **136**, 271–6.
Guyer, R. B. & Holmquist, J. W. (1954). Enzyme regeneration in high temperature-short time sterilized canned foods. *Food Technol.* (*Chicago*), **8**, 547–50.
Haard, N. F. (1977). Physiological roles of peroxidase in post-harvest fruits and vegetables. In *Enzymes in Food and Beverage Processing*, ed. R. L. Ory & A. J. St. Angelo. *ACS Symposium Series no. 47*, 143–71.
Halliwell, B. (1977). Generation of hydrogen peroxide, superoxide and hydroxyl radicals during the oxidation of dihydroxyfumaric acid by peroxidase. *Biochem. J.*, **163**, 441–8.
Halliwell, B. & Rycker, J. de (1978). Superoxide and peroxidase-catalysed reactions. Oxidation of dihydroxyfuromate, NADH and dithiothreitol by horseradish peroxidase. *Photochem. Photobiol.*, **28**, 757–63.
Halpin, B. E. & Lee, C. Y. (1987). Effect of blanching on enzymic activity and quality changes in green peas. *J. Food Sci.*, **52**, 1002–5.
Halpin, B., Pressey, R., Jen, J. & Mondy, N. (1989). Purification and characterisation of peroxidase isoenzymes from green peas. *J. Food Sci.*, **54**, 644–9.
Hendriks, T., Berg, B. M. van den & Schram, A. W. (1985). Cellular location of peroxidase isoenzymes in leaf tissue of Petunia and their affinity for concanavalin A-sepharose. *Planta*, **164**, 89–95.
Henley, J. P. & Sadana, A. (1985). Categorization of enzyme deactivations using a series-type mechanism. *Enz. Microbial Technol.*, **7**, 50–60.
Hirasawa, M., Gray, K. A., Shaw, R. W. & Knaff, D. (1987). Spectroscopic studies of spinach catalase. *Biochim. Biophys. Acta*, **911**, 37–44.
Hu, C. & Huystee, R. B. van (1989*a*). Immunochemical relatedness of two peroxidase isozymes from peanut cell culture. *Biochem. Cell Biol.*, **67**, 371–6.
Hu, C. & Huystee, R. B. van (1989*b*). Immunological similarity of the cationic and the anionic peanut peroxidase isozymes. *Biochem. Biophys. Res. Communs.*, **163**, 689–94.
Hu, C., Smith, R. & Huystee, R. B. van (1989). Biosynthesis and localisation of peanut peroxidases. A comparison of the cationic and the anionic isozymes. *J. Plant Physiol.*, **135**, 391–7.

Inamine, G. S. & Baker, J. E. (1989). A catalase from tomato fruit. *Phytochemistry*, **28,** 345–8.
Jacob, G. S. & Orme-Johnson, W. H. (1979). Catalase of *Neurospora crassa*. 2 Electron paramagnetic resonance and chemical properties of the prosthetic group. *Biochemistry*, **18,** 2975–80.
Joslyn, M. A. (1949). Enzyme activity in frozen vegetable tissue. *Adv. Enzymol.*, **9,** 613–52.
Joslyn, M. A. & Bedford, C. L. (1940). Frozen activity in frozen vegetables. *Ind. Eng. Chem.*, **32,** 702–6.
Joslyn, M. A. & David, J. J. (1952). Acetaldehyde and alcohol in raw or underblanched peas. *Quick Frozen Foods*, **15,** 51–3.
Khan, A. A. & Robinson, D. S. (1991). Purification of an anionic peroxidase isoenzyme from mango. (In press).
Keilin, D. & Mann, T. (1937). On the haematin compound of peroxidase. *Proc. Royal Society, London*, **127B,** 119–33.
Kermasha, S. & Metche, M. (1988). Studies on seed peroxidase *Phaseolus vulgaris* cv, haricot. *J. Food Sci.*, **53,** 247–52.
Kermasha, S., Alli, I. & Metche, M. (1988). Changes in peroxidase activity during the development and processing of *Phaseolus vulgaris* cv, Haricot seed. *J. Food Sci.*, **53,** 1753–5.
Labbee, M. D. & Esselen, W. B. (1954). Effect of peroxidase concentration, acidity and storage temperature on the development of off-flavors in fresh pack pickles. *Food Technol. (Chicago)*, **8,** 50–54.
Lagrimini, L. M., Burkhart, W., Moyer, M. & Rothstein, S. (1987). Molecular cloning of complementary DNA encoding the lignin-forming peroxidase from tobacco: Molecular analysis and tissue-specific expression. *Proc. Natl. Acad. Sci. USA*, **84,** 7542–6.
Lambien, A. M., Dunford, H. B., Huystee, R. B. van & Lobarzewski, J. (1985). Spectral and kinetic properties of cationic peroxidase secreted by cultured peanut cells. *Can. J. Biochem. Cell. Biol.*, **63,** 1086–92.
Lauriere, C. (1983). Enzymes and leaf senescence. *Physiol. Veg.*, **21,** 1159–77.
Lee, C. Y. & Hammes, J. K. (1979). Heat inactivation of peroxidase in corn-on-the-cob. *J. Food Sci.*, **44,** 785–7.
Lee, C. Y. & Pennesi, A. P. (1984). Isolation and further characterisation of a heat resistant peroxidase isoenzyme from cauliflower. *J. Food Sci.*, **49,** 1616–17.
Lee, C. Y., Pennesi, A. P. & Dickson, M. H. (1984). Characterisation of the cauliflower peroxidase isoenzyme. *J. Agric. Food Chem.*, **32,** 18–21.
Ling, A. C. & Lund, D. B. (1978). Determining kinetic parameters for thermal inactivation of heat-resistant and heat-labile isozymes from thermal destruction curves. *J. Food Sci.*, **43,** 1307–10.
Lu, A. T. & Whitaker, J. R. (1974). Some factors affecting rates of heat inactivation and reactivation of horseradish peroxidase. *J. Food Sci.*, **39,** 1173–8.
MacDonald, I. D. & Dunford, H. B. (1989). Mechanism of horseradish peroxidase-catalysed oxidation of malonaldehyde. *Archs Biochem. Biophys.*, **272,** 185–93.
McLellan, K. M. & Robinson, D. S. (1981). The effect of heat on cabbage and

Brussels sprout peroxidase enzymes. *Food Chem.*, **7**, 257–66.

McLellan, K. M. & Robinson, D. S. (1983). Cabbage and Brussels sprout peroxidase isoenzymes separated by isoelectric focusing. *Phytochemistry*, **22**, 645–7.

McLellan, K. M. & Robinson, D. S. (1984). Heat stability of peroxidases from orange. *Food Chem.*, **13**, 139–47.

McLellan, K. M. & Robinson, D. S. (1987*a*). Purification and heat stability of Brussels sprout peroxidase isoenzymes. *Food Chem.*, **23**, 305–19.

McLellan, K. M. & Robinson, D. S. (1987*b*). The heat stability of purified spring cabbage peroxidase isoenzymes. *Food Chem.*, **26**, 97–107.

Magnusson, R. P., Gostantas, J., Taurog, A. & Rapoport, B. (1987). Molecular cloning of the structural gene for procine thyroid peroxidase. *J. Biol. Chem.*, **262**, 13885–8.

Mason, H. S., Onopryenko, I. & Buhler, D. (1957). Hydroxylation: the activation of oxygen by peroxidase. *Biochim. Biophys. Acta*, **24**, 225–6.

Matile, P., Ginsburg, S., Schellenberg, M. & Thomas, H. (1987). Catabolites of chlorophyll in senescent leaves. *J. Plant Physiol.*, **129**, 219–28.

Merkler, D. J., Srikumar, K., Marchese-Ragona, S. P. & Wedler, F. C. (1988). Aggregation and thermo-inactivation of glutamine synthetase from an extreme thermophyle, *Baccilus caldolyticus*. *Biochim. Biophys. Acta*, **953**, 101–14.

Metodiewa, D., Gebicka, L. & Bachman, S. (1987). The radiation-induced inactivation and post irradiation effects of horseradish peroxidase solution. *J. Radioanal. Nucl. Chem., Letters*, **119**, 87–93.

Millis, C. R., Cai, D., Stankovich, M. T. & Tien, M. (1989). Oxidation–reduction potentials and ionisation states of extracellular peroxidases from the lignin-degrading fungus *Phanerochaete chrysosporium*. *Biochemistry*, **28**, 8484–9.

Monselise, S. P., Riov, J. & Kahan, R. S. (1969). Some changes in enzymatic activities of citrus peel tissues after fruit irradiation. In *Enzymological Aspects of Food Irradiation*. International Atomic Energy Agency, Vienna, pp. 71–81.

Moulding, P. H., Grant, H. F., McLellan, K. M. & Robinson, D. S. (1987). Heat stability of soluble and ionically bound peroxidase extracted from apples. *Int. J. Food Sci. Technol.*, **22**, 391–7.

Moulding, P. H., Singleton, D. E., McLellan, K. M. & Robinson, D. S. (1988). Purification and heat stability of Cox's apple pulp peroxidase isoenzymes. *Int. J. Food Sci. Technol.*, **23**, 343–51.

Moulding, P. H., Goodfellow, J., McLellan, K. M. & Robinson, D. S. (1989). The occurrence of isoperoxidases in Conference pears. *Int. J. Food Sci. Technol.*, **24**, 269–75.

Nanjundaswamy, A. M., Saroja, S. & Ranganna, S. (1973). Determination of thermal process for canned mango products. *Indian Food Packer*, **27**, 5–13.

Naveh, D., Mizrahi, S. & Kopelman, I. J. (1981). Quantitative determination of peroxidase in sweet corn by chemiluminescence. *J. Agric. Food Chem.*, **29**, 36–8.

Naveh, D., Mizrahi, S. & Kopelman, I. J. (1982). Kinetics of peroxidase deactivation in blanching of corn-on-the-cob. *J. Agric. Food Chem.*, **30,** 967–70.

Nebesky, E. A., Esselen, W. B. & Fellers, C. P. (1951). Studies on the peroxidase in pickles and pears. *Food Technol. (Chicago)*, **5,** 110–13.

O'Fagain, C., Sheehan, H., O'Kennedy, R. & Kilty, C. (1988). Maintenance of enzyme structure. Possible methods for enhancing stability. *Process Biochem.*, **23,** 166–71.

Onslow, M. W. (1920). Oxidising enzymes. II. The nature of the enzymes associated with certain direct oxidising systems in plants. *Biochem. J.*, **14,** 535–40.

Pang, An, Catesson, A., Francesh, C., Rolando, C. & Goldberg, R. (1989). On substrate specificity of peroxidases involved in the lignification process. *J. Plant Physiol.*, **135,** 325–9.

Paul, K. G. (1987). Peroxidases: past and present. *J. Oral Pathol.*, **16,** 409–11.

Pedreno, M. A., Ros Barcelo, A., Sabater, F. & Munoz, R. (1989). Control of pH of cell wall peroxidase activity involved in lignification. *Plant Cell Physiol.*, **30,** 237–41.

Penel, C. (1986). The role of calcium in the control of peroxidase activity. In *Molecular and Physiological Aspects of Plant Peroxidases*, ed. H. Greppin, C. Penel & Th. Gaspar. University of Geneva, Switzerland, pp. 155–64.

Pinsent, B. R. W. (1962). Peroxidase regeneration and its effect on quality in frozen peas and thawed peas. *J. Food Sci.*, **27,** 120–6.

Redinbaugh, M. G., Sabre, M. & Scandalios, J. G. (1990). The distribution of catalase activity isozyme protein, and transcript in the tissues of the developing maize seedling. *Plant Physiol.*, **92,** 375–80.

Reinecke, D. M. & Bandurski, R. S. (1988). Oxidation of indole-3-acetic acid to oxindole-3-acetic acid by an enzyme preparation from *Zea mays*. *Plant Physiol.*, **86,** 868–72.

Riov, J., Monselise, S. P. & Goren, R. (1972). Stimulation of phenolic biosynthesis in citrus fruit peel by gamma radiation. *Radiat. Res. Rev.*, **3,** 417–27.

Robinson, D. S. (1987). *Food—Biochemistry and Nutritional Value*. Longman Group UK, with John Wiley & Sons, New York.

Robinson, D. S., Bretherick, M. & Donnelly, J. K. (1990). The heat stability and isoenzyme composition of peroxidases in Ohane grapes. *Int. J. Food Sci. Technol.*, **24,** 613–18.

Ros Barcelo, A., Munoz, R. & Sabater, F. (1987). Lupin peroxidases. I. Isolation and characterization of cell wall-bound isoperoxidase activity. *Physiol. Planta.*, **71,** 448–54.

Ros Barcelo, A., Munoz, R. & Sabater, F. (1988*a*). Lupin peroxidases. III. Subcellular location of membrane-bound acidic isoperoxidases. *Plant Physiol. Biochem.*, **26,** 575–83.

Ros Barcelo, A., Pedreno, M. A., Munoz, R. & Sabater, F. (1988*b*). Lupin peroxidases II. Binding of acidic isoperoxidases to cell walls. *Physiologia Plantarum*, **73,** 238–44.

Schalch, H., Gaskell, J., Smith, T. L. & Cullen, D. (1989). Molecular cloning and sequences of lignin peroxidase genes of *Phanerochaete chrysosporium.*, *Mol. & Cell. Biol.*, **9,** 2743–7.

Schloss, P., Walter, C. & Mader, M. (1987). Basic peroxidases in isolated vacuoles of *Nicotiana tabacum* L. *Planta,* **170,** 225–9.

Schreier, B. & Miller, E. (1985). Studies on flavonol degradation by peroxidase (Donor: H_2O_2-oxidoreductase, EC 1.1.1.7): Part 2—Quercetin. *Food Chem.,* **18,** 301–17.

Servant, I., Gilot, B. & Gbert, H. J. (1986). Comportement thermique des peroxydases de legumes. I. Etude experimentale. *J. Food Engng,* **5,** 287–300.

Sessa, D. J. & Anderson, R. L. (1981). Soybean peroxidases: Purification and some properties. *J. Agric. Food Chem.,* **29,** 960–5.

Shannon, L. M. (1968). Plant isoenzymes. *Ann. Rev. Plant. Physiol.,* **19,** 187–210.

Smith, A. M., Morrison, W. L. & Milham, P. J. (1982). Oxidation of indole-3-acetic acid by peroxidase: Involvement of reduced peroxidase and compound III with superoxide as a product. *Biochemistry,* **21,** 4414–19.

Smith, J. L. & Stanley, D. (1989). Toughening in blanched asparagus: identification of phenolic compounds and evidence for a free radical mechanism. *J. Food Biochem.,* **13,** 271–87.

Srivastava, Om. P. & Huystee, R. B. van (1977). IAA oxidase and polyphenol oxidase activities of peanut peroxidase isozymes. *Phytochemistry,* **16,** 1527–30.

Svensson, B. E. (1988). Abilities of peroxidases to catalyse peroxidase-oxidase oxidation of thiols. *Biochem. J.,* **256,** 757–62.

Tamura, Y. & Morita, Y. (1975). Thermal denaturation and regeneration of Japanese-radish peroxidase. *J. Biochem.,* **78,** 561–71.

Thanabal, V., Ropp, J. S. de & La Mar, G. N. (1987). Identification of the catalytically important amino acid residue resonances in ferric low-spin horseradish peroxidase with nuclear overhauser effect measurement. *J. Am. Chem. Soc.,* **109,** 7516–25.

Thomas, P. (1987). Radiation preservation of foods of plant origin. Part IV. Subtropical fruits: citrus, grapes and avocados. *CRC Crit. Rev. Food Sci. Nutr.,* **24,** 53–89.

Thomas, R. L., Jen, J. J. & Morr, C. V. (1981). Changes in soluble and bound peroxidase-IAA oxidase during tomato fruit development. *J. Food Sci.,* **47,** 158–61.

Vamos-Vigyazo, L., Gajzago, I., Nadudvari-Markus, V. & Kerek, M. M. (1985). Changes in some enzyme activities and browning rate of apricots during storage. *Qual. Plant Plant Foods Hum. Nutr.,* **35,** 139–43.

Wagenknecht, A. C. & Lee, F. A. (1958). Enzyme action and off-flavor in frozen peas. *Food Res.,* **23,** 25–31.

Wariishi, H., Dunford, B., MacDonald, I. D. & Gold, M. H. (1989). Manganese peroxidase from the lignin-degrading Basidomycete *Phanerochaete chrysosporium. J. Biol. Chem.,* **264,** 3335–40.

Welinder, K. G. (1985). Plant peroxidases. Their primary, secondary and tertiary structures, and relation to cytochrome *c* peroxidase. *Eur. J. Biochem.,* **151,** 497–504.

Welinder, K. G., Smillie, L. B. & Schonbaum, G. R. (1972). Amino acid sequence studies of horseradish peroxidase. I. Tryptic peptides. *Can. J. Biochem.,* **50,** 44–62.

Willimott, S. G. & Wokes, F. (1926). Oxidising enzymes in the peel of citrus fruits. *Biochem. J.*, **20,** 1008–12.
Winter, E. (1971). Hitzebestandigkeit der peroxydase. *Z. Lebensm.-unters. Forsch.*, **145,** 3–6.
Yamamoto, H. Y., Steinberg, M. P. & Nelson, A. I. (1962). Kinetic studies in the heat inactivation of peroxidase in sweet corn. *J. Food Sci.*, **27,** 113–19.
Yamazaki, I., Tamura, M. & Nakajima, R. (1981). Horseradish peroxidase C. *Mol. Cell. Biochem.*, **40,** 143–53.
Yoon, J. R. & Park, K. H. (1982). Isolation and thermal inactivation of horseradish peroxidase isoenzymes. *Korean J. Fd Sci. & Technol.*, **14,** 125–9.
Zauberman, G., Fuchs, Y., Rot, I. & Wexler, A. (1988). Chilling injury, peroxidase, and cellulase activities in the peel of mango fruit at low temperature. *Hort. Sci.*, **23,** 732–3.
Zoueil, M. E. & Esselen, W. B. (1959). Thermal destruction rates and regeneration of peroxidases in green beans and turnips. *Food Res.*, **24,** 119–33.

2

Superoxide Dismutase

Judith K. Donnelly & David S. Robinson
Procter Department of Food Science, The University of Leeds, Leeds, UK

INTRODUCTION

In fresh and processed foods oxygen is the main causative agent of well-known post-mortem and post-harvest changes like oxidative rancidity in fats and oils and the discoloration of meat, fruits and vegetables as well as many other less well defined biochemical changes which are responsible for the loss of flavour and colour (Robinson, 1987). Oxidation of food constituents (e.g. lipids, vitamins and flavour compounds) is a major problem affecting all aspects of food quality, including nutritional value; however, the possibility that active oxygen species such as superoxide $O_2^{\cdot -}$ and hydroxl radical $OH^{\cdot}$, may be responsible for the initiation of oxidation in foods has not been thoroughly investigated (Lingnert *et al.*, 1989). In living systems, several theories have been proposed to explain the deleterious consequences of elevated concentrations of oxygen, the most popular being the superoxide theory of oxygen toxicity. According to this theory the undesirable and potentially toxic effects of oxygen to aerobic organisms are thought to be due in part to an increase in the formation of $O_2^{\cdot -}$ radicals. Superoxide dismutase (SOD, EC 1.15.1.1) is present in almost all living cells exposed to oxygen and is considered to constitute an essential defence against oxygen toxicity (McCord *et al.*, 1971).

A free radical ($R^{\cdot}$) may be defined as any species, capable of independent existence, that contains one or more unpaired electrons (Halliwell & Gutteridge, 1989), i.e. electrons which are present alone in atomic or molecular orbitals. Usually the presence of unpaired electrons is associated with increased chemical reactivity (Halliwell, 1987), since electrons have an inbuilt tendency to associate in pairs; the presence of a single, unpaired electron can be thought of as generating a certain 'tension' within free radicals to find a suitable partner (Slater, 1981). Indeed, within aqueous systems such as living cells and foods, unpaired electrons may afford some radical species a significant amount of chemical reactivity, ultimately causing extensive cell damage in living systems and flavour, colour and nutritional changes in food.

Much of the early research on free radicals was carried out by food scientists. As Willson (1987) pointed out, by the mid-1940s unsaturated fats were known to be particularly susceptible to oxidation by metal-catalysed, free radical chain reactions. Vitamins E and C were by then already known to delay the onset of rancidity in fatty foods.

THE FUNCTION OF SOD

In 1954 it was proposed that the most damaging effects of oxygen in the cellular environment could be attributed to the formation of oxygen free radicals (Gershman *et al.*, 1954); but it was not until the discovery of the enzymic function of superoxide dismutase, first described by McCord & Fridovich (1969), that many workers accepted the existence of oxygen radicals within the cellular environment. This change of view was forced upon researchers when McCord & Fridovich (1969) demonstrated that the protein previously known as cuprein, catalysed the dismutation of superoxide radicals (eqn (1)). McCord & Fridovich (1969) isolated SOD of bovine erythrocytes and obtained a blue–green copper-containing enzyme, which was recognised as being identical to previously identified haemocuprein and erythrocuprein. Cuprein had been isolated and studied by many workers since it was first discovered by Mann & Keilin in 1938, but without recognition of its enzymic activity (Fridovich, 1982). The discovery of the catalytic function of SOD, i.e. the removal or dismutation of superoxide radicals according to eqn (1), was of great significance to free radical research and the subsequent studies of reactive oxygen species in biological systems. In the past few years more papers have been published on SOD than any other type of enzyme (Halliwell & Gutteridge, 1989).

$$2O_2^{\cdot-} + 2H^+ \rightarrow H_2O_2 + O_2 \qquad (1)$$

In the 1990s the significance of the role of free radical reactions provides the basis of much research in both medicine and food science in relation to damage attributed to the manifestations of free radical activity. The study of natural defence mechanisms is of interest to those concerned with human diseases and toxicology and those concerned with food preservation alike.

Although the more reactive radicals only have a fleeting existence, their capacity to trigger chain reactions which greatly amplify the damage caused by the initiating radical necessitates efficient and safe non-radical producing mechanisms for their removal. In this chapter we consider current knowledge of oxy-radical reactions with regard to food systems, in which control over metabolic activities is diminished. The generation of oxygen-derived radicals and pathways for the reactions of free radicals with unsaturated fatty acids are described in order to emphasise the 'snowballing' effect of chain reactions and the far-reaching consequences of destructive, uncontrolled free radical activity. Within post-mortem and post-harvest foods free radicals may be removed by the action of indigenous free radical scavengers and antioxidants (e.g. phenolics, ascorbate and vitamin E) and SOD and by radical–radical interactions. Among the agents of antioxidant protection, SOD may form the first line of defence against oxygen toxicity by converting $O_2^{\cdot -}$ to hydrogen peroxide which can then be removed by catalases or peroxidases.

Following McCord & Fridovich's discovery of the catalytic function of SOD in 1969, SOD was subsequently found to be a universal enzyme which exists in three different metalloforms where each form incorporates a transition metal ion at the active site. SOD is not known to react catalytically with any substrate other than the superoxide radical.

SOD exists as a metalloprotein which incorporates either a copper or iron or manganese ion at the active site. The properties which characterise and differentiate the three forms of SOD are summarised in Table 1. The isoenzymes of SOD may be separated and detected on gels following electrophoresis (Almansa *et al.*, 1989). The metalloform of individual isoenzymes can be discovered by the use of selective inhibition by agents such as cyanide and hydrogen peroxide (Table 1). In their general chemical properties, Mn-SOD and Fe-SOD differ only in their response to H_2O_2. It is of interest that Fe-SOD (Asada *et al.*, 1975) and Cu/Zn-SOD (Hodgson & Fridovich, 1975) are slowly but

Table 1. Properties of the Metalloforms of SOD

Property	*Cu/Zn-SOD*	*Mn-SOD*	*Fe-SOD*
Molecular form	Dimer identical subunits	Prokaryotic dimer Eukaryotic tetramer Identical subunits	Dimer, occasionally tetramer. Identical subunits
Metal ion prosthetic group	One Cu^{2+} One Zn^{2+} per subunit	One or two Mn^{2+} per dimer	One Fe^{3+} per subunit
Molecular weight	32 000	42–46 000 (dimer)	40–46 000 (dimer)
Inhibition by Cn^-	At 1 mM CN^-	No effect	No effect
Inhibition by H_2O_2	At 0·5 mM H_2O_2	No effect	At 0·5 M H_2O_2 Inhibited
Inhibition by chloroform/ethanol	No effect	Inhibited	Inhibited
Effect of 10 mM azide	10% Inhibition	30% Inhibition	70% Inhibition
Effect of SDS	Stable to 4% solution SDS	Unstable to 1% solution SDS	Unstable to 1% solution SDS
Effect of pH on enzyme activity	No effect pH 5–10	Activity decreases at pH > 8	Activity decreases at pH > 8

(Adapted from Donnelly *et al.*, 1989.)

irreversibly inactivated by one of the products of the enzyme-catalysed reaction, namely hydrogen peroxide. The enzyme is thought to catalyse the same reaction whatever the metal present.

Although there is strong evidence to support the biological importance of dismutation of $O_2^{\cdot-}$ and the role of SOD, there remain a number of puzzling aspects to the picture we have to date. Gartner & Weser (1986) highlighted some of these aspects: one unresolved problem is why nature has not chosen a more simple mechanism for the removal of $O_2^{\cdot-}$ by the oxidation of $O_2^{\cdot-}$ to O_2, omitting the production of H_2O_2 (in that hypothetical case, the reactivity of metal ions would cause no further problems). Another problem is whether $O_2^{\cdot-}$ can migrate fast enough to SOD to prevent any side reactions. Also, the fact that cytosolic concentrations of SOD are so high is puzzling in view of the fact that $O_2^{\cdot-}$ is predominantly found in lipophilic environments (especially membranes) where relatively small concentrations of SOD are detected. The superoxide theory of oxygen toxicity does not adequately explain the presence of SOD in anaerobic bacteria, or the ability of some aerobes to survive without SOD in an oxygen-rich environment. Bors (1989) has recently proposed that since CuZn-SOD reacts with peroxyl (ROO.) radicals with rate constants comparable to its value for dismutating $O_2^{\cdot-}$, then the function of SOD may be expanded to include the scavenging of other radicals beside $O_2^{\cdot-}$.

The high dismutase activity of Cu^{2+} and its complexes *in vitro* has led some authors, notably Fee (1982), to propose that the biological role of SOD is not dismutation of $O_2^{\cdot-}$, but some other activity not yet determined and that the dismutase activity of the enzyme exists because of the fact that it contains a transition metal. However, Cu^{2+} aq. functions poorly as a dismutation catalyst at physiological pH, compared to SOD. If the catalytic activity of Cu/Zn SODs is fortuitous then other Cu-proteins, e.g. cytochrome *c* oxidase and ceruloplasmin, might be expected to perform the same function. None do so catalytically though it is possible that they do scavenge $O_2^{\cdot-}$ in some non-catalytic fashion (see Goldstein *et al.*, 1979 and Bannister *et al.*, 1980). Also SODs with iron or manganese as the prosthetic group are much more effective catalysts than free Mn or Fe ions at similar concentrations. The true biological function of SOD is almost certainly dismutation: SOD is the only efficient catalyst of this reaction *in vivo* (Czapski & Goldstein, 1988) and many authors have provided evidence for the electrostatic facilitation of superoxide binding to the active sites of SOD. Although SOD is well characterised, the biological function remains controversial—there is plenty of evidence which can be used to support the dismutation role of SOD (e.g. from amino-acid sequence invariance in a Cu/Zn-SOD and molecular genetic studies by Touati (1988) and others. However, there are some features of SOD activity which cannot be explained by the present superoxide theory of toxicity. Until the true nature of the physiological role of SOD in living systems becomes unambiguous and beyond dispute, the precise contribution of this enzyme to food quality cannot be firmly established.

THE BIOLOGICAL DISTRIBUTION OF METALLOFORMS OF SOD

Surveys of SODs from many different organisms has led to some generalisation and classification of the enzymes relating to their biological distributions (Steinman, 1982). Whereas Cu/Zn-SOD is the major enzyme in animals, plants and fungi, the prokaryotes, (i.e. blue-green algae and bacteria) and simple eukaryotes (algae and protozoa) contain Fe-SOD and Mn-SOD, sometimes both enzymes occurring together in the same organism. Research by Martin *et al.* (1986) has provided evidence that the bacterium *Streptococcus mutans* uses the same apoenzyme to form either Fe-SOD or Mn-SOD depending on the metal availability. These workers have commented that 'cambialistic' enzymes (i.e. those capable of making a co-factor substitution) may represent a previously unrecognised family of SODs. Other workers including Stallings *et al.* (1984) and Gregory (1985) have also cited recent reports of cases where the selectivity of dismutase apoproteins for metals is not absolute, at least for enzymes that usually contain iron or manganese. Fe-SOD, found in prokaryotes, is the most frequently occurring form of SOD in obligate anaerobes, although most of these organisms contain no SOD at all. Mn-SOD has been found to be present in eukaryotes, usually localised within the mitochondria (Rabinowitch & Fridovich, 1983). The occurrence of Mn-SOD rather than Cu/Zn-SOD in the mitochondria of eukaryotes is consistent with the endosymbiont theory which suggests that mitochondria (and chloroplasts) evolved from independent organisms which originally invaded the host cell and set up an endosymbiotic relationship with it. Cu/Zn-SOD has been found in the

Table 2. General Pattern of Distribution of SOD

Type of enzyme	*Usual source*
Fe-SOD	Prokaryotes
Mn-SOD	Prokaryotes + eukaryotes (mitochondria + peroxisomes)
Cu/Zn-SOD	Eukaryotes cytosol + chloroplasts + mitochondria)

cytosol of all eukaryotic cells, but is generally absent from prokaryotic organisms (see Table 2).

Although this is still the general pattern of distribution of SODs, exceptions have since been found which break most of these rules. Eukaryotic organisms have been found which lack Cu/Zn-SOD, from primitive eukaryotes such as green algae to higher organisms such as water lilies (Bridges & Salin, 1981), and the fungus, *Pleurotus olearius* (Steinman, 1982). Four bacterial species have been found to contain Cu/Zn-SOD. The Cu/Zn-SOD has been purified from two of these bacterial species and it is known that the majority of amino-acid residues which directly participate in enzyme activity and stability are conserved (Redford *et al.*, 1990). In the case of *Photobacterium leiognathi* the presence of Cu/Zn-SOD has been proposed to be the result of a symbiotic host–gene transfer from the ponyfish (Martin & Fridovich, 1981) rather than independent evolution, but this explanation does not suffice for all the other cases. For example, *Caulobacter crescentis* are free-living, non-symbiotic bacteria; and the more recent evidence from amino-acid sequence data of *P. leiognathi* SOD suggest significant structural differences from the bovine SOD model (Steffens *et al.*, 1983; Steinman, 1987). Redford *et al.* (1990) have recently suggested that as Cu/Zn-SOD of *C. crescentis* is known to be periplasmic, supporting a periplasmic location for *P. leiognathi,* then the structure of Cu/Zn-SOD from the latter bacterium could provide insight into structural differences of enzymes in different intracellular locations. The occurrence of Cu/Zn-SOD in a limited number of prokaryotes plus differences in amino acid sequences from chloroplast and cytosolic eukaryote sources had led to the proposal that genes for Cu/Zn-SOD are contained in prokaryotes and eukaryotic algae and protozoa, although their expression is limited for unknown reasons (Kanematsu & Asada, 1989, 1990).

Fe-SOD has been found in primitive eukaryotes which lack Cu/ZN-SOD, such as *Euglena gracilis* (Kanematsu & Asada, 1979), and although no animal tissues have been found to contain Fe-SOD it has been detected in several families of higher plants although only in chloroplasts (see Almansa *et al.*, 1989). Water lilies, mustard, ginko trees (a primitive fern) (Bridges & Salin, 1981), tomato plants (Kwiatowski *et al.*, 1985) and lemon trees (Sevilla *et al.*, 1984) are plants which contain an Fe-SOD.

The question of why some plants have expressed an iron SOD instead of enhancing the production of an already present chloroplastic Cu/Zn-SOD remains unanswered, although organisms which contain two metalloforms of SODs in chloroplasts may be at an advantage—for example in periods of nutritional stress (see Almansa, *et al.*, 1989). Garcia *et al.* (1981) have stated that the metalloenzyme system of SOD may be a useful tool in plant nutrition for studying interactions between, and imbalances, of micronutrients. Eukaryotic Mn-SOD is not confined to mitochondria, it has been claimed to be present in the cytosol of baboon liver and bakers' yeast (Steinman, 1982). Several reports of Mn-SOD in chloroplast fractions have also been published, for example by Hayakawa *et al.* (1985). Mn-SOD has also been found in peroxisomes of pea plants (Del Rio *et al.*, 1983; Palma *et al.*, 1987; Sandalio *et al.*, 1987) and in glyoxysomes from water-melon cotyledons (Sandalio & Del Rio, 1987).

THE PROTECTIVE ROLE OF SOD IN PLANTS

The SOD content of plants varies greatly with the age and the part of the plant examined (it is present in roots, stems, leaves and seeds). Cu/Zn-SOD in leaves is mainly associated with the chloroplasts although cytosolic isoenzymes may also be present. Oxygen, a product of photosynthesis, is produced in high concentration in the chloroplasts, where the presence of many strong reductants makes the formation of $O_2^{\cdot -}$ unavoidable. Chloroplasts therefore have a higher concentration of $O_2^{\cdot -}$ (approximately 6×10^{-9} M, Asada *et al.*, 1977) than the other parts of chlorophyll-containing cells and SOD is one of several enzymes of the chloroplast's protective mechanisms (see Schoner & Krause, 1990).

The SOD content of seeds increases dramatically with germination and the early stages of growth (Giannopolitis & Ries, 1977*b*;

Matkovics, 1977). In the dry seed of mung bean, one of the SOD isoenzymes (SOD 1) is undetectable and as the seedling grows the level increases; it then becomes the major SOD of the plant (Reddy & Venkaiah, 1984*a*). Bhattacharjee & Choudhuri (1986) have shown that pretreatment of jute seeds with chemicals such as cinnamic acid stimulates the activities of free-radical scavenging enzymes including SOD. This treatment may enhance the storage potential of seeds and render the seedlings tolerant against water deficit stress.

Since physiological stress in plants can result in oxidative injury, the capacity to increase the levels of antioxidants such as SOD in response is of great significance. The role of SOD in plant tissues has been the subject of a number of studies concerning environmental stress (Monk *et al.*, 1989). For example, Burke *et al.* (1985) have demonstrated increased SOD activity in drought-tolerant plants. Air pollution stress may be mediated by toxic oxygen species and may cause induction of SOD activity in plants. For example, the work of Lee & Bennett (1982) has provided evidence for an association between the greater SOD levels in younger leaves of the snap bean plant and lower ozone sensitivity than that of older leaves. Studies on Loblolly Pine and other tree species have shown that SOD activity increases in response to increased $O_2^{\cdot-}$ production by decomposition of ozone and oxidation of sulphur dioxide (see Tandy *et al.*, 1989).

Monk & Davies (1989) recently attributed resistance to a physiological disorder of potatoes (brought about by calcium deficiency) to a higher antioxidant status which featured increased levels of SOD. Monk *et al.* (1987) have suggested that SOD has a role in anoxia-tolerant plants, during and after anoxic stress. These workers proposed that SOD could be an anaerobic polypeptide in flood-/anoxia-tolerant species which survive in waterlogged habitats (since flooding conditions impose oxygen shortages on submerged organs). They point out that the identification of specific physiological adaptations to impart tolerance to flooding (post-anoxic injury) is of importance to agriculture in most parts of the world. On harvest, a high antioxidant status specific to a given cultivar could make a positive contribution to the quality of seeds, fruits and tubers (Lesham *et al.*, 1986).

Response to chilling injury has also been reported to relate to increased levels of SOD activity in photosynthetic systems (Clare *et al.*, 1984; Schoner & Krause, 1990). Gillham & Dodge (1987) have used pea plants to show that although the levels of ascorbate and other antioxidants within chloroplasts display seasonal variation when grown in glasshouse conditions, SOD activities vary little throughout the year. It seems that SOD activity in chloroplasts may be independent of variations in seasonal light intensity in temperate climates. However, both high and low temperatures can cause more severe damage to photosynthetic plants in the light than they do in the dark (Halliwell & Gutteridge, 1989). For example, the susceptibility of a fruit to sunscald, a form of photo-oxidative injury resulting in damage to the skin of the ripe fruit, was found to be related to SOD concentration in the fruit. When the SOD content was low, sunscald tolerance was low, and vice versa, suggesting a protective role for SOD against insolation (heat and high light intensity) damage in green tissues (Rabinowitch *et al.*, 1982). At low temperatures the rate of photosynthesis may be depressed, especially if the plants are well illuminated; enhanced production of oxygen species and chloroplastic injury may result. Resistance to chilling injury has also been reported to relate to increased levels of SOD activity in algae (Clare *et al.*, 1984) and spinach leaves (Schoner & Krause, 1990). Chilling can induce photoinhibition even at moderate light intensities, but Schoner & Krause (1990) have proposed that in cold-acclimated spinach plants increased levels of SOD activity contribute to protection against light at low temperatures; these workers reported an increase of almost 83% in SOD activity levels of plants acclimatised to 1°C over 10 days over control plants which were acclimatised to 18°C.

The distribution of free radicals in maize plants was found to correlate with metabolic activity, thus the young and actively dividing parts of the plants had the highest concentration of radicals (Michalov & Placek, 1982). This was parallelled by findings of higher SOD activity in young leaves than in mature and senescing leaves (Dhindsa *et al.*, 1982; Rabinowitch & Fridovich, 1983) and of high SOD content in other metabolically active parts of plants such as root nodule mitochondria in soybeans (Puppo *et al.*, 1987). As the patterns of SOD isoenzymes change during the life cycle of the plant and also in response to a variety of environmental effects, according to Rabinowitch & Fridovich (1983), these phenomena indicate a delicate genetic control of SOD.

An early survey of the SOD content during apple, banana, avocado and tomato fruit ripening found little difference between pre- and post-climacteric fruit (Baker, 1976). Later studies on tomatoes (Rabinowitch & Sklan, 1980) and peppers and cucumbers (Rabinowitch & Sklan, 1981) found the SOD content to vary with the different stages of ripening in the same way in all three fruits. Rabinowitch & Sklan (1981) found that the level of SOD activity was high in the immature green fruits they examined and that SOD activity passed through a minimum at the mature green stage in cucumbers and the mature green and breaker stages in peppers. SOD activity then increased during subsequent ripening and decreased again at the yellow or red-ripe stages. The authors showed that changes in SOD activity paralleled susceptibility to sunscald, i.e. susceptibility to sunscald was maximal when the SOD content was minimal. They propose that damage may be due to the photochemical generation of superoxide in chlorophyll: chlorophyll degraded (with maturation) and when poorly organised may play a role in photodynamic reactions involving superoxide radicals.

Fruit ripening involves many complex metabolic changes including lipid peroxidation, hydrogen peroxide formation and the breakdown of cell membranes and cell walls. Superoxide is involved in many of these processes, and SOD might therefore have an important role, the nature of which has yet to be determined, in the control of fruit ripening.

Plant polysaccharides have been found to be depolymerised as the result of a system producing superoxide and hydrogen peroxide and their subsequent interaction to produce damaging free radical species. Kon & Schwimmer (1977) showed that SOD and catalase gave complete protection from this system. SOD may also function in the control of lignification of plant tissues (Rabinowitch & Fridovich, 1983). Protection of plants against toxic concentrations of heavy metals (in particular copper) has been suggested as the role of Mn-SOD which together with catalase is more active in peroxisomes of Cu-tolerant than Cu-sensitive pea plants (Palma *et al.*, 1987). SOD may also have a role in response to wounding of tissue. Work on pea plants by Olsen & Cook (1987) has indicated the swift protein turnover of this enzyme. These researchers found evidence for a significant loss and dramatic recovery of activity of SOD, within 1 h, following wounding of etiolated stem tissue; catalase and peroxidase do not seem to follow such a rapid pattern of loss and recovery.

For the food scientist the presence of superoxide dismutase in foods has so far been generally ignored. This is partly due to the fact that superoxide dismutase is a newly discovered enzyme and also because the enzyme seems only to catalyse the removal of a reactive oxygen species: the enzyme does not seem to directly bring about the oxidation or reduction of naturally important substances present in foods, although as we have seen it may do so indirectly. However, because of the ubiquitous nature of superoxide dismutase enzymes, which are known to be present in most eukaryotic organisms with few exceptions, the possible role for the enzyme during redox reactions in food systems cannot be assumed to be negligible or unimportant.

Earlier in this chapter it was pointed out that according to the superoxide theory of O_2 toxicity, the undesirable and potentially toxic effects of oxygen are thought to be in part due to the formation of $O_2^{\cdot-}$ radicals and that SOD is considered to constitute an essential part of a defence mechanism against radical damage. The true biological function of SOD remains controversial, although we already know a great deal about the structure, character and ubiquitous nature of this enzyme. The majority of scientists in the field now believe that SOD plays an essential role in the dismutation of oxygen radicals—there is strong evidence, but no final proof of the biological importance of the dismutation of superoxide and the physiological role of SOD. In order to fully appreciate the debate and controversy surrounding arguments for and against the biochemical function of SOD, it is necessary to review aspects of superoxide chemistry and the widely varying consequences of formation of superoxide in biological systems.

OCCURRENCE OF SOD IN FOODS

Superoxide dismutases (SOD) are ubiquitous in the plants and animals that comprise our food supply. In Table 3 examples are given of reports where SOD has been detected in, or purified from, food sources, particularly foods of plant origin. It may be presumed that superoxide dismutase is present in all fresh food sources and yet its activity and influence upon the preservation and quality of foodstuffs is largely unknown.

Some work has been carried out on mammalian sources of the SOD enzyme, and it has been shown that bovine milk SOD is resistant to pasteurisation at 71·7°C for 15 s (Hicks *et al.*, 1979), and may protect the milk from rancidity by removing superoxide radicals which initiate lipid peroxidation (see Korycka-Dahl & Richardson, 1978). In milk $O_2^{\cdot-}$ is produced by enzymic and non-enzymic pathways, for example, by xanthine oxidase activity, or by reoxidation of photoreduced flavins. Superoxide dismutase may contribute to cheese flavour de-

Table 3. Types of SOD in Human Food Sources

	Reference
1. Cu/Zn-SOD	
Apple	Baker, 1976
Avocado	Baker, 1976
Banana	Baker, 1976
Barley	Jaaska & Jaaska, 1982
Cabbage	Steffens *et al.*, 1986
	Walker *et al.*, 1987
Chicken	Lonnerdal *et al.*, 1979
Cow	Steinman, 1982
Cucumber	Rabinowitch & Sklan, 1981
Jerusalem artichoke	Arron *et al.*, 1976
Lentils	Federico *et al.*, 1985
Maize/corn	Giannopolitis, & Ries, 1977*a*
Milk (bovine)	Hicks *et al.*, 1979
	Hicks, 1980
Mung bean seeds	Reddy & Venkaiah, 1984*a*
Oats	Giannopolitis & Ries, 1977*a*
Peas	Sawada *et al.*, 1972
	Giannopolitis & Ries, 1977*a*
Pepper	Rabinowitch & Sklan, 1981
Pig	Steinman, 1982
Rice	Kanematsu & Asada, 1989, 1990
Salmon	Sutton, *et al.*, 1983
Sheep	Steinman, 1982
Soya beans	Clarkson *et al.*, 1989
Spinach leaves	Asada *et al.*, 1973
	Kitagawa *et al.*, 1986
	Kanematsu & Asada, 1990
Tomato fruit	Baker, 1976
Trout	Sutton, *et al.*, 1983
Wheat	Neumann & Hart, 1986
Wheatgerm	Beauchamp & Fridovich, 1973
Yeast	Weser *et al.*, 1972
	Weisiger & Fridovich, 1973
	Steinman, 1980
2. Mn-SOD	
Chicken	Lonnerdal *et al.*, 1979
Chicken liver	Weisiger & Fridovich, 1973
Jerusalem artichoke	Asada *et al.*, 1977
Maize/corn	Giannopolitis & Ries, 1977*a*
	Baum & Scandalios, 1981
Mung beans	Reddy & Venkaiah, 1984*b*
Spinach leaves	Hayakawa *et al.*, 1985
Wheat	Neumann & Hart, 1986
Wheatgerm	Beauchamp & Fridovich, 1973
Yeast	Ravindranath & Fridovich, 1975
3. Fe-SOD	
Spinach leaves	Kanematsu & Asada, 1990

velopment, by controlling the concentration of superoxide (Schwimmer, 1981).

Loss of the red colour in meat may occur through the following reaction:

$$\underset{\text{(red)}}{\text{Oxymyoglobin}} \underset{O_2^{\overline{\cdot}}}{\rightleftharpoons} \underset{\text{(brown)}}{\text{metmyoglobin}}$$

Autoxidation of myoglobin and haemoglobin releases $O_2^{\overline{\cdot}}$ which, in turn, causes the oxidation of these pigments to the corresponding met states (Misra & Fridovich, 1972; Lynch *et al.*, 1976). However, superoxide dismutase may have a role in suppressing the autoxidation (see Rabinowitch & Fridovich, 1983). In spite of the high enzymic activity of SOD in muscle tissue the $O_2^{\overline{\cdot}}$ may have time to initiate the chain reactions of fatty acid autoxidation that lead to rancidity of the fatty parts of the meat.

Free radicals, e.g. $O_2^{\overline{\cdot}}$ which are potential precursors of off-flavour and off-colour compounds can be formed under some circumstances in the oxidatic reaction catalysed by peroxidase. Peroxidase activity is generally treated as deleterious to quality of plant foodstuffs and SOD may have a role in preventing loss of quality due to peroxidase activity. The role of SOD in the preservation of quality of harvested fruit and vegetables is unclear.

Lipid peroxidation, initiated either by a xanthine oxidase system producing $O_2^{\overline{\cdot}}$ or by lipoxygenase has been claimed to be inhibited by SOD (Kellogg & Fridovich, 1975; Richter *et al.*, 1975; Dhindsa *et al.*, 1981). The autoxidation of anchovy oil was found to be delayed by the presence of SOD (Michelson & Monod, 1975).

The biochemical role of the enzyme is still the subject of debate and therefore it is not surprising that there is little information on the importance and interactions of SOD in food. Although a brief comparison of SOD and peroxidase levels in fruit and vegetables has been carried out (Matkovics *et al.*, 1981), the interactive effects of SOD and peroxidase during the spoilage of fruits and vegetables is unknown. It remains to be determined whether SOD plays a protective or a reinforcing role in spoilage attributable to peroxidases but a product of SOD activity, H_2O_2, is required for peroxidase activity.

Peroxidase can react with $O_2^{\overline{\cdot}}$ to form the relatively inactive form of the enzyme, compound III or oxyperoxidase. Compound III slowly decomposes to release $O_2^{\overline{\cdot}}$. It appears that superoxide radicals could play a dual role in the decomposition of a harvested plant, on the one hand aiding the production of undesirable compounds from the action of peroxidase, and on the other hand inactivating the peroxidase itself. The role of SOD in preventing or aiding post-harvest plant degradation with respect to peroxidase activity is unclear.

THERMOSTABILITY OF SOD IN FOODS

Little information is available on the heat stability of SODs or their fate during thermal processing although bovine Cu/Zn-SOD is considered to be among the most heat stable of globular proteins, including those from extreme thermophiles (Tainer *et al.*, 1982). Indeed, Gartner *et al.* (1984) devised a new method for the isolation of Cu/Zn-SODs from erythrocytes which exploits the high thermal stability of mammalian Cu/Zn SOD. Yeast SOD is known to be heat stable with more than 10% of the original activity remaining after 10 min at 100°C. Cu/Zn-SOD in bovine milk can withstand pasteurisation at 63°C (Korycka-Dahl *et al.*, 1979), but at temperatures higher than 70°C the degree of enzyme inactivation increases rapidly with temperature. Raw milk undergoing HTST pasteurisation at 80°C loses approximately half of its SOD activity (Hicks *et al.*, 1979). In crude extracts of corn and oats the SOD activity has been found to be more resistant to boiling than peroxidase activity (Giannapolitis & Ries, 1977*a*). The SOD activities of crude extracts of corn and peas (Giannapolitis & Ries, 1977*a*) and cabbage (Walker, 1987) are known to be more thermostable than that of partially purified enzymes, indicating that cellular components may protect the protein from thermal denaturation.

The SOD activities found in cabbage extracts and sprout extracts are stable to 45°C, and at higher temperatures the cabbage SOD is the more heat sensitive; heating at 70°C for 2 min destroys virtually all the SOD activity in crude extract of cabbage (Walker *et al.*, 1987). In contrast with cabbage, one of the two main isoenzymes in Brussels sprout crude extract was found to be highly heat stable (demonstrated by the activity remaining after 30 min at 80°C; Walker & Robinson, 1991). The differential stabilities of the isoenzymes may account for the non-linear character of heat inactivation plots for SOD in extracts of *Brassica* species.

At present there is little evidence to support the possibility of SOD activity regeneration following thermal denaturation, as is the case with peroxidases (Moulding *et al.*, 1988) and it seems unlikely that SOD will be active in the majority of foods that have been heat-treated for food processing; however, Arnold & Lepock (1982) and Lingnert *et al.* (1989) have claimed that yeast Cu/Zn-SOD exhibits partially reversible heat denaturation.

ASSAYS OF SOD ACTIVITY

In general, the assay systems are based on coupled enzymic test systems. Such assay systems are not always exactly reproducible because of interferences between components. Exact quantitative measurements require the use of more direct assay techniques which are not usually convenient for routine laboratory purposes. The more commonly used indirect assays depend on a system to generate the $O_2^{\cdot-}$ radicals plus a detection system which involves an oxygen radical scavenging indicator. The more commonly used indirect assays are the xanthine oxidase/cytochrome *c* assay (McCord & Fridovich, 1969) and the nitroblue tetrazolium (NBT) assay in which $O_2^{\cdot-}$ is generated either by the photoreduction of riboflavin or by the xanthine oxidase/xanthine system (Beauchamp & Fridovich, 1971). The possible sources of interference in either of these assays must be considered prior to use; some of them are described by Beyer & Fridovich (1987).

STRUCTURE OF SOD

Cu/Zn-SOD

The Cu/Zn-SOD enzymes are mainly eukaryotic cytoplasmic or chloroplast enzymes (mol. wt 32 000). Each molecule is a homodimer containing about 153 amino-acids, one disulphide bridge and one Cu^{2+} and one Zn^{2+} ion per subunit. Using X-ray crystallography the 3-D structure of bovine Cu/Zn SOD (Fig. 1) has been refined to a resolution of 2A (Tainer *et al.*, 1982). Each subunit is composed of eight antiparallel beta-strands (containing 46% of the amino-acid residues) that form a flattened cylinder or beta-barrel, plus three external loops which contain a further 48% of the amino-acid residues.

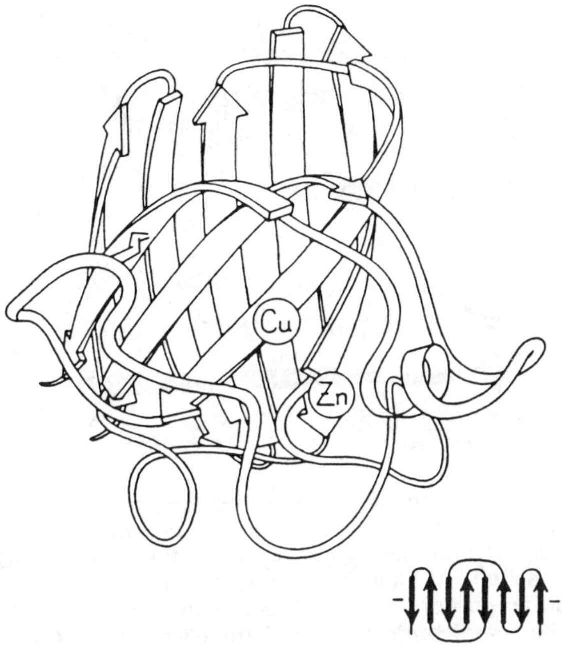

Fig. 1. Schematic backbone drawing of the SOD subunit. The β strands are shown as arrows and the disulfide as a zig-zag. Two long loops extend forward from pairs of β strands to form the upper and lower sides of the active-site channel. The Cu and Zn lie at the bottom of that channel, with the Cu accessible to solvent from the viewing direction. The small diagram (lower right) illustrates the Greek key topology of the β barrel, spread open and viewed from the outside. From Tainer *et al.*, 1982, by courtesy of Academic Press Inc (London) Ltd, London.

The antiparallel bands form the staves of the beta-barrel, with two strands connected across the top and another two connected across the bottom of the barrel. The remaining connections occur between adjacent β-strands. The Cu^{2+} ion is ligated at the surface of the barrel, at the base of a channel enclosed by the two largest of the three external loops. Both of these loops are functionally important; one of them contains residues which are important in precollision electrostatic guidance of the substrate (Getzoff *et al.*, 1983) and is called the 'electrostatic channel loop'. The second loop is subdivided into a Zn^{2+} binding region or subloop, plus a region which contributes to dimer contact and forms the second side of the active-site channel. The latter

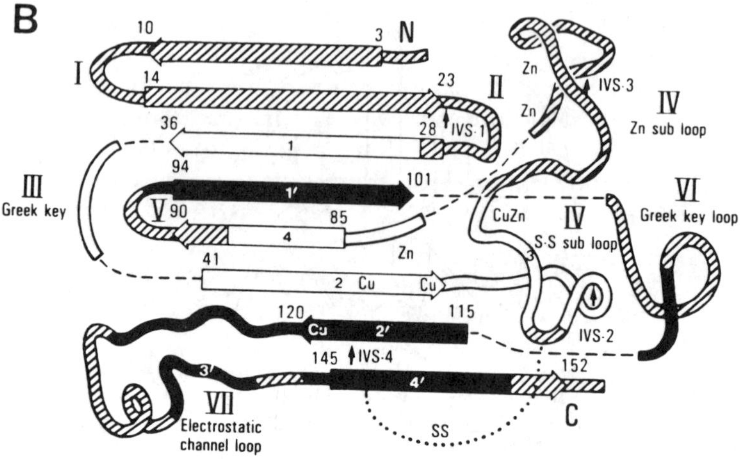

Fig. 2. Schematic diagram of β strand and loop secondary structure of SOD labelled to show loops and symmetrical regions. The β strands are shown as arrows and numbered (according to the human SOD sequence of 153 amino-acids) at their N- and C-terminal residues. Loops are labelled by roman numerals. Loop IV is subdivided into the S–S and Zn subloops at the metal bridging ligand His-63 (labelled CuZn). Twofold symmetry regions are coded in white (with black labels 1–4) and black (with white labels 1′–4′), the asymmetic regions are striped. Getzoff *et al.*, 1989. Reprinted by permission from J. Mol. Biol., vol 160, © Academic Press Inc. (London) Ltd.

region or subloop is stabilised by a disulphide bridge to the β-barrel and is therefore termed the disulphide subloop. Figure 2 shows the beta-barrel and loops spread out flat.

The great stability of Cu/Zn-SODs is considered to be due to the close packing in the hydrophobic interfaces between the dimeric subunits and between the two halves of the barrel. Mammalian Cu/Zn-SODs are highly thermostable, resistant to protease attack and tolerant to organic solvents including chloroform and ethanol. They also resist denaturation by solvents such as sodium dodecyl sulphate (SDS) and urea. Bovine SOD has been shown to remain active in 4% SDS (Forman & Fridovich, 1973) and 8 M urea (Malinowski & Fridovich, 1979).

The complete amino-acid sequences of Cu/Zn-SOD from a variety of organisms have been determined. Recently Getzoff *et al.* (1989) aligned amino-acid sequences of Cu/Zn-SOD from 15 species including animals, plants, fruit fly, fungi and photobacteria. These workers demonstrated the presence of 23 invariant amino-acids. They found that these residues could be divided into three structure–function categories: active site (15 residues), dimer-interface (4 residues) and β-barrel (4 residues). Glycine was the most commonly occurring invariant residue, it was presumed that no other amino-acid can satisfy the spatial or conformational constraints at a given position in the 3-D structure. Invariance of amino-acids at the active site of the subunit has particular significance in view of the ongoing debate about the true biological function of SOD. There is complete invariance of the metal liganding residues and the importance of their spatial configuration is indicated by the number of invariant residues which orientate them.

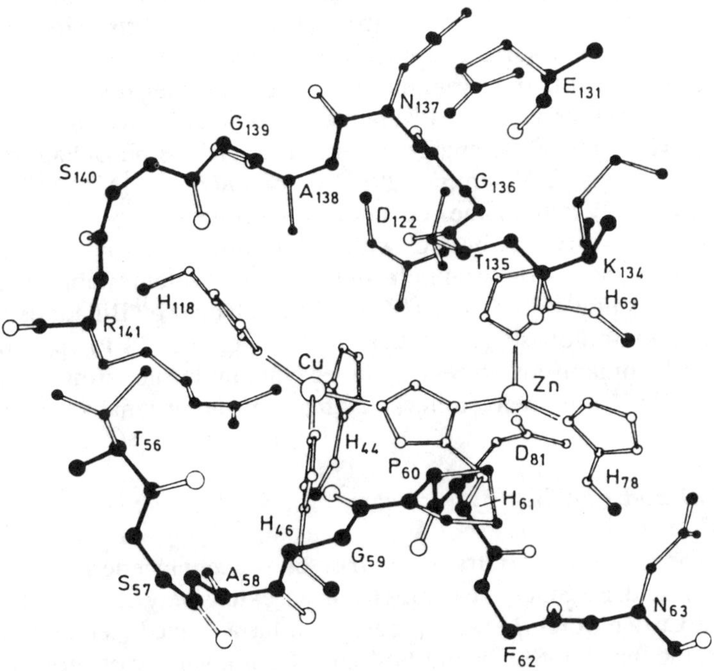

Fig. 3. Drawing of the active-site channel of bovine erythrocyte Cu_2Zn_2 superoxide dismutase, as viewed from the solvent. The copper is liganded by His 44, 46, 61 and 118. The Zn is liganded by His 61, 69, 78 and Asp 81. The geometry of its ligands is tetrahedral. The imidazolate ring of His 61 is the bridging ligand between Cu and Zn. Tainer *et al.*, 1983. Reprinted by permission from Nature Vol 306, © 1983 Macmillan Magazines Ltd.

The Cu^{2+} ion is ligated to four histidine residues (bovine SOD nos 44, 46, 61 and 118) in the amino acid sequence in an uneven tetrahedral distortion from a square plane (Fig. 3). The Zn^{2+} ion has one aspartic acid (bovine SOD no. 81) and three histidine ligands, (bovine SOD nos 61, 69 and 78), one of which (bovine SOD His 61) it shares with the Cu^{2+}, in a tetrahedral formation which is distorted towards a trigonal pyramid. The common histidine ligand is referred to as the imidazole bridge. Histidine 61, which is involved in the positioning of both metals, may also be involved in the supply of protons required for dismutation.

At the active site, Cu^{2+} is at the bottom of the 'channel', and is partially exposed to the solvent. An arginine residue (bovine SOD Arg 141, Fig. 1) at the outer end of the 'channel' appears to assist in guiding the substrate, $O_2^{\cdot-}$, down to the Cu^{2+} at the active site. Whilst most of the subunit surface is negatively charged and repels $O_2^{\cdot-}$, the positive charge of the arginine stabilises the negatively charged substrate as well as assisting in the electrostatic guidance. Solvents or reagents which affect the charge on arginine, such as phosphate buffer (Mota de Freitas & Valentine, 1984) or butanedione (Malinowski & Fridovich, 1979) cause some reduction of enzyme activity.

The Zn^{2+} is wholly buried within the protein structure and is thought to play a structural role only, aiding protein stability. In a calorimetric study of human Cu/Zn-SOD conducted by Biliaderis *et al.* (1987), these authors suggested that the binding of Zn^{2+} to apoprotein initiated the organising process of the enzyme molecule, although both Cu^{2+} and Zn^{2+} are involved in the stability of the enzyme structure.

Mn-SOD and Fe-SOD

The secondary and tertiary structures of the phylogenetically older Mn-SOD and Fe-SOD have not been as extensively investigated as those of Cu/Zn-SOD. However, results so far achieved (see Steinman, 1982) show that Mn-SODs and Fe-SODs from a variety of prokaryotic sources are structurally similar to each other. Both have extensive areas of α-helix in the tertiary structure, but have little β-sheet. The folded proteins are roughly rectangular in shape, with the metal binding site close to, or even part of, the subunit interface (Stallings *et al.*, 1984). For both Mn-SOD and Fe-SOD there is usually one metal binding site per subunit.

Mn-SOD

At pH 7·0 the rate constants for Mn-SOD are similar to Cu/Zn-SOD. However, unlike the Cu/Zn enzyme at alkaline pH the Mn-SOD rate constant decreases. Mn-SOD is also more susceptible to denaturation by heat or by chemicals (such as detergents) than the Cu/Zn-SOD, but like Fe-SOD resistant to attack by cyanide. No crystal structure has been reported for Mn-SOD, but the circular dichroism (Cd)-spectra and amino-acid sequences are similar for all Mn-SODs studied to date. Due to the relatively low binding constants compared to Cu/Zn-SODs, the metal contents vary between 1 and 2 g atoms per dimer. The molecular weights reported for Mn-SODs vary between 39 500 and 94 000. The subunit molecular weights are approximately 20 000. Mn-SOD probably exists as a dimer and as a tetramer.

Fe-SOD

Fe-SOD also exists as a dimer or tetramer. Each subunit seems to have a functional metal ion, with a molecular weight around 22 000. In common with Mn-SOD the ground state of the metal atoms of iron enzyme is the tervalent state. As with Cu/Zn-SODs, hydrogen peroxide destroys Fe-SOD activity (whereas Mn-SOD is unaffected), and in common with Mn-SOD (unlike Cu/Zn-SOD), Fe-SOD is not affected by cyanide. X-ray crystallography of Fe-SOD from *Pseudomonas ovalis* and *Escherichia coli* reveals highly helical foldings. The monomer contains six major helical segments and three strands of antiparallel β-sheet. The iron is ligated to four protein side chains and the active centre is exposed to the solvent (see Gartner & Weser, 1986).

SEQUENCE HOMOLOGY OF SOD

Great sequence homologies are notable between Mn-SODs and Fe-SODs from different sources, including Mn-SOD from eukaryotic mitochondria. However, these are different from those of Cu/Zn-SODs. The lack of sequence homology between Mn and Fe and Cu/Zn-SOD and the differences between the 3-D structure of Cu/Zn-SOD and the secondary structure predicted for known sequences of Mn-SOD suggests that Cu/Zn and Mn/Fe families of SODs have evolved independently and do not share a common ancestral protein (Steinman, 1982).

Extensive sequence homology, between such disparate sources of Cu/Zn-SOD as human and bovine erythrocytes, cabbage and yeast, indicates a high degree of evolutionary conservation of the secondary structure of the enzyme (see Steffens *et al.*, 1986; Getzoff *et al.*, 1989).

MECHANISM OF SUPEROXIDE DISMUTATION

In all SODs the transition metal is vital for catalytic activity. In Cu/Zn-SOD the Cu^{2+} ion is responsible for catalysis and the role of zinc is probably structural. Removal of Cu^{2+} from Cu/Zn-SOD causes loss of enzymic activity which is restored only by its replacement. Zn^{2+} can be replaced in Cu/Zn-SOD by Ag^{2+}, Co^{2+} or Cd^{2+} with no loss of enzyme activity (Forman & Fridovich, 1973).

Since $O_2^{\cdot-}$ is unstable with respect to O_2 and H_2O_2 it spontaneously dismutates to these products. $O_2^{\cdot-}$ is the conjugate base of a weak acid, the hydroperoxyl radical ($OOH^{\cdot}$) which has a pK_a value of 4·8 and so there are three possible dismutation reactions:

$$OOH^{\cdot} + OOH^{\cdot} \rightarrow H_2O_2 + O_2 \qquad K_2 = 8 \times 10^5\ \text{M}^{-1}\ \text{s}^{-1} \qquad (2)$$

$$OOH^{\cdot} + O_2^{\cdot-} + H^+ \rightarrow H_2O_2 + O_2 \qquad K_2 = 8 \times 10^7\ \text{M}^{-1}\ \text{s}^{-1} \qquad (3)$$

$$O_2^{\cdot-} + O_2^{\cdot-} + 2H^+ \rightarrow H_2O_2 + O_2 \qquad K_2 < 0{\cdot}3\ \text{M}^{-1}\ \text{s}^{-1} \qquad (4)$$

Dismutation is thus most rapid at acidic pH values (pH 4·8) and becomes slower with increasing pH values as the concentration of $OOH^{\cdot}$ decreases, for example at pH 7 the overall rate constant is approximately $5 \times 10^5\ \text{M}^{-1}\ \text{s}^{-1}$.

The reaction of one $O_2^{\cdot-}$ with another is probably hindered by electrostatic repulsion. At physiological pH superoxide exists predominantly as $O_2^{\cdot-}$ (rather than the perhydroxyl radical $OOH^{\cdot}$), and so catalysis of dismutation must overcome electrostatic repulsion between the negatively charged radicals and consequent slow rate of reaction

$$O_2^{\cdot-} + O_2^{\cdot-} + 2H^+ \rightarrow H_2O_2 + O_2$$

At the active site of the enzyme, bound copper acts as an electron carrier between two superoxide radicals in the two successive half-reactions, the two radicals never actually meet. A cyclical change in SOD brought about by $O_2^{\cdot-}$, in the first half-reaction (eqn (5)) Cu^{2+} is reduced to Cu^+, and $O_2^{\cdot-}$ oxidised to molecular oxygen. In the second half-reaction (eqn (6)) Cu^+ is oxidised to Cu^{2+} and a second $O_2^{\cdot-}$ is simultaneously reduced and protonated to form hydrogen peroxide. A detailed explanation of the catalytic mechanism can be found in the review by Gartner & Weser (1986).

$$E - Cu^{2+} + O_2^{\cdot-} \rightarrow E - Cu^+ + O_2 \quad (5)$$

$$E - Cu^+ + O_2^{\cdot-} + 2H^+ \rightarrow E - Cu^{2+} + H_2O_2 \quad (6)$$

(E = one SOD subunit)

The two half-reactions have rate constants identical to each other and hence to the overall rate constant of $2 \times 10^9\ \mathrm{M}^{-1}\ \mathrm{s}^{-1}$. The large rate constant for catalysis, and the dependence of the reaction rate on viscosity (Rotilio *et al.*, 1972) suggest that the reaction rate is diffusion limited even at alkaline pH values. Thus it is one of the fastest enzymic reactions known.

SOD is powerful enough to increase the rate of reaction by several orders of magnitude at physiological pH. The catalytic activity of Cu/Zn-SOD has been shown by direct assay to be invariant over the pH range 5–9·5 (Klug *et al.*, 1972; Rotilio *et al.*, 1972) and to decrease rapidly above pH 10. Above pH 10 positive charges are lost from the lysine residues on the enzyme surface. The positive charge on the surface, in combination with the electrostatic repulsion by negatively charged areas on the surface, serves to actively guide $O_2^{\cdot-}$ anions to the active site channel. Since conserved residues in sequence alignment reflect biological selection for function, together with evidence of SOD protection against $O_2^{\cdot-}$ mediated stress, amino acid sequence conservation both at the active site of SOD and in the electrostatic channel loop indicates that the primary biological function of SOD is to remove $O_2^{\cdot-}$.

FORMATION OF OXYGEN RADICALS

Molecular or triplet O_2 has two unpaired electrons, each located in a different π^* orbital. These two electrons have the parallel spins and so if O_2 attempts to oxidise another atom or molecule by accepting a pair of electrons from it, both new electrons must be of antiparallel spin so as to fit the vacant spaces in the π^* orbitals. This imposes a spin restriction on oxidation which tends to make O_2 accept electrons one at a time (to allow time for spin inversions between collisional events). This makes O_2 poorly reactive towards non-radical species, and thus permits the existence of many organic substances in aerobic environments.

Transition metals are found at the active sites of many oxidases and oxygenase enzymes, including xanthine oxidase and SOD: their ability to accept and donate single electrons can overcome the spin restriction. If a single electron is added to the ground-state O_2 molecule (univalent reduction), it must enter one of the π^* orbitals, producing the superoxide radical:

$$\pi^*2p(\uparrow)(\uparrow) \xrightarrow{\bar{e}} (\uparrow\downarrow)(\uparrow)$$

ground state O_2 superoxide $O_2^{\cdot-}$

In biological materials the univalent pathway for the reduction of O_2 is a commonplace biochemical event; this pathway being the most facile. Some examples of biochemical systems which produce $O_2^{\cdot-}$ are shown in Table 4.

Although $O_2^{\cdot-}$ is produced in all aerobically respiring cells, quantitative measurements are difficult because of its intrinsic instability and the ubiquity of SOD which catalyses its dismutation. In respiring cells most of the molecular O_2 consumed is tetravalently reduced to water by the cytochrome oxidase and other water-producing oxidases without release of the intermediates such as $O_2^{\cdot-}$ and H_2O_2. However,

Table 4. Examples of Biochemical Systems which may Produce $O_2^{\cdot-}$ *in vivo*

Sources of $O_2^{\cdot-}$	*Examples*
Small molecules	Hydroquinones, leucoflavins, adrenalin, thiols (by autoxidation).
Enzymes and proteins	Xanthine oxidase (degradation of purines to nitrogenous excretory products), aldehyde oxidase, peroxidase (oxidative reaction), haemoglobin, myoglobin. Iron-sulphur proteins.
Cell organelles	Chloroplasts, mitochondria, peroxisomes, glyoxysomes.
Whole cells	Polymorphonuclear leukocytes (during the oxygen burst which leads to the destruction of invading bacteria and viruses).

Information from Fee, 1982 used by permission of Pergamon Press.

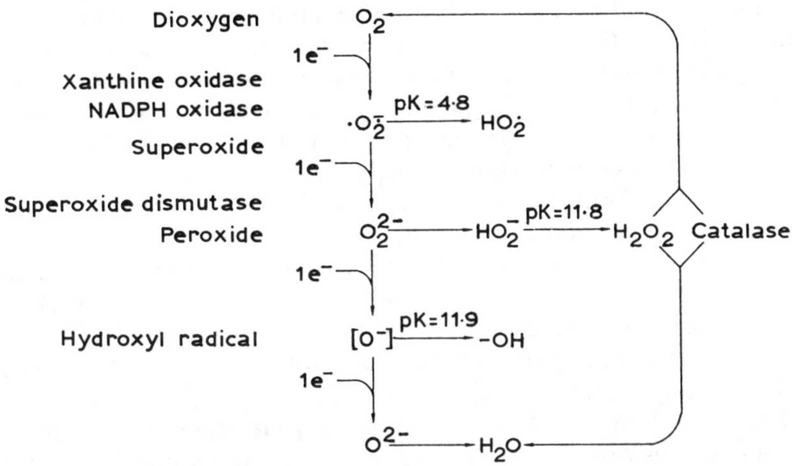

Fig. 4. Proposed mechanism for the step-wise reduction of dioxygen, including the enzymes, which are involved in the removal of reactive oxygen species. From Gartner & Weser, 1986, by courtesy of Springer-Verlag-Berlin.

Table 5. Reduction Products of Molecular Oxygen

No. of ē	*Products*	*Stability*	*Reactivity*	*Sources*
1	$O_2^{\cdot-}$	Unstable	Fair oxidant, Good reductant, Free radical	Autoxidations Some oxidases Leakage from electron transport
2	H_2O_2	Stable	Good oxidant	Many oxidases
3	$OH^{\cdot}(+H_2O_2)$	Very unstable	Powerful oxidant	Haber-Weiss reaction (metal catalysed)
4	H_2O	Stable	Non-toxic	Cytochrome oxidase and other Cu-oxidases

From McCord, 1979.

Britton *et al.* (1978) estimated that $O_2^{\cdot-}$ production accounted for as much as 17% of the O_2 uptake in the bacterium *Streptococcus faecalis*. During the course of normal mammalian metabolism, approximately 98% of molecular O_2 undergoes complete oxidation: Cadenas (1989) has more recently cited an upper limit value of about 5% for the fraction of total O_2 reduced only to $O_2^{\cdot-}$.

Once $O_2^{\cdot-}$ is formed within the system, it can undergo a variety of chemical and metabolic reactions yielding other reactive oxygen species (Fig. 4): for example, it may become protonated to form the more powerful reducing agent, the hydroperoxy radical ($OOH^{\cdot}$). Addition of further electrons may form hydrogen peroxide (H_2O_2) and the hydroxyl radical ($OH^{\cdot}$, one of the most reactive species known) and ultimately, water.

The production of $OH^{\cdot}$ may result in the generation of other radical species such as organic peroxyl radicals ($ROO^{\cdot}$), because $OH^{\cdot}$ reacts with cellular components to produce organic radicals ($R^{\cdot}$) which in turn react with O_2.

CHEMICAL REACTIVITY OF SUPEROXIDE

In aqueous solution, superoxide is a moderately strong reductant and in comparison with other reduced products of molecular oxygen it is a weak oxidant (see Table 5). It will often participate in rapid, one-electron, reductions, but will only oxidise strong reducing agents such as ascorbic acid, or α-tocopherol. Although $O_2^{\cdot-}$ may seem unreactive in comparison with other free radicals, its relative stability may contribute to its toxicity because it can diffuse and strike specific targets.

As superoxide radicals are poorly reactive towards organic molecules in aqueous solutions (such as biological systems), at least as an oxidant (Larson, 1988), $O_2^{\cdot-}$ may not be reactive enough to cause cell damage in living systems. However, biochemical systems generate $O_2^{\cdot-}$ in all cells utilising oxygen (see Table 1) and fluxes of $O_2^{\cdot-}$ generated *in vitro* have been shown to damage cellular components in several instances (Fridovich, 1979). In many of these actions $O_2^{\cdot-}$ is considered to be the precursor of the more reactive attacking agent ($OH^{\cdot}$) (Table 5). In post-harvest and post-mortem foods compartmentalisation of various substrates and enzymes is gradually lost and under these conditions chemical reactive species may be released or accumulate in

Table 6. Examples of Damage Caused by $O_2^{\cdot-}$ and the More Reactive Radicals Derived from it

1. Induction of lipid peroxidation	(Kellogg & Fridovich, 1975; Dhindsa *et al.*, 1981)
2. Depolymerisation of polysaccharides	(Kon & Schwimmer, 1977)
3. Killing of bacteria and inactivation of viruses	(Michelson, 1977)
4. Hydroxylation of aromatic compounds	(Aruoma *et al.*, 1989)
5. Bleaching of pigments	(Halliwell, 1987)
6. Calcium protein (enzyme) degradation	(Davies, 1986)

aqueous and non-aqueous environments, thus damage caused by the production of free radicals may be extensive.

Table 6 shows some examples of damage to biological material caused by O_2 and its more reactive species.

THE ROLE OF SUPEROXIDE IN PRODUCTION OF A MORE REACTIVE SPECIES

Superoxide may contribute to oxygen toxicity within an aqueous environment in either living cells or foods through its role as the precursor of a more reactive species, i.e. a reduced form of superoxide, the hydroxyl radical. Thus formation of $OH^{\cdot}$ can usually be prevented by SOD.

Equations (7) and (8) describe the possible mechanism for the production of the hydroxyl radicals in the presence of $O_2^{\cdot-}$, H_2O_2 and 'free' transition metal ions (i.e. not protein-bound). Fridovich (1979) considers that the greatest biological danger posed by $O_2^{\cdot-}$ and H_2O_2 is not due to their individual reactivities, but to their cooperation in the production of $OH^{\cdot}$ in the presence of transition metal ions. Thus oxygen toxicity is considered, at least partially, to be due to synergism between $O_2^{\cdot-}$ and H_2O_2.

It may be the case that in undamaged living cells, control is maintained over availability of intracellular metal ions and that the reaction between peroxide and metal ions is not allowed to occur unnecessarily. In both living cells and hence foods the most abundant transition metals are iron and copper. Contamination by metals, especially iron and copper has long been known to accelerate the deterioration of foods (Willson, 1987). In foods either of these elements and possibly traces of other transition metals may react with $O_2^{\overline{\cdot}}$ and H_2O_2 to produce more reactive species.

$$O_2^{\overline{\cdot}} + M^{n+} \rightleftharpoons O_2 + M^{(n-1)+} \quad (7)$$

Superoxide acting as a metal ion reductant

$$M^{(n-1)+} + H_2O_2 \rightleftharpoons M^{n+} + OH^{\cdot} + OH^{-} \quad (8)$$

Fenton reaction

The net reaction is sometimes referred to as a metal-catalysed Haber–Weiss reaction.

$$O_2^{\overline{\cdot}} + H_2O_2 \xrightleftharpoons{\text{Fe-salt catalyst}} O_2 + OH^{\cdot} + OH^{-} \quad (9)$$

In living cells, Fee and others have argued that $O_2^{\overline{\cdot}}$ can be replaced, in its role as generator of $OH^{\cdot}$ in the presence of H_2O_2, by other compounds and they claim that it is unlikely that the O_2 makes a significant contribution to oxygen toxicity (see Halliwell, 1984). Good reducing agents such as glutathione and ascorbate may be present at greater steady state concentrations than free $O_2^{\overline{\cdot}}$; so the role of $O_2^{\overline{\cdot}}$ in eqn (7) may be replaced by other metal ion reductants within the cell. However, under these circumstances ascorbate and glutathione could be acting as pro-oxidants rather than antioxidants. Studies which have been reviewed by Fee (1980) show that ascorbic acid levels comparable to or below those found in tissues can effectively out-compete $O_2^{\overline{\cdot}}$ being formed by the action of the enzyme xanthine oxidase. However, since ascorbic acid and other metal ion reductants are themselves rapidly oxidised by $O_2^{\overline{\cdot}}$ (and by $OH^{\cdot}$), $OH^{\cdot}$ production may eventually become $O_2^{\overline{\cdot}}$-dependent. The in-vivo situation probably depends on the concentration of ascorbic acid (Halliwell & Gutteridge, 1989).

However, harvested foods are non-living systems and therefore in foods there is only a limited ability to generate the natural reducing agents such as ascorbate and glutathione. Thus the alternative metal ion reducing agents (e.g. ascorbate) would rapidly be consumed by such a system, whereas it is reasonable to assume that production of intracellular $O_2^{\overline{\cdot}}$ is continuous and would allow the production of $OH^{\cdot}$ to continue. Nevertheless, the reduced form of the transition metals may be generated by oxidation of ascorbate and glutathione and like substances, or by the oxidation of superoxide radical; and in the presence of accumulating amounts of hydrogen peroxide they would be responsible for the production of hydroxyl radicals (eqn (8)). This factor may also be particularly significant in processed foods in which the amounts of active SOD and catalase may decrease.

Gutteridge & Quinlan (1986) have found evidence for an alternative mechanism for the production of $OH^{\cdot}$ in foods by the addition of carminic acid which is a major component of the cochineal red food colouring. In the scheme, $O_2^{\cdot -}$ plays no part in the Fenton reaction (eqn (8)) except as a precursor of H_2O_2.

The radical $OOH^{\cdot}$ is considered to be a more reactive species than $O_2^{\cdot -}$ in the cellular environment (see Table 5), but it comprises only a tiny proportion of the physiological O_2 concentration in many living systems and is more likely to be present within systems in organelles where the pH is low (see Halliwell, 1984). However, in food systems where the physiological pH is low, e.g. citrus and other fruits, reactions involving $OOH^{\cdot}$ may be of greater significance.

THE ROLE OF OXYGEN RADICALS IN LIPID PEROXIDATION

It has been known for a long time that oxygen free radicals and trace metals are involved in the oxidative process in fats which results in the development of a rancid or 'off' flavour. This knowledge is now more widely known because of new medical research on free radical chemistry and more awareness of the living cell's antioxidant protective mechanisms. The oxidation of unsaturated fatty acids is perhaps the most significant factor affecting the stability of nutritional value of many foods (Chan, 1987) and is of increasing significance in medical research because of the current preference for polyunsaturated fatty acids (PUFA) in the diet. Hammer & Wills (1978) and others have provided evidence of a dietary regulation of membrane phospholipid composition in animal studies. PUFAs are particularly susceptible to free radical attack and lipid peroxidation (a free-radical-mediated process that affects PUFA) in biomembranes which leads to loss of membrane function. Lipid peroxidation is a process which involves initiation, propagation and/or termination phases, each of which is influenced by a wide range of factors including the structure of the lipid, the presence of antioxidants and the mechanism of initiation

(Aust, 1985). Peroxidation can be initiated either by lipoxygenases which catalyse the formation of hydroperoxides, or by other substances which give rise to free radicals. Many chemicals and enzyme systems within living organisms and foods can reduce iron (either directly or indirectly) and so initiate the peroxidation by the formation of free radicals.

Usually, living cells are protected against damage due to iron because the ions are protein-bound and compartmentalised. However, if the cell structure is damaged, for example, when tissue is injured by infection, mechanical damage or action of toxin, cellular organisation becomes disrupted. Within a processed food, treatments and the addition of other ingredients compound the problems of exposure to trace elements created by the cellular disruption. When the cell structure is damaged in such a way that the Fe^{2+} is decompartmentalised access to organelles allows rapid and extensive peroxidation (Slater *et al.*, 1987). Free Fe^{2+} can stimulate free radical reactions in various ways:

(i) by production of $O_2^{\cdot-}$ through catalysis of a redox shuttle in which electrons pass from NAD(P)H to oxygen:

$$Fe^{2+}\text{-chelate} + O_2 \rightleftharpoons Fe^{3+}\text{-chelate} + O_2^{\cdot-} \qquad (10)$$

$$\text{NADPH} \rightleftharpoons \text{Flavoprotein} \xrightleftharpoons{Fe^{3+}\text{-chelate}} Fe^{2+}\text{-chelate} \qquad (11)$$

(ii) by production of $OH^{\cdot}$ from $O_2^{\cdot-}$ and H_2O_2 via the Fenton reaction (eqns (8) and (9));

(iii) by production of reactive lipid alkoxy species by reacting with lipid hydroperoxide (eqn (16)).

The Initiation Phase of Lipid Peroxidation

Whilst superoxide is a relatively poor oxidant species, $OH^{\cdot}$ radicals which may be derived from $O_2^{\cdot-}$, are strongly oxidising and may initiate further damage in living tissues and foods by abstracting an hydrogen atom or by adding to an ethylenic double bond of a polyunsaturated fatty acid. The initiation of the process of peroxidation of the lipid thus occurs through the production of another radical species, the carbon-centred radical ($L^{\cdot}$),

$$LH + OH^{\cdot} \rightleftharpoons L^{\cdot} + H_2O \text{ or } LHOH^{\cdot} \qquad (12)$$

Other free radical species may also participate in the abstraction of hydrogen atoms of unsaturated fatty acids:

$$LH + X^{\cdot} \rightleftharpoons L^{\cdot} + XH \tag{13}$$

The exact nature of $X^{\cdot}$ in foods may be difficult to determine because of its low concentration (Chan, 1987), but it could be generated either by redox reactions, or as a consequence of high energy irradiation. Alternatively, in foods a photo-oxidation process in natural sunlight, which involves the formation of an excited dye, such as riboflavin (vitamin B_2) or erythrosine, may be responsible for the first step in the generation of free radicals.

The Propagation Phase

Oxygen may react with free radicals produced as a result of the initial reaction, yielding more free radicals and the hydroperoxide (LOOH). Whilst the initiation step may be relatively slow, the propagation phase may be rapid. For every free radical ($L^{\cdot}$) used, another is generated.

$$L^{\cdot} + O_2 \rightleftharpoons LOO^{\cdot} \tag{14}$$

$$LOO^{\cdot} + LH \rightleftharpoons L^{\cdot} + LOOH \tag{15}$$

The hydroperoxide may also participate in metal catalysed initiation reactions (eqns (16) and (17)) so there is autocatalysis or positive feedback generating reactive peroxyl-($LOO^{\cdot}$) and alkoxyl ($LO^{\cdot}$) radicals.

$$LOOH + M^{n} \rightleftharpoons LO^{\cdot} + OH^{-} + M^{n+} \tag{16}$$

$$LOOH + M^{n+} \rightleftharpoons LOO^{\cdot} + H^{+} + M^{n} \tag{17}$$

The generation of radicals may therefore be maintained, resulting in a self-perpetuating or chain reaction which, once started, may turn over many times, until two radicals combine to terminate the reaction.

The Termination Phase

When radical concentration is sufficiently high they may react together (eqns (18)–(20)) (or with other compounds) to yield stable end-products. As hydroperoxides accumulate, their breakdown products (e.g. alkoxyl radicals) become increasingly important. For example, alkoxyl radicals give rise to aldehydes which are the source of the rancid fat's characteristic odour. Several aldehydes produced by metal ion catalysed lipid peroxide decomposition are highly cytotoxic, even in minute amounts (Halliwell, 1987).

$$\left.\begin{array}{l} L^{\cdot} + L^{\cdot} \rightleftharpoons L{-}L \\ LOO^{\cdot} + LOO^{\cdot} \rightleftharpoons L{-}O{-}O{-}L + O_2 \\ LOO^{\cdot} + L^{\cdot} \rightleftharpoons L{-}O{-}O{-}L \end{array}\right\} \text{stable end-products} \quad \begin{array}{r} (18) \\ (19) \\ (20) \end{array}$$

THE ROLE OF FREE RADICALS IN FOOD IRRADIATION

Although the usual sources of free radicals *in vivo* are univalent biochemical redox reactions involving oxygen, UV light and other types of high energy irradiation are among many exogeneous initiators of free radical products derived from water both in living systems and foods. The major constituent of living cells and hence our food supply is water. Exposure of water to high energy ionizing radiation such as gamma-rays or X-rays results in the radiolytic production of reactive $OH^{\cdot}$ radicals, which may cause damage to living cells where DNA seems particularly susceptible to disruption, possibly by hydroxylation (Aruoma *et al.*, 1989). Dissolved oxygen also may be reduced to superoxide by hydrated electrons (produced by the radiolysis of water) and in the presence of metal ions, radicals may, in turn, produce more $OH^{\cdot}$ radicals (eqns (7) and (8)). The presence of oxygen may further increase injury to living cells because, in the absence of antioxidants, organic peroxyl radicals ($ROO^{\cdot}$) are produced from the reaction between O_2 and the organic free radicals ($R^{\cdot}$) arising from the action of $OH^{\cdot}$ in cell constituents. Thus irradiation, due to its ability to generate reactive oxygen species and disrupt DNA, is a very effective pasteurisation technique. The mechanism by which irradiation pasteurises food, destroys pests and prevents sprouting in potatoes arises from the disruption of DNA by reactive radicals.

Food irradiation processes have been used in many countries for more than 30 years. The forms of ionizing energy used include gamma-rays, X-rays and accelerated electrons. These processes usually occur in an aerobic environment and have a wide range of uses, including delay of ripening of fruits, elimination of pathogens and the production of shelf-stable products and sterile ingredients. An expert committee of the World Health Organisation/Food and Agricultural Organisation in 1981 concluded that the preservation of food by radiation doses of up to 10 kGY presented no toxicological hazard for human consumption (WHO/FAO, 1988). Even 'low' levels of irradiation, up to 1 kGY, used to destroy insects and parasites and inhibit sprouting and delay ripening of food plants, produce some reactive oxygen radicals in the foodstuff which might cause the destruction of small amounts of some vitamins and other micronutrients. At low doses, up to 1 kGy, the loss of nutrients from food is insignificant, although in the range 1–10 kGy some vitamin losses may occur in food exposed to air during irradiation and further storage (WHO/FAO, 1988).

Foods with a high fat content, for example dairy foods, are currently unsuitable for radiation treatment, because of the adverse flavours and tastes produced by the subsequent lipid oxidation. The $OH^{\cdot}$ radicals which are generated by irradiation of the food are able to generate other free radical species and initiate the chain reactions of lipid peroxidation. It is possible that this problem could be overcome by the use of added antioxidants. As in healthy living systems, protection against damage by free radical reactions in irradiated fresh foods is afforded by the indigenous antioxidants which minimise oxidative damage.

THE PREVENTION OF FREE RADICAL DAMAGE

Living cells, tissues and fresh foods are protected against inorganic intermediates of biological oxygen reduction ($O_2^{\cdot-}$ and H_2O_2) produced by redox reactions or the effects of irradiation, using a variety of defence mechanisms which inhibit or remove them. Apart from radical–radical interactions, radicals may also be inhibited or removed by scavengers or antioxidants including metalloenzymes such as SOD which catalyse highly efficient scavenging reactions. In healthy living systems there exists a sophisticated range of antioxidant mechanisms to prevent oxidative damage, and synergistic (cooperative) interactions are known to occur between antioxidant scavengers; this synergism may be significantly reduced in damaged cells.

Antioxidants in our foods are either indigenous or added during processing. In fresh foods, free radicals are held in check by the indigenous antioxidants, but in stored foods increased chemical reactivity allowing increased production of oxy-radicals may parallel a decline in the antioxidant action. Antioxidants in foods provide protection against colour deterioration in addition to their function in the retardation of lipid peroxidation. In addition to SOD, indigenous antioxidants in foods include catalase and peroxidases which remove H_2O_2. Non-enzymic antioxidant protection is provided by the actions of free-radical scavenging or chain-breaking susbstances such as ascorbate, glutathione, α-tocopherol and β-carotene. These scavengers of free radicals and reactive species are not continuously regenerated in foods and therefore may be differentially lost during processing and storage.

INDUSTRIAL APPLICATIONS OF SOD

SOD and synthetic SOD 'mimics' are of increasing interest to the pharmaceutical industry because of their potential therapeutic use as antioxidant drugs to reduce tissue damage on re-oxygenation following local ischaemia and also in the healing of wounds and burns. The administration of SOD has potential benefit following heart attacks and during cardiac surgery and organ transplants. Thus superoxide dismutase is an enzyme which might have potential use as a natural antioxidant in foods. Certainly it is known that the superoxide radical will rapidly oxidise substances like ascorbic acid and α-tocopherol. There is evidence that SOD has a role in suppressing autoxidation (Rabinowitch & Fridovich, 1983; Lingnert *et al.*, 1989). Lipid peroxidation, initiated either by a xanthine oxidase system producing $O_2^{\cdot-}$ or by lipoxygenase, has been claimed to be inhibited in the presence of SOD (Kellogg & Fridovich, 1975; Richter *et al.*, 1975; Dhindsa *et al.*, 1981). Our own studies on aqueous emulsified linoleic acid (unpublished data) indicate that SOD can prolong the initiation step of the autoxidation process. The SOD enzyme may reduce the damaging action of some organic peroxy radicals ($ROO^{\cdot}$) (Gee *et al.*, 1985; Willson, 1987). Although SOD has been patented as an antioxidant in foods (Michelson & Monod, 1975), definite proof of inhibition of both oxidative enzymes and formation of autoxidation products has not yet been given. Furthermore the effectiveness of SOD as an antioxidant in foods is untried and may in any case be strictly limited by the conditions occurring in the food (Korycka-Dahl & Richardson, 1978; Lingnert *et al.*, 1989).

Indigenous SOD may have a role in the control of fruit ripening and in preventing the loss of quality due to peroxidase activity in harvested fruits and vegetables (Donnelly *et al.*, 1989). $O_2^{\cdot -}$ can rapidly oxidise important nutrients such as ascorbic acid (vitamin C) and α-tocopherol (vitamin E) and so SOD may provide protection, but the enzyme's antioxidant role may be destroyed in foods that have been severely heat treated during processing (Walker *et al.*, 1987). Non-enzymic oxidation by superoxide radicals probably occurs throughout the various stages of beer production and also in the finished product and it has therefore been thought (Bamforth, 1983) that since barley SOD may be destroyed in processing, the addition of a heat-stable SOD may be desirable. A method describing the partial purification of a thermostable soya bean SOD has recently been published (Clarkson *et al.*, 1989). Such a source of SOD may in future be incorporated into other beverages and foods that need to be pasteurised, although catalase may also be required to remove the generated H_2O_2.

Lingnert *et al.* (1989) have also reported on the potential use of SOD in food products, drawing attention to the influence of pH, water activity, storage temperature and heat treatment on the antioxidative properties of SOD in model systems. These authors concluded that whilst yeast SOD could act as an antioxidant in various model systems, including an emulsified linoleic acid system, catalase in fact had a pro-oxidative effect in the linoleic acid system.

REFERENCES

Almansa, M. S., Del Rio, L. A., Alcaraz, C. F. & Sevilla, F. (1989). Isoenzyme pattern of superoxide dismutase in different varieties of citrus plants. *Physiologica Plantanum,* **76,** 563–8.

Arnold, L. D. & Lepock, J. R. (1982). Reversibility of the thermal denaturation of yeast superoxide dismutase. *FEBS Letts,* **146,** 302–6.

Arron, G., Henry, L., Palmer, J. M. & Hall, D. O. (1976). Superoxide dismutases in mitochondria from *Helianthus tuberosus* and *Neurospora crassa. Biochem. Soc. Trans.,* **4,** 618–20.

Aruoma, O. I., Halliwell, B. & Dizdaroglu, M. (1989). Iron ion-dependent modification of bases in DNA by the superoxide radical-generating system hypoxanthine/xanthine oxidase. *J. Biol. Chem.,* **264,** 13024–8.

Asada, K., Urano, M. & Takahashi, M.-A. (1973). Subcellular location of superoxide dismutase in spinach leaves and preparation of properties of crystalline spinach superoxide dismutase. *Eur. J. Biochem.,* **36,** 257–66.

Asada, K., Yoshikawa, K., Takahashi, M.-A., Maeda, Y. & Enmaryi, K. (1975). Superoxide dismutases from a blue-green alga, *Plectonema boryanum*. *J. Biol. Chem.*, **250,** 2801–7.

Asada, K., Takahashi, M.-A., Tanaka, K. & Natano, Y. (1977). Formation of active oxygen and its fate in chloroplasts. In *Biochemical and Medical Aspects of Active Oxygen,* ed. O. Hayaisha & K. Asada. University of Tokyo Press, Tokyo, pp. 45–63.

Aust, S. D. (1985). Lipid peroxidation. In *Handbook of Methods for Oxygen Research,* ed. R. A. Greenwald. CRC Press, Boca Raton, Florida, pp. 203–7.

Baker, J. E. (1976). Superoxide dismutase in ripening fruits. *Plant Physiol.*, **58,** 644–7.

Bamforth, C. W. (1983). Superoxide dismutase in barley. *J. Inst. Brew.*, **89,** 420–3.

Bannister, J. V., Bannister, W. H., Hill, H. A. O., Mahood, J. F., Willson, R. L. & Wolfenden, B. S. (1980). Does caeruloplasmin dismute superoxide? No. *FEBS Lett.*, **118,** 127–9.

Baum, J. A. & Scandalios, J. G. (1981). Isolation and characterization of the cytosolic and mitochondrial superoxide dismutases of maize. *Archs Biochem. Biophys.*, **206,** 249–64.

Beauchamp, C. & Fridovich, I. (1971). Superoxide dismutase: Improved assays and an assay applicable to acrylamide gels. *Analyt. Biochem.*, **44,** 276–87.

Beauchamp, C. & Fridovich, I. (1973). Isoenzymes of superoxide dismutase from wheatgerm. *Biochem. Biophys. Acta.*, **317,** 50–64.

Beyer, W. F. & Fridovich, I. (1987). Assaying for superoxide dismutase activity: some large consequences of minor changes in conditions. *Analyt. Biochem.*, **161,** 559–66.

Bhattacharjee, A. & Choudhuri, M. A. (1986). Chemical manipulation of seed longevity and stress tolerance capacity of seedlings of *Corchorus capsularis* and *C. olitorius*. *J. Plant Physiol.*, **125,** 391–400.

Biliaderis, C. G., Weselake, R. J., Petkau, A. & Friesen, A. D. (1987). A calorimetric study of human Cu/Zn superoxide dismutase. *Biochem. J.*, **248,** 981–4.

Bors, W. (1989). An expanded function for superoxide dismutase. In Progam and Abstracts of Fifth Conference on Superoxide Dismutase, Department of Physical Chemistry. Hebrew University of Jerusalem, Jerusalem, Israel, p. 45.

Bridges, S. M. & Salin, M. L. (1981). Distribution of iron-containing superoxide dismutase in vascular plants. *Plant Physiol.*, **68,** 275–8.

Britton, L., Malinowski, D. P. & Fridovich, I. (1978). Superoxide dismutase and oxygen metabolism in *Streptococcus faecalis,* and comparisons with other organisms. *J. Bacteriol.*, **134,** 229–36.

Burke, J. J., Gamble, P. E., Hathfield, J. L. & Quisenberry, J. E. (1985). Plant morphological and biochemical responses to field water deficits. I. Response of glutathione reductase activity and paraquat sensitivity. *Plant Physiol.*, **79,** 415–9.

Cadenas, E. (1989). Biochemistry of oxygen toxicity. *Ann. Rev. Biochem.*, **58,** 79–110.

Chan, H. W.-S. (1987). Oxygen free radicals in food. *Proc. of the Nutrition Society,* **46,** 35–41.

Clare, D. A., Rabinowitch, H. D. & Fridovich, I. (1984). Superoxide dismutase and chilling injury in *Chlorella ellipsoidea. Arch. Biochem. Biophys.*, **231,** 158–63.

Clarkson, S. P., Large, P. J. & Bamforth, C. W. (1989). Purification of a cyanide-sensitive superoxide dismutase from soya beans: A food-compatible enzyme preparation. *J. Sci. Food Agric.*, **48,** 87–97.

Czapski, G. & Goldstein, S. (1988). The uniqueness of superoxide dismutase (SOD)—why cannot most copper compounds substitute SOD *in vivo*? *Free Rad. Res. Comms.*, **4,** 225–9.

Davies, K. J. A. (1986). Intracellular proteolytic systems may function as secondary antioxidant defences: An hypothesis. *J. Free Radicals in Biol. Med.*, **2,** 155–73.

Del Rio, L. A., Lyon, D. S., Olah, I., Glick, B. & Salin, M. L. (1983). Immunocytochemical evidence for a peroxisomal location of Mn-SOD in leaf protoplasts from a higher plant. *Planta,* **158,** 216–24.

Dhindsa, R. S., Plumb-Dhindsa, P. L. & Thorpe, T. A. (1981). Leaf senescence. Correlated with increased levels of membrane permeability and lipid peroxidation, and decreased levels of superoxide dismutase and catalase. *J. Exp. Bot.*, **32,** 93–101.

Dhindsa, R. S., Plumb-Dhindsa, P. L. & Reid, D. M. (1982). Leaf senescence and lipid peroxidation. Effects of some phytohormones and scavengers of free radicals and singlet oxygen. *Physiol. Plant,* **56,** 543–7.

Donnelly, J. K., McLellan, K. M., Walker, J. L. & Robinson, D. S. (1989). Superoxide dismutases in foods. A review. *Food Chem.*, **33,** 243–70.

Federico, R., Medda, R. & Floris, G. (1985). Superoxide dismutase from *Lens esculenta* purification and properties. *Plant Physiol.*, **78,** 357–8.

Fee, J. A. (1980). A comment on the hypothesis that oxygen toxicity is mediated by superoxide. In *Oxygen and Life. Second BOC Priestley Conference. Royal Soc. Chem.*, London, pp. 77–97.

Fee, J. A. (1982). On the question of superoxide toxicity and the biological function of superoxide dismutases. In *Oxidases and Related Redox Systems* (*Proceedings of the Third International Symposium*), ed. T. E. King, H. S. Mason & M. Morrison. Pergamon Press, Oxford, pp. 101–49.

Forman, H. J. & Fridovich, I. (1973). On the stability of bovine superoxide dismutase. The effect of metals. *J. Biol. Chem.*, **248,** 2645–9.

Fridovich, I. (1979). Superoxide and superoxide dismutases. In *Advances in Inorganic Biochemistry 1,* ed. G. L. Eichhorn & L. G. Marzilli. Elsevier North Holland, pp. 67–90.

Fridovich, I. (1982). The discovery of superoxide dismutases: A history. In *Superoxide Dismutase,* Vol. I, ed. L. W. Oberley. CRC Press, Boca Raton, Florida, pp. 1–9.

Garcia, J. E., Gomez, M., Yanez, J., Lopez-Gorge, J. & Del Rio, L. A. (1981). Isozyme pattern of the metalloenzyme system superoxide dismutase during growth of peas (*Pisum sativum* L.) under different iron nutrient concentrations. *Z. Pflanzenphysiol. Bot.*, **105S,** 21–9.

Gartner, A. & Weser, U. (1986). Molecular and functional aspects of superoxide dismutases. In *Biomimetic and Bioorganic Chemistry II.* (*Topics in Current Chemistry* 132), ed. F. Vogtle & E. Weber. Springer, Berlin, pp. 1–61.

Gartner, A., Hartmann, H.-J. & Weser, U. (1984). A simple, rapid and efficient isolation of erythrocyte Cu_2/Zn_2-superoxide dismutase. *Biochem. J.*, **221,** 549–51.

Gee, C. A., Kittridge, K. J. & Willson, R. L. (1985). Peroxy free radicals, enzymes and radiation damage: sensitisation by oxygen and protection by superoxide dismutase and antioxidants. *Br. J. Radiol.*, **58,** 251–6.

Gershman, R., Gilbert, D. L., Nye, S. W., Dwyer, P. & Fenn, W. O. (1954). Oxygen poisoning and x-irradiation: A mechanism in common. *Science,* **119,** 623–6.

Getzoff, E. D., Tainer, J. A., Weiner, P. K., Kollman, P. A., Richardson, J. S. & Richardson, D. C. (1983). Electrostatic interaction between superoxide and copper-zinc superoxide dismutase. *Nature,* **306,** 287–90.

Getzoff, E. D., Tainer, J. A., Stempien, M. M., Bell, G. I. & Hallewell, R. A. (1989). Evolution of Cu/Zn-superoxide dismutase and the Greek key β-barrel structural motif. *Proteins: Structure, Function, Genetics,* **5,** 322–36.

Giannopolitis, C. N. & Ries, S. K. (1977*a*). Superoxide dismutases: occurrence in higher plants. *Plant Physiol.*, **59,** 309–14.

Giannopolitis, C. N. & Ries, S. K. (1977*b*). Superoxide dismutases: purification and quantitative relationship with water-soluble protein in seedlings. *Plant Physiol.*, **59,** 315–18.

Gillham, D. J. & Dodge, A. A. (1987). Chloroplast superoxide and hydrogen peroxide scavenging systems from pea leaves: seasonal variations. *Plant Sci.*, **50,** 105–9.

Goldstein, I. M., Kaplan, H. B. Edelson, H. S. & Weissman, G. (1979). Ceruloplasmin. A scavenger of superoxide anion radicals. *J. Biol. Chem.*, **254,** 4040–5.

Gregory, E. M. (1985). Characterization of the O_2-induced manganese-containing superoxide dismutase from *Bacteroides fragilis. Arch. Biochem. Biophys.*, **238,** 83–9.

Gutteridge, J. M. C. & Quinlan, G. J. (1986). Carminic acid-promoted oxygen radical damage to lipid and carbohydrate. *Food Additives and Contaminants,* **3,** 289–93.

Halliwell, B. (1984). Superoxide dismutase and the superoxide theory of oxygen toxicity. A critical appraisal. In *Copper Proteins and Copper Enzymes,* Vol. II, ed. R. Lonti. CRC Press, Boca Raton, Florida, pp. 63–102.

Halliwell, B. (1987). Free radicals and metal ions in health and disease. *Proc. of the Nutrition Society,* **46,** 13–26.

Halliwell, B. & Gutteridge, J. M. C. (1989). *Free Radicals in Biology and Medicine,* 2nd edn. Clarendon Press, Oxford.

Hammer, C. T. & Wills, E. D. (1978). The role of lipid components of the diet in the regulation of the fatty acid composition of the rat liver endoplasmic reticulum and lipid peroxidation. *Biochem. J.*, **174**, 585–93.

Hayakawa, T., Kanematsu, S. & Asada, K. (1985). Purification and characterisation of thylakoid-bound Mn-superoxide dismutase in spinach chloroplasts. *Planta*, **166**, 111–16.

Hicks, C. L. (1980). Occurrence and consequence of superoxide dismutase in milk products. A review. *J. Dairy Sci.*, **63**, 1199–204.

Hicks, C. L., Bucy, J. & Stofer, W. (1979). Heat inactivation of superoxide dismutase in bovine milk. *J. Dairy Sci.*, **62**, 529–32.

Hodgson, E. K. & Fridovich, I. (1975). Interaction of bovine erythrocyte superoxide dismutase with H_2O_2. Inactivation of the enzyme. *Biochemistry*, **14**, 5294–9.

Jaaska, Vilve & Jaaska, Vello (1982). Isoenzymes of superoxide dismutase in barley. *Biochem. Physiol. Pflanzen*, **177**, 375–86.

Kanematsu, S. & Asada, K. (1979). Ferric and manganic superoxide dismutases in *Euglena gracilis*. *Archs Biochem. Biophys.*, **195**, 535–45.

Kanematsu, S. & Asada, K. (1989). Cu/Zn-superoxide dismutases in rice: occurrence of an active, monomeric enzyme and two types of isozyme in leaf and non-photosynthetic tissues. *Plant Cell Physiol.*, **30**, 381–91.

Kanematsu, S. & Asada, K. (1990). Characteristic amino acid sequences of chloroplast and cytosol isozymes of Cu/Zn-superoxide dismutase in spinach, rice and horsetail. *Plant Cell Physiol.*, **31**, 99–112.

Kellogg, E. W. & Fridovich, I. (1975). Superoxide, hydrogen peroxide and singlet oxygen in lipid peroxidation by a xanthine oxidase system. *J. Biol. Chem.*, **250**, 8812–17.

Kitagawa, Y., Tsunasawa, S., Tanaka, N., Katsube, Y., Sakiyama, F. & Asada, K. (1986). Amino acid sequence of copper, zinc superoxide dismutase from spinach leaves. *J. Biochem.*, **99**, 1289–98.

Klug, C., Rabani, J. & Fridovich, I. (1972). A direct demonstration of the catalytic action of superoxide dismutase through the use of pulse radiolysis. *J. Biol. Chem.*, **247**, 4839–42.

Kon, S. & Schwimmer, S. (1977). Depolymerization of polysaccharides by active oxygen species derived from a xanthine oxidase system. *J. Food Biochem.*, **1**, 141–52.

Korycka-Dahl, M. & Richardson, T. (1978). Activated oxygen species and oxidation of food constituents. *CRC Crit. Rev. Fd. Sci. & Nutr.*, **10**, 209–41.

Korycka-Dahl, M., Richardson, T. & Hicks, C. L. (1979). Superoxide dismutase activity in bovine milk serum. *J. Food Protection*, **42**, 867–71.

Kwiatowski, J., Safianowska, A. & Kaniuga, Z. (1985). Isolation and characterization of an iron containing superoxide dismutase from tomato leaves, *Lycopersicon esculentum*. *Eur. J. Biochem.*, **146**, 459–66.

Larson, R. A. (1988). The antioxidants of higher plants. *Phytochemistry*, **27**, 969–78.

Lee, E. H. & Bennett, J. H. (1982). Superoxide dismutase. A possible protective enzyme against ozone injury in snap beans (*Phaseolus vulgaris* L.). *Plant Physiol.*, **69**, 1444–9.

Lesham, Y. Y., Halevy, A. H. & Frenkel, C. (1986). *Processes and Control of*

Plant Senescence (*Developments in Crop Science 8*). Elsevier, Amsterdam, pp. 162–99.

Lingnert, H., Akesson, G. & Eriksson, C. E. (1989). Antioxidative effect of superoxide dismutase from *Saccharomyces cerevisiae* in model systems. *J. Agric. Food Chem.*, **37,** 23–8.

Lonnerdal, B., Keen, C. L. & Hurley, L. S. (1979). Isoelectric focusing of superoxide dismutase isozymes. *FEBS Letts,* **108,** 51–5.

Lynch, R. A., Lee, G. R. & Cartwright, G. E. (1976). Inhibition by superoxide dismutase of methemoglobin formation from oxyhaemoglobin. *J. Biol. Chem.*, **251,** 6049–55.

McCord, J. M. & Fridovich, I. (1969). Superoxide dismutase. An enzymatic function for erythrocuprein (haemocuprien). *J. Biol. Chem.*, **244,** 6049–55.

McCord, J. M. (1979). Superoxide dismutases: Occurrence, structure, function and evolution. In *Isozymes: Current Topics in Biological and Medical Research,* **3,** 1–21.

McCord, J. M., Keele, B. B. Jr & Fridovich, I. (1971). An enzyme-based theory of obligate anaerobisis: The physiological function of superoxide dismutase. *Proc. Natl. Acad. Sci. USA,* **68,** 1024–7.

Malinowski, D. P. & Fridovich, I. (1979). Chemical modifications of arginine at the active site of the bovine erythrocyte superoxide dismutase. *Biochemistry,* **18,** 5909–17.

Mann, T. & Keilin, D. (1938). Haemocuprien, a copper-protein compound of red blood corpuscles (Letter to the Editor). *Nature,* **142,** 148.

Martin, J. P. Jr & Fridovich, I. (1981). Evidence for a natural gene transfer from the ponyfish to its bioluminescent bacterial symbiont *Photobacter leiognathi.* The close relationship between bacteriocuprein and the copper-zinc superoxide dismutase of teleost fishes. *J. Biol. Chem.*, **256,** 6080–9.

Martin, M. E., Byers, B. R., Olson, M. O. J., Salin, M. L., Arceneaux, J. E. L. & Tolbert, C. (1986). A streptococcus mutans superoxide dismutase that is active with either manganese or iron as a cofactor. *J. Biol. Chem.*, **261,** 9361–7.

Matkovics, B. (1977). Effects of plant and animal tissue lesions on superoxide dismutase activities. In *Superoxide and Superoxide Dismutase,* ed. A. M. Michelson, J. M. McCord & I. Fridovich. Academic Press, London, pp. 501–15.

Matkovics, B., Varga, Sz. I. & Matkovics, I. (1981). Some data on the superoxide dismutase and peroxidase contents of fruits, seeds and different parts of plants. *Acta Biologica Szeged.*, **27,** 25–31.

Michalov, J. & Placek, J. (1982). Free radicals in the root system and stem of *Zea mays. J. Exp. Bot.*, **33,** 511–14.

Michelson, A. M. (1977). Toxic effects of active oxygen. In *Biochemical and Medical Aspects of Active Oxygen,* ed. O. Hayaisho & K. Asada. University of Tokyo Press, Japan.

Michelson, A. M. & Monod, J. (1975). Superoxide dismutases and their applications as oxidation inhibitors. *US Pat. No. 3 920 521.* Nov. 18.

Misra, H. P. & Fridovich, I. (1972). The generation of superoxide radical during the autoxidation of haemoglobin. *J. Biol. Chem.*, **247,** 6960–2.

Monk, L. S. & Davies, H. V. (1989). Antioxidant status of the potato tuber and Ca^{2+} deficiency as a physiological stress. *Physiol. Plant,* **75,** 411–16.
Monk, L. S., Fagerstedt, K. V. & Crawford, R. M. M. (1987). Superoxide dismutase as an anaerobic polypeptide—a key factor in recovery from oxygen deprivation in *Iris pseudacorus*? *Plant Physiol.,* **85,** 1016–20.
Monk, L. S., Fagerstedt, K. V. & Crawford, R. M. M. (1989). Oxygen toxicity and superoxide dismutase as an antioxidant in physiological stress. *Physiol. Planta.,* **76,** 456–9.
Mota de Freitas, D. & Valentine, J. S. (1984). Phosphate is an inhibitor of copper-zinc superoxide dismutase. *Biochemistry,* **23,** 2079–82.
Moulding, P. H., Singleton, D. E., McLellan, K. M. & Robinson, D. S. (1988). Purification and heat stability of Cox's apple pulp peroxidase isoenzymes. *Int. J. Food Sci. Tech.,* **23,** 343–51.
Neumann, P. R. & Hart, G. E. (1986). Genetic control of the mitochondrial form of superoxide dismutase in hexaploid wheat. *Biochemical Genetics,* **24,** 435–46.
Olsen, J. A. & Cook, C. R. (1987). Superoxide dismutase in wounded etiolated pea seedlings. *Phytochemistry,* **26,** 71–3.
Palma, J. M., Gomez, M., Yanez, J. & Del Rio, L. A. (1987). Increased levels of peroxisomal active oxygen-related enzymes in copper-tolerant pea plants. *Plant Physiol.,* **85,** 570–4.
Puppo, A., Dimitrijevic, L. & Rigaud, J. (1987). O_2 consumption and superoxide dismutase content in purified mitochondria from soybean root nodules. *Plant Sci.,* **50,** 3–11.
Rabinowitch, H. D. & Fridovich, I. (1983). Superoxide radicals, superoxide dismutases and oxygen toxicity in plants. *Phytochem. Phytobiol.,* **37,** 679–90.
Rabinowitch, H. D. & Sklan, D. (1980). Superoxide dismutase. A possible protective agent against sunscald in tomatoes. *Planta,* **148,** 162–7.
Rabinowitch, H. D. & Sklan, D. (1981). Superoxide dismutase activity in ripening cucumber and pepper fruit. *Physiol. Plant,* **52,** 380–4.
Rabinowitch, H. D., Sklan, D. & Budowski, P. (1982). Photo-oxidative damage in the ripening tomato fruit: protective role of superoxide dismutase. *Physiol. Plant,* **54,** 369–74.
Ravindranath, S. D. & Fridovich, I. (1975). Isolation and characterisation of a manganese-containing superoxide dismutase from yeast. *J. Biol. Chem.,* **256,** 6107–12.
Reddy, C. D. & Venkaiah, B. (1984*a*). Studies on isoenzymes of superoxide dismutase from mung bean (*Vigna radiata*) seedlings. *J. Plant Physiol.,* **116,** 279–84.
Reddy, C. D. & Venkaiah, B. (1984*b*). The purification and characterisation of a manganese superoxide dismutase from mung bean. *Biochem Int.,* **8,** 707–14.
Redford, S. M., McRee, D. E., Getzoff, E. D., Steinman, H. M. & Tainer, J. A. (1990). Crystallographic characterization of a Cu/Zn-superoxide dismutase from *Photobacterium leiognathi. J. Mol. Biol.,* **212,** 449–51.
Richter, C., Wendel, A., Weser, U. & Azzi, A. (1975). Inhibition by

superoxide dismutase of linoleic acid peroxidation induced by lipoxidase. *FEBS Letts,* **51,** 300–3.

Robinson, D. S. (1987). *Food: Biochemistry and Nutritional Value.* Longman, Harlow, pp. 446–7.

Rotilio, G., Bray, R. C. & Fielden, E. M. (1972). A pulse radiolysis study of superoxide dismutase. *Biochim. Biophys. Acta,* **268,** 605–9.

Sandalio, L. M. & Del Rio, L. A. (1987). Localisation of superoxide dismutase in glyoxysomes from *Citrullus vulgaris*. Functional implications in cellular metabolism. *J. Plant Physiol.,* **127,** 395–409.

Sandalio, L. M., Palma, J. M. & Del Rio, L. A. (1987). Localization of manganese superoxide dismutase in peroxisomes isolated from *Pisum sativum* L. *Plant Sci.,* **51,** 1–8.

Sawada, Y., Ohyama, T. & Yamazaki, I. (1972). Preparation and physicochemical properties of green pea superoxide dismutase. *Biochim. Biophys. Acta,* **268,** 305–12.

Schoner, S. & Krause, G. H. (1990). Protective systems against active oxygen species in spinach: response to cold acclimation in excess light. *Planta,* **180,** 383–9.

Schwimmer, S. (1981). Other roles of enzyme action on phenols as determinants of food colour quantity. *Source Book of Enzymology.* AVI, Westport, Connecticut, USA, pp. 284–97.

Sevilla, F., Del Rio, L. A. & Hellin, A. (1984). Superoxide dismutases from a citrus plant: presence of two iron-containing isoenzymes in leaves of lemon trees (*Citrus limonum* L.). *J. Plant Physiol.,* **116,** 381–7.

Slater, T. F. (1981). Free radical scavengers. In *Cyanidanol-3 in Diseases of the Liver, International Workshop,* ed. H. O. Conn. *Royal Society of Medicine International Congress and Symposium Series no. 47*, Crans-Montana, Switzerland. Academic Press, London and the Royal Society of Medicine, London, pp. 11–15.

Slater, T. F., Cheeseman, K. M., Davies, M. J., Proudfoot, K. & Zin, W. (1987). Free radical mechanisms in relation to tissue injury. *Proc. of the Nutrition Society,* **46,** 1–12.

Stallings, W. C., Pattridge, K. A., Strong, R. K. & Ludwig, M. L. (1984). Manganese and iron superoxide dismutases are structural homologs. *J. Biol. Chem.,* **259,** 10695–9.

Steffens, G. J., Bannister, J. V., Bannister, W. H., Flohe, L., Guenzler, W. A., Kim, S.-M. A. & Oetting, F. (1983). The primary structure of Cu/Zn-superoxide dismutase from *Photobacterium leiognathi*: Evidence for a separate evolution of Cu/Zn-superoxide dismutase in bacteria. *Hoppe-Seyler's Z. Physiol. Chem.,* **364,** 675–90.

Steffens, G. J., Michelson, A. M., Otting, F., Puget, K., Strassburger, W. & Flohe, L. (1986). Primary structure of Cu/Zn-superoxide dismutase of *Brassica oleracea* proves homology with corresponding enzymes of animals, fungi and prokaryotes. *Biol. Chem. Hoppe-Seyler,* **367,** 1007–16.

Steinman, H. M. (1980). The amino acid sequence of copper-zinc superoxide dismutase from bakers' yeast. *J. Biol. Chem.,* **255,** 6758–65.

Steinman, H. M. (1982). Superoxide dismutases: protein chemistry and structure–function relationships. In *Superoxide Dismutase,* Vol. I, ed. L. W. Oberley. CRC Press, Boca Raton, Florida, pp. 11–18.

Steinman, H. M. (1987). Bacteriocuprein superoxide dismutase of *Photobacterium leiognathi*. Isolation and sequence of the gene and evidence for a precursor form. *J. Biol. Chem.*, **262,** 1882–7.

Sutton, J. G., Goodwin, J., Horscroft, G., Stockdale, R. E. & Frake, A. (1983). Identification of trout and salmon bloods by simple immunological technique and by electrofocusing patterns of red cell enzyme superoxide dismutase. *J. Assoc. Off. Anal. Chem.*, **66,** 1164–74.

Tainer, J. A., Getzoff, E. D., Beem, K. M., Richardson, J. S. & Richardson, D. C. (1982). Determination and analysis of the 2A structure of copper–zinc superoxide dismutase. *J. Mol. Biol.*, **160,** 181–217.

Tainer, J. A., Getzoff, E. D., Richardson, J. S. & Richardson, D. C. (1983). Structure and mechanism of copper–zinc superoxide dismutase. *Nature,* **306,** 284–7.

Tandy, N. E., Giulio, R. T. & Richardson, C. T. (1989). Assay and electrophoresis of superoxide dismutase from Red Spruce (*Picea rubens* Sarg.), Loblolly Pine (*Pinus taeda* L.) and Scotch Pine (*Pinus sylvestris* L.). *Plant Physiol.*, **90,** 742–8.

Touati, D. (1988). Molecular genetics of superoxide dismutases. *Free Radical Biology & Medicine,* **5,** 393–402.

Walker, J. L. (1987). Superoxide dismutase in Brassicas. PhD thesis, University of Leeds.

Walker, J. L., McLellan, K. M. & Robinson, D. S. (1987). Heat stability of superoxide dismutase in cabbage. *Food Chem.*, **23,** 245–56.

Walker, J. L. & Robinson, D. S. (1991). Isolation and purification of superoxide dismutase purified from Brussel sprouts (*Brassica* oleracea. L var. bullata sub. var. gemmifera). *Food Chem.*, **41,** 1–9.

Weisiger, R. A. & Fridovich, I. (1973). Superoxide dismutase. Organelle specificity. *J. Biol. Chem.*, **248,** 3582–92.

Weser, U., Fretzdorf, A. & Prinz, R. (1972). Superoxide dismutase in bakers' yeast. FEBS *Letts,* **27,** 267–9.

WHO/FAO (1988). *Food Irradiation. A Technique for Preserving and Improving the Safety of Food.* World Health Organisation, Geneva.

Willson, R. L. (1987). Vitamin, selenium, zinc and copper interactions in free radical protection against ill-placed iron. *Proc. of the Nutrition Society,* **46,** 27–34.

3

Amine Oxidases and Amino Acid Oxidases

J. R. Burt

Torry Research Station, Aberdeen, UK

INTRODUCTION

The amino acid and amine oxidases constitute two quite disparate groups of enzymes. Occurring widely throughout the plant and animal

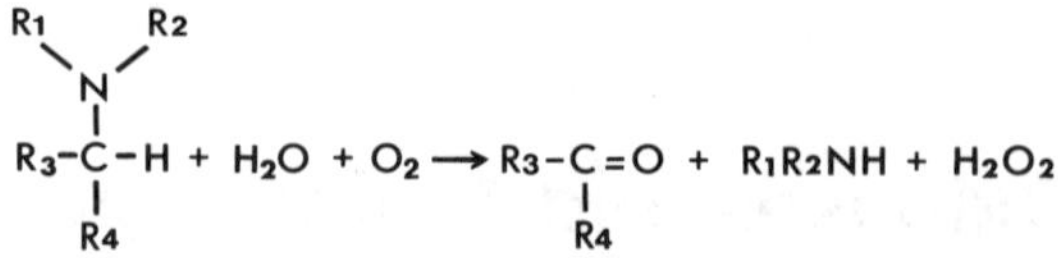

Fig. 1. Reaction catalysed by amine oxidases and amino acid oxidases.

kingdoms they are also to be found in many micro-organisms. During their action, in most cases, electrons are transferred directly to molecular oxygen, hydrogen peroxide is produced and the substrate is deaminated as it is oxidized. With only one or two important exceptions, the reactions catalysed by enzymes of both these groups can be summarized as shown in Fig. 1. Table 1 illustrates the range of N- and C-substituents that have been found in the different substrates for these enzymes.

The oxidases have been classified systematically on the basis of the chemical nature of their prosthetic groups. These prosthetic group classes have then been subdivided according both to the types of reaction they catalyse and to their occurrence (Keevil & Mason, 1978). Table 2 lists the enzymes which could be regarded as coming under the scope of this chapter; the properties of a number of representative enzymes will be discussed in detail, emphasis being placed on those which have, or which are likely to have, greater importance in foods.

AMINE OXIDASES

The different enzymes which metabolize amines have attracted much research attention over the years. Several excellent reviews have been published since the first amine oxidase, tyramine oxidase, was discovered (Hare, 1928). These reviews depict not only the accumulation

Table 1. Possible Substituents for the General Substrate, $R_1R_2N—CHR_3R_4$ for Amine and Amino Acid Oxidases

Substituent	*Amine oxidases*	*Amino acid oxidases*
R_1	H or alkyl	H
R_2	H or alkyl	H or CH_3
R_3	H or alkyl or aralkyl	COOH
R_4	H or alkyl or aralkyl	H or alkyl or aralkyl

Table 2. Classification of Amine and Amino Acid Oxidases (after Keevil & Mason, 1978)

I. Copper-containing oxidases		
C. Cuproprotein-catalysed electron transfer to oxygen		
4. Two electron donor/H_2O_2 produced		
d. Amine donors		
EC 1.4.3.6	Amine oxidase	Leguminous plants (Keevil & Mason, 1978)
EC 1.4.3.6	Diamine oxidase	Mammalian kidney (Keevil & Mason, 1978)
		Pea seedlings (Hill, 1971; Yanigasawa *et al.*, 1981)
		Lentil, bean, *Euphorbia* (Floris *et al.*, 1983)
		Human intestine (Bieganski *et al.*, 1980)
		Human pregnancy serum (Gahl *et al.*, 1982)
		Fish (Matsumiya & Otake, 1981, 1983, 1984, 1985, 1986)
EC 1.4.3.13	Protein-lysine 6-oxidase	Vertebrates, yeast (Duine, 1989)
	Amine oxidase (PLP)	Micro-organisms, fungi (Keevil & Mason, 1978)
	Amine oxidase	Tea leaf (Tsushida & Takeo, 1985)
		Aspergillus, yeasts, plants, mammals (Duine, 1989)
		Ruminant plasma (Mondovì *et al.*, 1983)
	Benzylamine oxidase (PLP)	Blood plasma (Keevil & Mason, 1978)
	Diamine oxidase	Human placenta (Crabbe *et al.*, 1976)
	Lysyl protocollagen oxidase	Mammalian connective tissue (Keevil & Mason, 1978)
	Methylamine oxidase	Micro-organisms (Kung & Wagner, 1970)
	Monoamine oxidase	Blood plasma (Keevil & Mason, 1978)
IV. Flavin-containing oxidases		
B. Flavoprotein mixed function oxidases		
2. External source of reducing equivalents		
f. Oxidation of amines		
	Dialkyl arylamine *N*-oxidase (FAD)	Liver microsomes (Keevil & Mason, 1978)

(*continued*)

Table 2.—*contd.*

C. Flavoprotein-catalysed electron transfer to oxygen		
2. Two-electron donor/two-electron acceptor forming H_2O_2		
e. Amine donors		
EC 1.4.3.1	D-Aspartate oxidase (FAD)	Kidney (Keevil & Mason, 1978)
EC 1.4.3.2	L-Amino acid oxidase (FAD or FMN)	Snake venom, kidney (Keevil & Mason, 1978)
EC 1.4.3.3	D-Amino acid oxidase (FAD)	Kidney (Keevil & Mason, 1978)
EC 1.4.3.4	Monoamine oxidase (peptidyl FAD)	Kidney, liver mitochondria, brain, skeletal muscles (Keevil & Mason, 1978)
EC 1.4.3.4	Monoamine oxidase	Vertebrates (Hall & Urueña, 1983)
		Fish (Syed & Coombs, 1982; Kinemuchi *et al.*, 1983)
		Echinoderms (Nicotra *et al.*, 1986)
		Fish eggs (Nicotra & Senatori, 1989)
EC 1.4.3.4	Tyramine oxidase (FAD)	*Sarcina* (Keevil & Mason, 1978)
EC 1.4.3.5	Pyridoxamine phosphate oxidase (FAD)	Liver (Keevil & Mason, 1978)
		Mammals, yeast, bacteria, animals, plants (McCormick & Merrill, 1980; Tsuge *et al.*, 1979, 1982*a,b,* 1986)
EC 1.4.3.7	D-Glutamate oxidase (flavin)	Crayfish (Keevil & Mason, 1978)
EC 1.4.3.8	Ethanolamine oxidase (flavin)	*Arthrobacter* (Keevil & Mason, 1978)
EC 1.4.3.10	Putrescine oxidase (FAD)	*Micrococcus rubens* (Keevil & Mason, 1978)
EC 1.4.3.11	L-Glutamate oxidase (flavin)	*Azotobacter* (Keevil & Mason, 1978)
		Streptomyces (Kusakabe *et al.*, 1983)
EC 1.4.3.12	Cyclohexylamine oxidase	*Pseudomonas* (Tokieda, 1977)
EC 1.4.3.14	L-Lysine oxidase (FAD)	*Trichoderma viride* (Kusakabe *et al.*, 1980)
EC 1.4.3.15	D-Glutamate (D-aspartate) oxidase (FAD)	*Aspergillus* (Mizushima, 1957)

EC 1.4.3.16	L-Aspartate oxidase (FAD)	Bacteria (Nasu *et al.*, 1982)
EC 1.5.3.1	Sarcosine oxidase (flavin)	Liver (Keevil & Mason, 1978)
EC 1.5.3.2	*N*-Methylamino acid oxidase (FAD)	Kidney (Keevil & Mason, 1978)
EC 1.5.3.5	(S)-6-Hydroxynicotine oxidase (peptidyl flavin)	*Arthrobacter* (Keevil & Mason, 1978)
EC 1.5.3.6	(R)-6-Hydroxynicotine oxidase (FMN)	*Arthrobacter* (Keevil & Mason, 1978)
	Pipecolate oxidase	*Pseudomonas* (Keevil & Mason, 1978)
	Polyamine oxidase	Catfish (Kumazawa & Suzuki, 1987)
		Rat liver (Hölttä, 1983)
		Graminae (Smith, 1975)
	Spermidine oxidase (haem) (FAD)	*Serratia marcescens* (Keevil & Mason, 1978)
	Spermidine oxidase	*Pseudomonas* (Padmanabhan & Kim, 1965)
V. Oxidases with uncertain prosthetic groups		
B. Mixed function oxidases		
2. External source of reducing equivalents		
f. Oxidation of nitrogen atoms		
	Alkyl hydrazine oxidase	Liver (Keevil & Mason, 1978)
	Trimethylamine oxidase	*Pseudomonas* (Keevil & Mason, 1978)
		Mammalian liver, fish, zooplankton, bacteria (Barrett & Kwan, 1985)
C. Electron transfer to oxygen		
4. Two-electron donor/H_2O_2 formed		
d. Amine donors		
	Allohydroxy-D-proline oxidase	Bacteria (Drawert & Barton, 1978)
	Methylamine oxidase	*Pseudomonas* (Keevil & Mason, 1978)
EC 1.5.3.4	N^6-Methyllysine oxidase	Kidney (Keevil & Mason, 1978)

of detailed knowledge but also the evolution of the schemes that were devised for classifying the large number of individual such enzymes described (Blaschko, 1963; Zeller, 1963; Kappeller-Adler, 1970, 1971; Bachrach, 1973; Smith, 1975; Keevil & Mason, 1978; Hall & Urueña, 1983; Mondovi & Riccio, 1984; Tabor & Tabor, 1984).

The first attempts at classifying the amine oxidases were described by Zeller in 1940 (see Kappeller-Adler, 1970) who divided them into two groups—the monoamine oxidases and the diamine oxidases—and these trivial names, along with others such as histaminase and tyramine oxidase, still persist to confuse the readers of the literature. These enzymes are now known not to possess the specificities apparently promised, or implied, by their nomenclature. There is extensive overlap, in terms of common substrates, between the monoamine oxidases and the diamine oxidases and also many instances of certain mono- and diamines not being oxidized by their supposedly appropriate enzymes. As more amine oxidases were discovered in different organs and organisms, the inadequacies of this system became apparent and Zeller (1963) concluded that '. . . it is best to characterize a given diamine oxidase by its origin and by one of its best substrates'.

In 1959, two similar classification systems based on differences in inhibitor specificities, rather than on differences in substrate specificities, were separately published by Blaschko and Zeller with their respective co-workers (see Kappeller-Adler, 1970). The Blaschko scheme was based on whether an amine oxidase was resistant to or inhibited by carbonyl reagents whilst the Zeller scheme was more specific since semicarbazide (and not carbonyl reagents in general) was the diagnostic inhibitor reagent used.

Information on many other inhibitors of amine oxidases accumulated. In particular, the actions of copper-specific reagents proved most interesting and from considerations of the inhibitors *per se* it was a logical step to move to a system based on the parts of the enzyme molecule affected by such inhibitors. A system for classifying amine oxidases could thus be regarded in more general terms and the scheme for all oxidases devised by Keevil & Mason (1978), which was based primarily on the nature of their prosthetic groups, was able to encompass the amine oxidases (see Table 2). Within these two groups the oxidases are further subdivided in terms of the types of their interactions with oxygen. Most of the amine oxidases appear to fall into one of two main groups depending on whether they contain copper or contain flavin.

Copper-Containing Amine Oxidases

Copper occurs in only five classes of enzymes, one of which comprises the copper-containing group of amine oxidases (Stewart & Klinman, 1988) and is quite a rare co-factor in mammalian enzymes. Having said that the copper-containing oxidases are very widespread in nature. Mondovì & Riccio (1984) and Finazzi-Agro (1989) have reviewed and summarized the structural and functional properties of the enzymes in this class. They point out that the copper-containing amine oxidases are soluble and are, in structural terms, remarkably similar. All contain carbohydrate (2–14%), have molecular weights of 120 000–250 000 (some values outside this range have, however, been reported; see Matsumiya & Otake, 1986) and are composed of two subunits which have, in some instances been shown to be linked by a disulphide bridge. The two subunits are not, necessarily, the same or even of identical molecular weights, but each subunit contains one firmly bound copper atom. There are exceptions to this pattern, though: some diamine oxidases contain both copper and manganese, presumably one atom of each per molecule (Crabbe *et al.*, 1976; Matsumiya & Otake, 1986).

Each enzyme molecule is associated with one molecule of a carbonylic compound which is now known to be pyrroloquinoline quinone (PQQ) (Fig. 2) (Amayama *et al.*, 1984; Lobenstein-Verbeek *et al.*, 1984). Pyridoxal phosphate, long suspected on circumstantial evidence to be the carbonylic co-factor, is possibly, for most copper-containing amine oxidases, involved only in the synthesis of PQQ. There is still evidence, however, that pyridoxal phosphate may be the co-factor for some (e.g. chick embryo lysyl oxidase; Levene, 1989). The sequence of amino acids in the region of the active site has been determined and the probable location of the peptide link between the ε-amino group of a lysine and one of the three carboxyl groups of PQQ identified (Fig. 3).

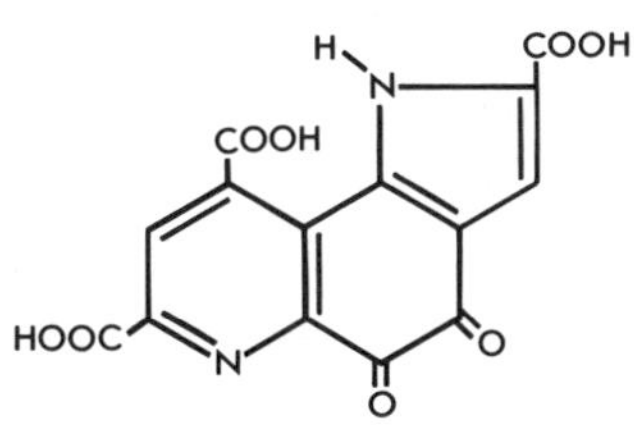

Fig. 2. Pyrroloquinoline quinone.

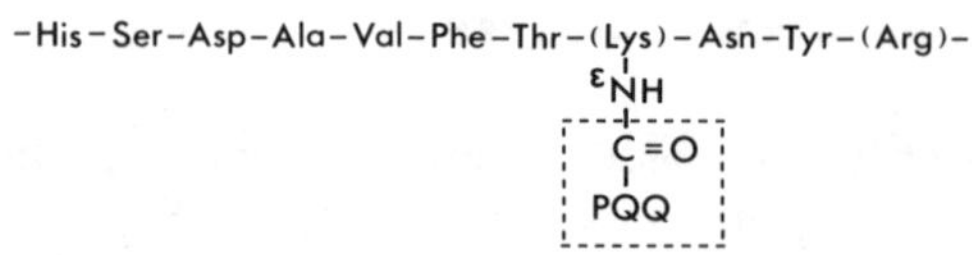

Fig. 3. Sequence of amino acids at the reactive site of porcine diamine oxidase (van der Meer *et al.*, 1989). The identity of the amino acids in brackets is only tentative.

The copper atoms are coordinately bound in the holoenzyme complex by more than one nitrogen atom per copper site; one of the PQQ nitrogens and an imidazole (histidine) nitrogen are probably involved in this way. Other possible copper ligands are water, two molecules per copper atom, at one site and protein at the other (Mondovì & Riccio, 1984; Dooley *et al.*, 1989). Copper is essential for catalysis and it may also fulfil a structural function (Mondovì & Riccio, 1989), but its exact rôle has not yet been elucidated; it possibly binds oxygen and superoxide and may be involved as a mediator in the transfer of electrons from the reduced co-factor to molecular oxygen via the metal coordinated ligand orbitals.

The copper-containing amine oxidases can act on a wide range of substrates and there are significant interspecies differences both in absolute terms as well as in terms of relative rates between enzymes from different sources. The diamine oxidases, such as those from pig kidney and pea and lentil seedlings, can oxidise many diamines, polyamines and histamine whilst the plasma amine oxidases can oxidise aliphatic and aromatic monoamines as well as spermine and spermidine.

Despite the complexity of the reactions catalysed by the copper-containing amine oxidases involving, as they do, three substrates and three products, a reasonable picture of their active sites and their probable mechanisms of action has emerged (Mondovì & Riccio, 1984). In the first step an amino group in the unprotonated form binds to a protonated group in the enzyme and the other amino or imidazole

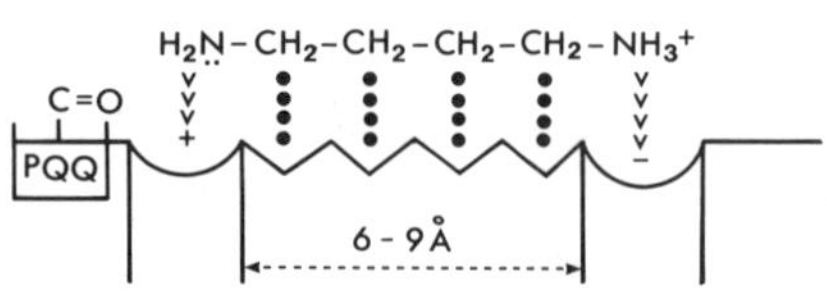

Fig. 4. Schematic model for enzyme substrate interaction at the active site of PQQ amine oxidases (after Mondovì & Riccio, 1984). Electrostatic forces ⩔; van der Waals forces ⁚.

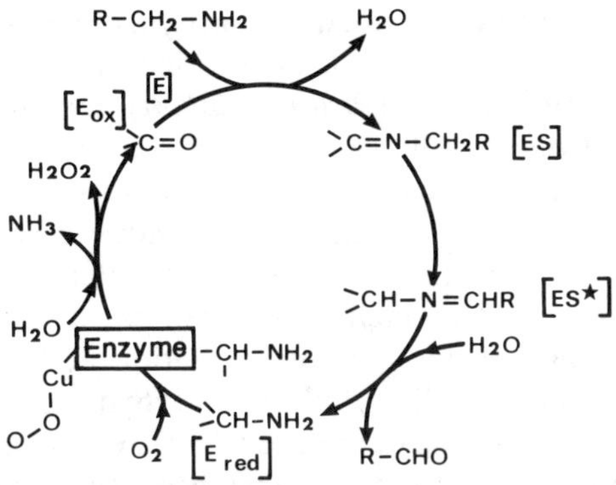

Fig. 5. Reaction mechanism for copper-PQQ-containing amine oxidases (after Mondovì & Riccio, 1984; Robinson, 1987).

group in the protonated form is attracted to a second, negatively charged, site. The distance between these two charged sites is 6–9 Å, there being a hydrophobic region between them which binds, by van der Waals forces, the hydrocarbon chain or aromatic part of the substrate (Fig. 4). The unprotonated nitrogen then forms a Schiff-base with the carbonyl group of PQQ which is close by. An activated enzyme–substrate complex is formed as the Schiff-base tautomerizes between the amino group bound to the enzyme and the aldehyde product. As the latter is split off, the enzyme in its reduced form reacts with dioxygen, and ammonia and hydrogen peroxide are released leaving the enzyme in the reoxidized form. Finally the oxidized enzyme is converted into the resting form. Figure 5 illustrates this possible ping-pong type, transaminase mechanism of action. Recent work by Klinman *et al.* (1989) has provided some additional fine details of this mechanism (at least so far as bovine plasma amine oxidase is concerned); ammonia is transferred quantitatively to the co-factor from the substrate during the reductive half reaction, and substrate oxidation occurs through a base-assisted proton abstraction mechanism.

Diamine Oxidase: EC 1.4.3.6

Recommended name; amine oxidase (copper-containing). Systematic name; amine:oxygen oxidoreductase (deaminating) (copper-containing). Other names; diamino oxhydrase, histaminase, amine oxidase (pyridoxal-containing) (IUB, 1984).

Procedures have been described for purifying and characterizing this enzyme from several sources, viz. pig kidney (Yamada *et al.*, 1967; Mondovì *et al.*, 1971), pea (*Pisum sativum*) seedlings (Hill, 1971), lentil (*Lens esculenta*) seedlings (Floris *et al.*, 1983), human intestine (Bieganski *et al.*, 1980), human pregnancy plasma (Baylin & Margolis, 1975), yellowtail (*Seriola quinqueradiata*) pyloric caeca (Matsumiya & Otake, 1986). These methods consist of various combinations of a variety of techniques which include ammonium sulphate fractionation, ethanol/chloroform precipitation, precipitation at low ionic strength, chromatography on DEAE cellulose and hydroxylapatite columns, affinity chromatography on derivatised Sepharoses, gel filtration and column electrophoresis. In some instances the enzyme has been crystallized. In addition to the sources mentioned above, the presence of diamine oxidase has been demonstrated in many other organisms and tissues.

Flavin-linked Amine Oxidases

If anything, the flavin-linked amine oxidases are even more diverse than the copper-containing ones. Stubbe (1989) has drawn attention to this and to the richness in chemistry exhibited by these enzymes. Flavin adenine dinucleotide (FAD) is the most commonly encountered prosthetic group but flavin mononucleotide (FMN) also occurs in this class of enzymes. Probably as a result of their diversity it has not proved possible to draw up as unified a scheme for the flavin-linked amine oxidases as it has been for the copper-linked ones—even if such a unity were to exist.

Monoamine Oxidase: EC 1.4.3.4

Recommended name; amine oxidase (flavin-containing). Systematic name; amine:oxygen oxidoreductase (deaminating) (flavin-containing). Other names; tyramine oxidase, tyraminase, amine oxidase, adrenalin oxidase (IUB, 1984).

As has already been mentioned, the title of this enzyme should not be taken too literally. Many diamines, particularly those which have the two amino groups well separated, can act as substrates but the most common substrates are primary monoamines and also secondary and tertiary amines with small substituents (IUB, 1984). The enzyme is located in the mitochondria and is associated with the outer membrane of that organelle. In fact, it can be used as a convenient marker enzyme, since it is stable and easy to assay, for mitochondrial outer membranes in tissue fractionation studies (Greenawalt, 1974, 1979; Comte & Gautheron, 1979). This is not, however, applicable to mitochondria from all sources (Greenawalt, 1974).

Preparation of the enzyme in purified forms involves, first, isolating mitochondria from the tissue concerned and then solubilizing the enzyme using sonication or treatment with non-ionic detergents or bile salts; treatment with phospholipase has also been employed. The main problems arise as a result of the tight attachment of the enzyme to the mitochondrial membrane and a fine balance has to be struck, during the solubilizing steps, between getting enough of the enzyme into solution and risking inactivation. Early techniques for isolating monoamine oxidases have been summarized by Kappeller-Adler (1971) and, more recently, procedures for purifying it from a wide variety of sources have been described; pig brain (Tipton, 1971; Salach *et al.*, 1976), beef liver (Yasunobu & Gomes, 1971; Salach, 1978; Salach & Weyler, 1987), rat liver (Youdim & Tenne, 1987), human liver and brain (Abell, 1987), human placenta (Salach & Weyler, 1987), *Sarcina lutea* (Yamada *et al.*, 1971). Various combinations of the following techniques have been utilized by the workers cited above to prepare purified preparations of monoamine oxidase: ammonium sulphate or ethanol fractionation; alumina C_{γ} adsorption; DEAE-cellulose, DEAE-Sephadex or hydroxylapatite column chromatography; sucrose gradient or gel separations; and immunoaffinity chromatography.

Values quoted for the molecular weight of monoamine oxidase range from 58 000 to 146 000 with one or two molecules of FAD per molecule of enzyme. There are 7 or 8 sulphydryl groups per mol of FAD, one of which is involved at the active site of the enzyme, the FAD being linked to the thioether group of cysteine by a covalent bond at the 8α position on the isoalloxazine ring. The amino acid sequence in the vicinity of the flavin–cysteine grouping is:

Ser–Gly–Gly–Cys(FAD)–Tyr

Histidine is also involved in the bonding of FAD to the apoenzyme (Kenney, 1980).

Monoamine oxidase exists in two, distinct, isoenzymic forms which have been designated Types A and B (Johnston, 1968; Hall & Urueña, 1983; Nicotra *et al.*, 1986; Abell, 1987; Salach & Weyler, 1987; Youdim & Tenne, 1987). These two types differ in substrate and

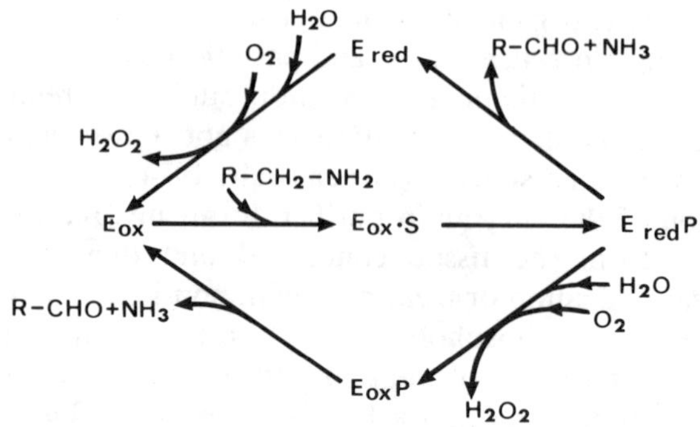

Fig. 6. Reaction mechanisms for liver monoamine oxidase (after Husain *et al.,* 1982).

inhibitor specificities but are, otherwise, remarkably similar; they have, however, been separated by sodium dodecylsulphate-polyacrylamide gel electrophoresis. It is postulated that they perform distinct physiological rôles, particularly in brain tissue. A third 'classical' type of monoamine oxidase has been described from carp liver (Kinemuchi *et al.*, 1983). Metal ions are not believed to be necessary for the catalytic action of monoamine oxidase. Findings that iron may be essential are probably due to the fact that it might somehow be involved in the synthesis of the enzyme molecule (Youdim, 1976; Dallman, 1986).

Various forms of ping-pong mechanisms have been proposed for monoamine oxidases and it may be that enzymes from different sources act in different ways (Oi *et al.*, 1970; Yasunobu *et al.*, 1971; Abell, 1987). Husain *et al.* (1982), however, concluded that, depending on the substrate, monoamine oxidase acts through either a binary (ping-pong) or a ternary (sequential) complex mechanism (Fig. 6). For example, benzylamine is oxidatively deaminated by the lower loop and β-phenylethylamine by the upper. Further detail of the mechanism has been supplied by compelling evidence (Silverman and co-workers, cited by Stubbe, 1989) that substrate radicals are involved and that thiyl radicals may participate in the reaction.

Pyridoxamine-Phosphate Oxidase: EC 1.4.3.5

Systematic name; pyridoxamine-5′-phosphate:oxygen oxidoreductase (deaminating) (IUB, 1984).

Pyridoxamine-phosphate oxidase is unusual in that it can accept as substrates both the amine, pyridoxamine-5′-phosphate and the alcohol, pyridoxine-5′-phosphate, and oxidize each to the same product, the coenzyme pyridoxal 5′-phosphate (Wada & Snell, 1961). It was later shown that the enzyme could also oxidize secondary amines belonging to the family of *N*-substituted pyridoxamine phosphates, the *N*-(5′-phospho-4′-pyridoxyl)amines (Kazarinoff & McCormick, 1975). A recent review (McCormick & Merrill, 1980) describes the history of the development of knowledge about this enzyme. Pogell (1958) was the first to describe it adequately and showed that its activity was present in extracts from several mammalian tissues and microbial cells. Wada & Snell (1961) also demonstrated that riboflavin 5′-phosphate (FMN) constitutes the prosthetic group. This enzyme is widely distributed in animals, plants and microbes (McCormick & Merrill, 1980; Tsuge *et al.*, 1980, 1982*a, b*; Langham *et al.*, 1982; Pflug & Lingens, 1983; Tsuge *et al.*, 1986; Fonda, 1988). Its relative activities towards the two main substrates (pyridoxamine 5′-phosphate and pyridoxine 5′-phosphate) vary from organism to organism and, within an organism, from organ to organ. It is now known to occur in two or three isoenzymic forms in certain tissues (Tsuge *et al.*, 1982*a, b,* 1986) each isoenzyme exhibiting different preferences for the two substrates.

Purified preparations of pyridoxamine-phosphate oxidase have been isolated from rabbit liver (Wada & Snell, 1961; Pogell, 1963; Kazarinoff & McCormick, 1975; Merrill *et al.*, 1979; Bowers-Komro *et al.*, 1986), rat brain (Cash *et al.*, 1980), rat liver (Nutter *et al.*, 1983), pig brain (Churchich, 1984), baker's yeast (Tsuge *et al.*, 1979, 1982*b*) and wheat seedlings (Tsuge *et al.*, 1982*a*). The enzyme is soluble, occurring in the cytosol fraction, and is readily extracted. After extraction various combinations of the following techniques have been employed to purify the enzyme to homogeneity: acid and ethanol precipitation; chromatography on Sephadex, DEAE-Sephadex, calcium phosphate gel and organomercurial agarose columns; and affinity chromatography.

Yeast and liver pyridoxamine-phosphate oxidase are similar in that their molecules are composed of two subunits with unit weights around 25 000–28 000 Daltons, there being one FMN molecule per 53 000–54 000 Daltons (Tsuge *et al.*, 1980; Nutter *et al.*, 1983; Bowers-Komro & McCormick, 1984). The enzyme from *Alcaligenes faecalis* is somewhat different in that it consists of a single 38 000 Daltons unit associated with one FMN molecule (Yamamoto & Nagasaki, 1969, cited by Tsuge *et al.*, 1980).

Kinetic mechanisms similar to those proposed for monoamine oxidase (Fig. 6) have also been suggested for pyridoxamine phosphate

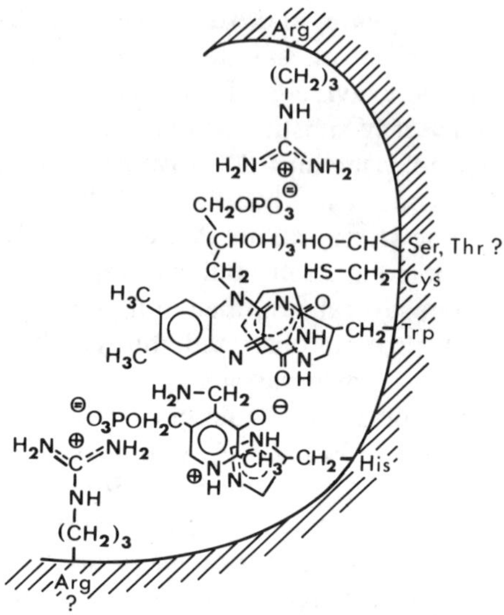

Fig. 7. Topography of the active site of pyridoxamine 5′-phosphate oxidase. (Reproduced by kind permission of Food and Nutrition Press Inc., Connecticut, from McCormick & Merrill, 1980.)

oxidase (Choi *et al.*, 1983). As with the former enzyme the reaction proceeds via a binary or a ternary complex depending on the substrate. In the case of pyridoxine phosphate oxidation a binary complex (ping-pong) mechanism operates whilst pyridoxamine phosphate is oxidized via a ternary complex (sequential) mechanism. All the evidence relating to the involvement of different amino acids with this enzyme's catalytic activity has been summarized by Bowers-Komro & McCormick (1984) and a plan of the active site drawn up (Fig. 7).

Polyamine Oxidase

Several of the mono- and diamine oxidases described will act on the polyamines spermine and spermidine but a group of enzymes, known collectively as polyamine oxidases and which have distinctly different properties from the enzymes in those two other groups, has recently been described (Tabor & Tabor, 1984). Members of this group of enzymes have been found in rat liver (Hölttä, 1977, 1983), in various members of the family Graminae (the grasses) (Smith, 1983), in a pseudomonad (Padmanabhan & Kim, 1965), in a catfish (*Parasilurus asotus*) (Kumazawa & Suzuki, 1987) and in various human tissues (Suzuki *et al.*, 1984). The rat liver polyamine oxidase occurs in the peroxisomes while the Graminae one is in the cell walls of vascular tissues.

Procedures for purifying the enzyme from rat liver and seedlings of oat, maize and barley have been published (Hölttä, 1983; Smith, 1983). These employ combinations of ammonium sulphate or acetone precipitation of extracts, column chromatography using DEAE-cellulose or -Sephadex and hydroxylapatite, gel filtration and polyacrylamide gel electrophoresis. The purified enzymes have molecular weights which range from 60 000 to 80 000 and contain tightly bound FAD. There is evidence that ferrous iron may also be a co-factor for the rat liver enzyme.

Polyamine oxidases can oxidise spermine and spermidine, as well as N^1-acetylspermidine and N^1-acetyl- and N^1,N^{12}-diacetylspermine. In fact the acetylated derivatives are better substrates than their respective parent compounds (Bolkenius & Seiler, 1981). The reactions proceed according to the scheme illustrated in Fig. 8. It is interesting to note that the enzymes fall into two subgroups according to which side of the N^4 atoms of spermine and spermidine scission takes place. Incidentally, a spermidine dehydrogenase from *Serratia marcescens* behaves in this regard like the polyamine oxidase from oat seedlings.

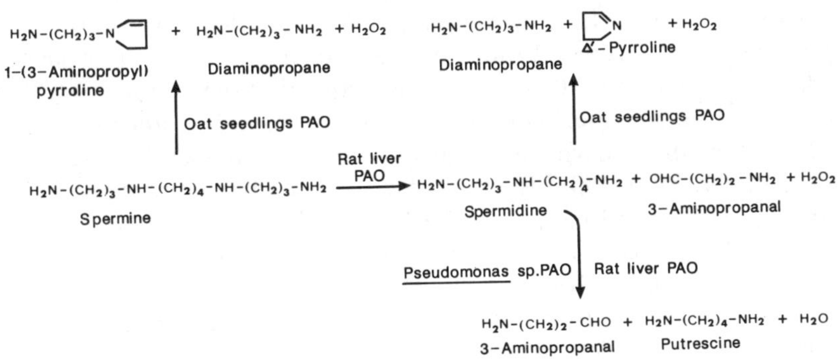

Fig. 8. The reactions catalysed by polyamine oxidases (PAOs) from oat seedlings, rat liver and *Pseudomonas* sp.

AMINO ACID OXIDASES

This group contains enzymes which have a wide range of specificities. At one end of the spectrum there are the L- and D-amino acid oxidases which can act on most amino acids with the appropriate stereochemistry or even on occasions, albeit at very low levels, exhibit activity towards certain enantiomers. On the other hand there are the specific amino acid oxidases, such as L-glutamate oxidase, which act exclusively on one enantiomer of a single amino acid.

As with the amine oxidases, the reaction of amino acid oxidases can be described as oxidative deamination. The α-amino acid is converted into the corresponding α-keto acid, ammonia is released and hydrogen peroxide formed.

$$\underset{\displaystyle NH_2}{R.\underset{|}{C}H.COOH} + H_2O + O_2 \rightarrow \underset{\displaystyle O}{R.\underset{\|}{C}.COOH} + NH_3 + H_2O_2$$

The hydrogen peroxide thus formed can, however, react non-enzymically with the α-keto acid to produce the next lower homologous carboxylic acid thus

$$\underset{\displaystyle O}{R.\underset{\|}{C}.COOH} + H_2O_2 \rightarrow R.COOH + CO_2 + H_2O$$

The actual reaction products will thus depend on whether, and how quickly, or not hydrogen peroxide is removed from the reactive milieu.

The subject of the oxidative deamination of amino acids has been reviewed by Meister (1965). Neubauer (1909) and Knoop (1910) (cited by Meister) first described the production of α-keto acids from amino acids but it was not until some 20 years later that the picture began to clarify and the two mechanisms, transamination and oxidative deamination, for the enzymic production of α-keto acids could be differentiated from each other. Krebs, in the early 1930s, showed that mammalian liver and kidney catalysed the oxidation of D- and L-amino acids (Krebs, 1935). D-Amino acid oxidase activity was separated from the system which catalysed L-amino acid oxidation and the latter was subsequently shown to be due in part to coupled dehydrogenase–transaminase activities as well as, more recently (see Nakano & Danowski, 1971), to a specific L-amino acid oxidase. All the amino acid oxidases described so far in the literature are flavoproteins (see

Table 2), FAD being far and away the most common prosthetic group.

D-Amino acid oxidases occur throughout the animal kingdom. Amongst the vertebrates the enzymes from mammalian liver and kidney have been the most studied but activities have been reported from mammalian brain as well as from birds, amphibians, reptiles and fish. Insects, molluscs, moulds and bacteria also contain specific or the general D-amino acid oxidases (Meister, 1965). The enzymes occur in the peroxisomal or glyoxysomal cellular fractions (Tolbert, 1978). The occurrence in mammalian kidney of an L-amino acid oxidase has already been mentioned but the most widely studied examples of this enzyme are those which occur in the venoms of certain snakes.

D-*Amino Acid Oxidase: EC 1.4.3.3*

Systematic name; D-amino acid:oxygen oxidoreductase (deaminating) (IUB, 1984).

The classical source for preparing D-amino acid oxidase is pig kidney. In purified or partially purified form the enzyme is unstable and a stabilizer (benzoate) is used. A pure, crystalline enzyme–benzoate complex can be prepared from extracts of kidney by employing heat treatment, ammonium sulphate fractionation, column chromatography on calcium phosphate gel-cellulose and on Sephadex G-75 (Yagi, 1971). The benzoate can be readily removed from the complex and the holoenzyme and apoenzyme prepared (Curti *et al.*, 1973; Tu *et al.*, 1973).

The molecular weight of D-amino acid oxidase monomer (per FAD molecule) is 36 000–40 000 (Tojo *et al.*, 1982); the monomer is known to associate and form di-, tri- and, perhaps, higher oligomers (Tojo *et al.*, 1984); the dimeric form is probably the one in which the holoenzyme exists and it is probable that FAD binding induces co-operative binding (Tanaka & Yagi, 1980). D-Amino acid oxidase has a high content of aromatic amino acids (Yagi, 1971) and there is strong evidence that arginine, tyrosine and histidine are in the active site and implicated in the catalytic activity of the enzyme. The arginine acts as a positively charged group near the flavin N(1)—C(2)═O grouping and interacts with the carboxylate group of the substrate (Nishino *et al.*, 1980; Ferti *et al.*, 1981; Fitzpatrick & Massey, 1983; Simonetta *et al.*, 1984). The tyrosine interacts electrostatically with the α-amino group of the substrate and probably has a rôle to play in reduction of the flavin (Nishino *et al.*, 1980; Ferti *et al.*, 1981), while the histidine functions as a proton abstractor from the α-carbon atom of the substrate, thus forming the substrate carbanion intermediate during the electron transfer step; the flavin thus forms an *N* (5)-substrate intermediate (Nishino *et al.*, 1980; Ferti *et al.*, 1981; Swenson *et al.*, 1984). Cysteine is the only other amino acid for which some evidence exists for involvement in the active site and it is probably involved in the binding of FAD (Ronchi *et al.*, 1976).

The mechanism for the reaction catalysed by D-amino acid oxidase has been reviewed by Yagi (1971) who outlined the history of its elucidation. The isolation and crystallization of an enzyme–substrate intermediate (the purple complex) for a neutral amino acid such as

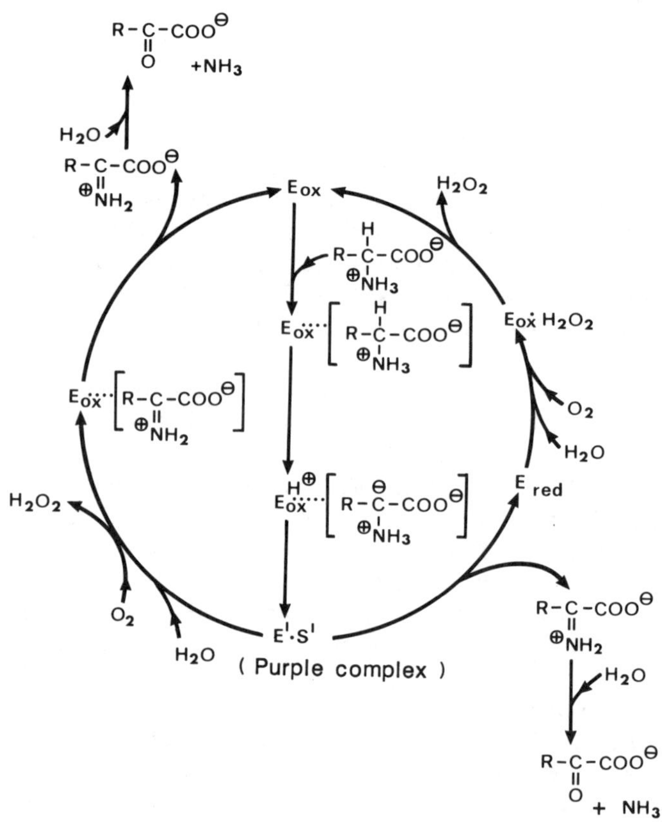

Fig. 9. Reaction mechanisms for D- (and L-) amino acid oxidases (after Yagi, 1971; Bright, 1976; Nishino *et al.*, 1980).

D-alanine has aided this exposition. The basis of this reaction mechanism is illustrated in Fig. 9; both left hand and right hand routes are utilized by this enzyme, as they also are by L-amino acid oxidase.

The enzyme is relatively non-specific being able to act on a wide range of D-α-amino acids and *N*-monoalkylated D-α-amino acids. Dicarboxylic amino acids are not oxidized whereas L-proline is, albeit at a low rate (Meister, 1965; Yagi, 1971).

FOOD AND NUTRITIONAL IMPLICATIONS

Pre-harvest

The biochemical importance of many amines has long been recognized. As a class they are pharmacologically and nutritionally important. They occur widely in nature in diverse molecular forms and exhibit a wide range of physiological activities which include regulation of enzyme mechanisms, of growth and of differentiation processes, interaction with polyanions (e.g. nucleic acids and heparin) and with biological membranes, bactericidal and pressor activities, neurotransmission and, also, toxicity.

Amine Oxidases

The considerable importance of the functions exerted by amine oxidases *in vivo* stems from the physiological importance of their substrates. These enzymes can be involved in regulating the cellular concentrations of biogenic amines and also, by influencing the relative concentrations of amines, they can affect the latters' competitive power for the physiological receptor. In plants, amine oxidases appear in the biosynthetic pathways of certain alkaloids (e.g. nicotine and strychnine) and are involved actively during periods of growth. Amine oxidases also fulfil a scavenging and detoxifying role protecting living organisms from amines that have been ingested or produced *in vivo*. The detoxification of histamine and the catabolism of the catecholamines (DOPamine, epinephrine and nor-epinephrine) are particularly important examples of these last functions.

Cheese Production

Many cheeses are rich in amines, tyramine and histamine in particular, and this can cause severe reactions in the consumer on occasion (see later, under 'Monoamine oxidase inhibitors'). The amines are produced by enzymic decarboxylation from the corresponding amino acids by bacteria in the starter cultures. Studies have shown that it is the availability of free amino acids which is the major limiting factor in the formation of biologically active amines, not only in cheese but also in other fermented foods. Of secondary importance are the activities of amine-producing or -reducing enzymes (amino acid decarboxylases and amine oxidases, respectively). Those cheeses which possess oxidase activity tend to contain lower concentrations of amines and it is possible that amine contents could be controlled by the judicious selection of dairy starters (Voigt & Eitenmiller, 1978; Sumner, 1987). The possible effects of such selections on the sensory properties of the cheese would have to be considered, however.

Vitamin B_6 Metabolism

The key position of pyridoxamine-phosphate oxidase in the control of vitamin B_6 metabolism has long been recognized. Not only does it have as substrates two separate B_6 vitamins (pyridoxamine 5′-phosphate and pyridoxine 5′-phosphate) but its activity is strongly affected, through a product inhibition mechanism, by the coenzyme pyridoxal 5′-phosphate.

Crabby Taint in Eggs

Another amine oxidase (trimethylamine oxidase), plays a central rôle in preventing a problem which has been described as crabby or fishy taint in the brown-shelled eggs laid by certain breeds of hen. Normally in the hen, trimethylamine, of dietary origin *per se* or produced from dietary trimethylamine oxide by reduction in the gut, is oxidised by the liver microsomal trimethylamine oxidase to give a tasteless and odourless product, trimethylamine oxide (Fig. 10). In the affected poultry strains, an inherited deficiency in trimethylamine oxidase activity reduces the extent of trimethylamine oxidation and allows some of the free base to become excreted in the eggs where, at concentrations above 0.8 μg per g the crabby taint is detectable. This relatively simple picture becomes complicated, however, when hens belonging to these susceptible strains are fed diets which include rapeseed meal. Sinapine (the choline ester of sinapic acid) present in the

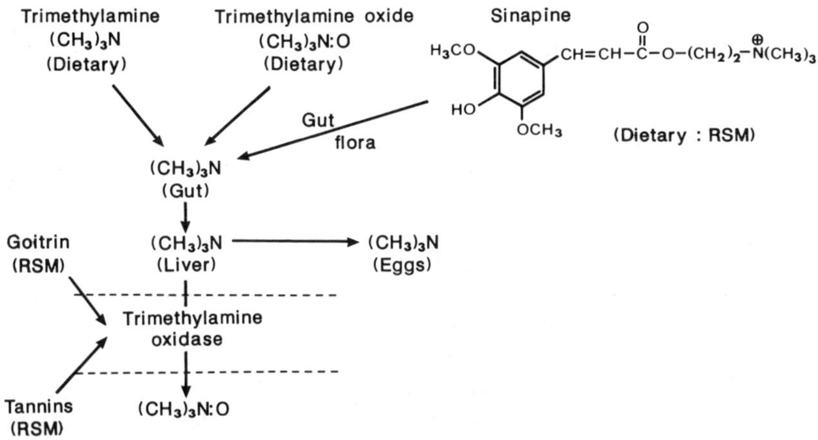

Fig. 10. The influence of rape seed meal (RSM) constituents on trimethylamine and trimethylamine oxide metabolism in susceptible hens.

meal is converted to trimethylamine by enteric bacteria in the caecum and can overload the hepatic trimethylamine oxidase system. Furthermore, rapeseed meal contains goitrin (5-vinyl-2-oxazolidinethione: OZT) and tannins, both of which act as inhibitors of trimethylamine oxidase thus increasing still further the tendency for trimethylamine to appear in the eggs (Fenwick & Curtis, 1980; Pearson *et al.*, 1980; Fenwick *et al.*, 1981; Butler *et al.*, 1984). The inherited enzymic insufficiency mentioned appears to be closely related to the ability to produce brown-shelled eggs but recent work has shown that the genetic factor which influences trimethylamine oxidase activity is not directly linked to shell pigmentation (Goh *et al.*, 1985). Thus selective breeding techniques would stand a chance of producing brown-shelled eggs without their being susceptible to crabby taint.

Lathyrism

Another, more serious, amine oxidase deficiency which can affect food production is lathyrism. This condition is probably caused by inhibition of lysyl oxidase, which catalyses the development of cross-links in collagen and elastin. Insufficiency will lead to impaired skeletal development in, for instance, young chicks, while ingestion of lysyl oxidase inhibitors, such as those contained in sweet pea (*Lathyris odoratus*) seeds, can cause spinal deformities, skin fragility, and rupture of the aorta and other aneurysms (Metzler, 1977; Kagan *et al.*,

1989). It is pertinent to mention at this stage the putative use of lysyl oxidase inhibitors in the chemotherapeutic treatment of cystic fibrosis (Kagan *et al.*, 1989).

Amino Acid Oxidases

The main physiological function of the amino acid oxidases is somewhat obscure but they remove the enantiomeric amino acids which might not otherwise be accepted by the organism. The corresponding keto acid can be produced and then be converted into the required stereoisomeric amino acid and brought into the metabolic pool by the action of the appropriate transaminase.

Production of Pyridoxal Phosphate

Pyridoxamine phosphate oxidase from bakers' yeast can be immobilized by binding to a commercial organomercurial agarose (Affi-Gel 501) (Tsuge *et al.*, 1978) or to relatively non-toxic halogenoacetyl polysaccharides (Tsuge & Okada, 1984). Such immobilized preparations lend themselves to the continuous production of the coenzyme pyridoxal 5′-phosphate from culture media containing pyridoxine 5′-phosphate, pyridoxamine 5′-phosphates or *N*-(5′-phosphopyridoxyl)-L-serine.

Post-harvest

The post-harvest importance of amine oxidases in foods is mainfest in a number of direct as well as indirect ways and it is, perhaps, appropriate to deal first with one of the indirect ones because of its dietary importance to the human consumer.

Monoamine Oxidase Inhibitors

The in-vivo activity of amine oxidases in general and monoamine oxidases in particular helps control the intracellular levels of amines and, in the case of monoamine oxidases in the intestine and liver, reduces the amounts which would otherwise be absorbed from the food. The action of monoamine oxidase-inhibiting drugs, which have been used in the treatment of tuberculosis and of certain conditions such as depressive illnesses, suppresses the activity of these control and scavenging mechanisms. In such cases, several patients developed severe headaches, hypertension, nausea and vomiting, and occasionally, brain haemorrhage or heart failure. These symptoms were caused by the vasoconstrictive and cardio-stimulatory actions of food-derived pressor amines directly, or indirectly as in the case of tyramine which induces the release of norepinephrine from the sympathetic nervous system. Many foods and drinks contain amines in amounts sufficient to trigger these reactions but cheeses can play a pre-eminent role in this respect; in fact the syndrome is popularly known as the 'cheese reaction'. Other foods which contain significant quantities of amines are several types of meat, poultry and fish products, yeast extract, bananas, citrus fruits, spinach, beer, red wines (chianti), chocolates, tea and coffee (Smith, 1981).

Scombrotoxic Fish Poisoning

Another unpleasant set of reactions (nausea, vomiting and diarrhoea in extreme instances) caused, allegedly, by ingested amines has been termed 'scombrotoxin poisoning' because of its association, but not invariably, with spoiled fish of the families *Scomberesocidae* and *Scombridae*; non-scombroid fish like sardines, anchovies and herring have also been implicated. Histamine, produced by bacterial decarboxylases from histidine, has been indicted but is now known to be not the only causative agent. Like histamine, scombrotoxin is stable and is produced at ambient temperatures of 10–20°C; but samples containing high concentrations of histamine are not necessarily scombrotoxic, whilst some scombrotoxic samples contain little if any histamine.

It is possible that a microbial toxin could produce these gastrointestinal symptoms, the histamine being present coincidentally, or other microbial metabolites may directly or indirectly behave synergistically with histamine. For instance, other biogenic amines could act as competitive or non-competitive inhibitors of monoamine oxidase thereby interfering with histamine detoxication (Clifford *et al.*, 1989; Ijomah *et al.*, 1991). It would be pleasant to speculate that the amounts of amines present could have been ameliorated to some extent by the amine oxidases present but this has not been demonstrated. Attempts, however, to reduce deliberately the amounts of histamine produced in minced mackerel flesh by the addition of plant or animal diamine oxidase preparations were not successful (Pavelka, 1982).

Sodium Cyclamate and Cyclohexylamine

Sodium cyclamate, a quondam permitted artificial sweetener, is metabolised in the gut to cyclohexylamine by a pseudomonad sulphamatase. Cyclohexylamine is known to be toxic but it appears to be readily eliminated from the body by excretion in the urine. A further metabolite, cyclohexanone, is also excreted in the urine and this has been shown to result from the action of cyclohexylamine oxidase, present in pseudomonads as well, on the cyclohexylamine (Tokieda *et al.*, 1977).

Catabolism of Lysinoalanine

The heat treatment of some proteins, particularly under alkaline conditions, can lead to the formation of lysinoalanine, a known inducer of kidney lesions in the rat (Annan & Manson, 1981). It is a moot point, however, if it is lysinoalanine itself which causes the lesions or if it is the products of its oxidation through the action of L-amino acid oxidase in the kidney. These have been shown to be enantiomers of 1,7-diazabicyclo[4.3.0]nonane-6,8-dicarboxylic acid when the hydrogen peroxide also produced is removed by the action of catalase known also to be in the kidney (Engelsma *et al.*, 1979; Leegwater & Tas, 1980). This is just one example of the somewhat ironic situations where it is the product of a detoxication process which actually causes pathological lesions to develop (Fig. 11).

Dimethylnitrosamine Production

Dimethylamine occurs at very low levels in fresh fish and is produced during spoilage in the unfrozen state (Huss, 1988); the levels produced, however, are low in comparison with those of trimethylamine, which is the major product of trimethylamine oxide (TMAO) degradation under such conditions where bacterial actions (TMAO reductases are the enzymes responsible) predominate (Ikeda 1980). In the frozen state, on the other hand, bacterial activity is suppressed and TMAO is degraded by the action of an intrinsic TMAO demethylase, the products in this case being dimethylamine and formaldehyde (Rehbein, 1988).

Dimethylamine can also be formed by another route which involves the action of trimethylamine oxidase on trimethylamine and this has been claimed to be of major significance in the production of relatively high concentrations (2043 μg/g) of dimethylamine in dried seafoods, particularly in dried squid products. Fresh squid tissue contains high trimethylamine oxidase activities and it has been pointed out that the potential for production in the human body (or during processing prior to ingestion) of *N,N*-dimethylnitrosamine could be sufficient to have

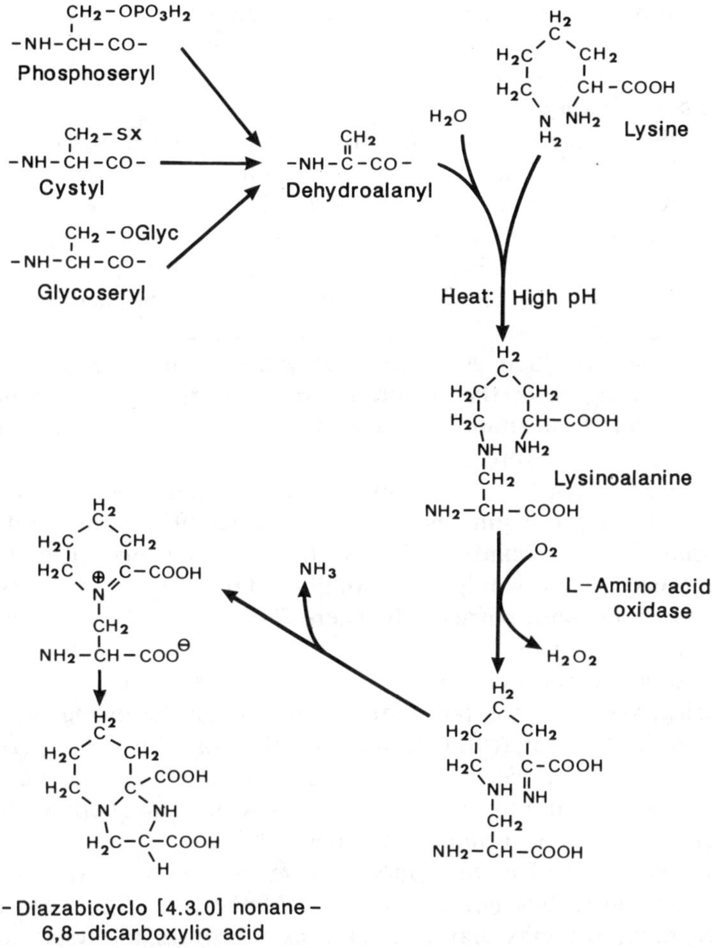

Fig. 11. The action of L-amino acid oxidase on lysinoalanine (after Leegwater & Tas, 1980; Annan & Manson, 1981).

carcinogenetic implications on an epidemiological basis in certain parts of the world (Lin *et al.*, 1984).

So far as the actual concentrations of *N,N*-dimethylnitrosamine that have been observed in dried and smoked seafood products are concerned, these appear to be several orders of magnitude less than those of the amine itself (Matsui *et al.*, 1984*a,b*,; Sikorski, 1988). The highest levels of *N*-nitroso compounds in general found in foods are in those products like bacon, which can contain high levels of nitrates and nitrites, and malt dried in contact with combustion gases.

Effects on Quality

In addition to the specific reactions described above, amine oxidases and amino acid oxidases will affect the quality of foods in more general ways. Amines present in foods will add to overall flavour be it desirable, as in some cheeses, or undesirable, as in spoiling fish. Similarly free amino acids are known to be partly responsible for the sweet and meaty notes of fresh flesh foods. The actions of the respective oxidases will in the first instance reduce the contributions to flavour made by their substrates but will also produce a secondary effect resulting from the flavorous characteristics of the carbonylic products and of the ammonia released. Furthermore, in the case of the amino acids, the α-keto acids produced will enter the pool of keto and amino acids which are readily interactive and interchangeable through the activity of transaminases (Murray & Burt, 1974). The production of carbonylic compounds could also have an influence on carbonyl–amino reactions which lead to Maillard browning, to reduction of available lysine and, particularly where formaldehyde is concerned, to toughening by reaction with proteins (Rehbein, 1988).

The other reaction product, hydrogen peroxide, is highly reactive in biological systems. It is postulated that hydrogen peroxide can affect the growth of bacteria, thus leading to retention of freshness (Burt & Hardy, 1991) but can also have deleterious effects in promoting oxidations of lipid and other constituents with consequential changes in flavour and colour (Harel & Kanner, 1985).

The possibility of textural modification by the use of lysyl oxidase has been raised but not yet studied (Matheis & Whitaker, 1987). Whilst many proteins may not serve as substrates for lysyl oxidase, collagen, elastin and soluble proteins from bovine aorta will. This enzyme catalyses the oxidative deamination of lysyl and δ-hydroxylysyl residues in the substrate to allysyl (α-amino adipic acid δ-semialdehyde) and hydroxyallysyl (δ-hydroxy α-amino adipic acid δ-semialdehyde) residues, respectively. The semialdehydes can react spontaneously together or with other aldehydes (aldol condensation) or with lysyl, δ-hydroxylysyl and histidyl residues (aldimine reaction) to form intra- and intermolecular cross-linkages. These cross-linkages could act as the bases for a texture-modifying matrix in certain formulated foods.

Specifically excluded from consideration here is the oxidation of tyrosine by tyrosinase (phenolase, phenoloxidase, catechol oxidase) which is involved in the formation of blackspot in crustaceans and mushrooms. This topic is discussed in Chapter 6.

Analytical Applications

Insoluble or immobilized enzymes have many advantages when applied in analytical studies; their specificity, sensitivity and re-use potential being the main ones. It is therefore not surprising that amine and amino acid oxidases have been studied in this context. The particular advantages, in terms of recording and automating, when immobilised enzymes are incorporated into electrode systems confer additional bonuses. Clark (1979) listed 15 amine and amino acid oxidases which had been, or could be, used in association with the hydrogen peroxide-sensing platinum electrode. He points out that applications exist both for those with wide and for those with narrow specificities. Electrodes other than the hydrogen peroxide-sensing platinum one have also been used; L-phenylalanine, L-leucine and L-methionine have been determined using L-amino acid oxidase in association with an ammonium ion electrode and L-amino acid oxidase plus horseradish peroxidase in association with an iodide-sensing electrode (Guilbault, 1976). D-Amino acids have also been determined using D-amino acid oxidase in conjunction with an ammonium ion electrode (Guilbault, 1976).

In the food field, monoamine oxidase immobilized in a collagen membrane along with an oxygen-sensing electrode has been used to determine amines in meat paste (Karube *et al.*, 1980; Karube & Suzuki, 1982). Putrescine oxidase but not in an immobilized state, has also been used in conjunction with an oxygen electrode to determine the putrescine, cadaverine and spermidine contents in mackerel (Todoriki *et al.*, 1988). Lentil diamine oxidase has been immobilized and used both in an automated electrode system and in test strips (in the latter case in association with horseradish peroxidase and leuco crystal violet) to assess the biogenic amine contents of spoiling fish; a fast plate reading system, using the same ingredients which were employed in the test strips, has also been developed and tested (Gouygou *et al.*, 1991). This dye-linked system, originally used by Lerke *et al.* (1983) to screen fish for histamine, has also been used by

Sumner & Taylor (1989) to screen dairy-related bacteria for their ability to form histamine. Somewhat earlier, however, Jahns & Rand (1977, 1978) used immobilized diamine oxidase and xanthine oxidase, coupled with *o*-tolidine and peroxidase in the one instance and with resazurin on its own in the other, to assess the freshness of fish. Such rapid screening techniques as these have to be employed with caution, particularly in instances where they may be applied to the native material or to undenatured extracts. As catalase is present in the tissue or in the bacterial flora, any hydrogen peroxide produced could be destroyed before it had a chance to oxidise the indicator dye; alternatively enzyme systems, other than the diamine oxidase added, could also produce hydrogen peroxide.

A system for detecting lysine produced by bacteria has been developed (LiMuti & Paulson, 1989). This also uses a peroxidase–chromogen (4-aminoantipyrene plus 3-aminophenol) combination to detect hydrogen peroxide, in this instance incorporated into an agar plate, so that lysine-producing bacteria can be readily identified as they grow.

The extent of amino acid racemization in alkali treated proteins has been measured with the aid of D-amino acid oxidase and catalase co-immobilized on porous succinamidopropyl glass beads. The amino acid mixture, prepared by acid hydrolysis, was passed repeatedly over the glass beads; comparison of the amino acid profiles, before and after such exposure, gave a measure of the extent of racemization for each amino acid (Chung *et al.*, 1985).

Finally, L-glutamate oxidase has been used along with an oxygen electrode to measure the concentrations in soy sauce of L-glutamate, an essential flavour constituent (Kusakabe *et al.*, 1984). The enzyme has also been immobilized by entrapment in gelatin and used in conjunction with a platinum hydrogen peroxide sensing electrode (Wollenberger *et al.*, 1989). This last mentioned application raises the possibility of automatic control, based on the production of an important constituent of the product, in the soya fermentation industry.

ACKNOWLEDGEMENTS

The author is indebted to several colleagues for their assistance and in particular to Dr D. M. Gibson and Mr C. K. Murray for their helpful suggestions.

REFERENCES

Abell, C. W. (1987). Monoamine oxidase A and B from human liver and brain. In *Methods in Enzymology,* Vol. 142, ed. S. Kaufman. Academic Press, London, pp. 638–50.

Amayama, M., Hayashi, M., Matsushita, K., Shinagawa, E. & Adachi, O. (1984). Microbial production of pyrroloquinoline quinone. *Agricultural and Biological Chemistry,* **84,** 561–5.

Annan, W. D. & Manson, W. (1981). The production of lysinoalanine and related substances during processing of proteins. *Food Chemistry,* **6**(3), 255–61.

Bachrach, U. (1973). *Function of Naturally Occurring Amines.* Academic Press, New York, London.

Barrett, E. L. & Kwan, H. S. (1985). Bacterial reduction of trimethylamine oxide. *Annual Review of Microbiology,* **39,** 131–49.

Baylin, S. B. & Margolis, S. (1975). Purification of histaminase (diamine oxidase) from human pregnancy plasma by affinity chromatography. *Biochimica et Biophysica Acta,* **397,** 294–306.

Bieganski, T., Kusche, J., Feussner, K.-D., Hesterberg, R., Richter, H. & Lorenz, W. (1980). Human intestinal diamine oxidase: substrate specificity and comparative inhibitor study. *Agents and Actions,* **10**(1/2), 108–10.

Blaschko, H. (1963). Amine oxidase. *The Enzymes,* **8,** 337–51.

Bolkenius, F. N. & Seiler, N. (1981). Acetylderivatives as intermediates in polyamine catabolism. *International Journal of Biochemistry,* **13,** 287–92.

Bowers-Komro, D. M. & McCormick, D. B. (1984). Steric restrictions in the active-site region of liver pyridoxamine (pyridoxine) 5′-phosphate oxidase. In *Progress in Clinical Biological Research. 144A. Chemical and Biological Aspects of Vitamin B_6 Catalysis: Part A.* Alan R. Liss Inc., New York, pp. 387–96.

Bowers-Komro, D. M., Hagen, T. M. & McCormick, D. B. (1986). Modified purification of pyridoxamine (pyridoxine) 5′-phosphate oxidase from rabbit liver by 5′-phosphopyridoxyl affinity chromatography. In *Methods in Enzymology,* Vol. 122, ed. F. Chytil & D. B. McCormick. Academic Press, London, pp. 116–20.

Bright, H. J. (1976). Flavoprotein oxidases. In *Food Related Enzymes,* ed. J. R. Whitaker. *Advances in Chemistry* series, American Chemical Society, Washington DC, pp. 305–23.

Burt, J. R. & Hardy, R. (In press). The composition and deterioration of pelagic fish. In *Pelagic Fish: the Resource and its Exploitation,* ed. J. R. Burt, R. Hardy & K. J. Whittle. Fishing News Books, Oxford, UK.

Butler, E. J., Pearson, A. W. & Greenwood, N. M. (1984). Trimethylamine taint in eggs: the occurrence of the causative metabolic defect in commercial hybrids and pure breeds in relation to shell colour. *Journal of the Science of Food and Agriculture,* **35,** 272–8.

Cash, C. D., Maitre, M., Rumigny, J. F. & Mandel, P. (1980). Rapid purification by affinity chromatography of rat brain pyridoxal kinase and pyridoxamine-5′-phosphate oxidase. *Biochemical Biophysical Research Communications,* **96**(4), 1755–60.

Choi, J. D., Bowers-Komro, D. M., Davis, M. D., Edmondson, D. E. & McCormick, D. B. (1983). Kinetic properties of pyridoxamine (pyridoxine)-5′-phosphate oxidase from rabbit liver. *Journal of Biological Chemistry*, **258**, 840–5.

Chung, S.-Y., Swaisgood, H. E. & Catignani, G. L. (1985). Measurement of amino acid racemization in alkali-treated proteins using an immobilized D-amino acid oxidase-catalase reactor. *Journal of Agricultural and Food Chemistry*, **133**, 201–4.

Churchich, J. E. (1984). Brain pyridoxine-5-phosphate oxidase. A dimeric enzyme containing one FMN site. *European Journal of Biochemistry*, **138**, 327–32.

Clark, L. C. J. Jr (1979). The hydrogen peroxide sensing platinum anode as an analytical enzyme electrode. In *Methods in Enzymology*, Vol. LVI, ed. S. Fleischer & L. Packer. Academic Press, London, pp. 448–79.

Clifford, M. N., Walker, R., Wright, J., Hardy, R. & Murray, C. K. (1989). Studies with volunteers on the role of histamine in suspected scombrotoxicosis. *Journal of the Science of Food and Agriculture*, **47**, 365–75.

Comte, J. & Gautheron, D. C. (1979). Preparation of outer membrane from pig heart mitochondria. In *Methods in Enzymology*, Vol. LV, ed. S. Fleischer & L. Packer. Academic Press, London, p. 101.

Crabbe, J. M. C., Waight, R. D., Bardsley, W. G., Barker, R. W., Kelly, I. D. & Knowles, P. F. (1976). Human placental diamine oxidase. Improved purification and characterization of a copper- and manganese-containing amine oxidase with novel substrate specificity. *Biochemical Journal*, **155**, 679–87.

Curti, B., Ronchi, S., Branzoli, U., Ferri, G. & Williams, C. H. Jr (1973). Improved purification, amino acid analysis and molecular weight of homogeneous D-amino acid oxidase from pig kidney. *Biochimica et Biophysica Acta*, **327**, 266–73.

Dallman, P. R. (1986). Biochemical basis for the manifestations of iron deficiency. *Annual Reviews of Nutrition*, **6**, 13–40.

Dooley, D. M., Cote, C. E., McGuirl, M. A., Bates, J. L., Perkins, J. B., Moog, R. S., Singh, I., Knowles, P. F. & McIntire, W. C. (1989). Copper-PQQ interactions in amine oxidases. In *PQQ and Quinoproteins*, ed. J. A. Jongejan & J. A. Duine. Kluwer Academic Publishers, Dordrecht/Boston/London, pp. 307–16.

Drawert, F. & Barton, H. (1978). On the enzymatic analysis of hydroxyproline. III. Hydroxyproline-2-epimerase and allohydroxy-D-proline oxidase of miscellaneous organisms. *Chemie Mikrobiologie Technologie der Lebensmittel*, **5**, 165–9.

Duine, J. A. (1989). PQQ and quinoproteins; an important novel field in enzymology. In *PQQ and Quinoproteins*, ed. J. A. Jongejan & J. A. Duine. Kluwer Academic Publishers, Dordrecht/Boston/London, pp. 351–60.

Engelsma, J. W., Meulen, J. D. v.d., Slump, P. & Haagsma, N. (1979). The oxidation of lysinoalanine by L-amino acid oxidase. Identification of products. *Lebensmittel-Wissenschaft &-Technologie*, **12**, 203–7.

Fenwick, G. R. & Curtis, R. F. (1980). Rapeseed meal in rations for laying chickens. A review of the effect on egg quality. *Journal of the Science of Food and Agricultre,* **31,** 515–25.

Fenwick, G. R., Pearson, A. W., Greenwood, N. M. & Butler, E. J. (1981). Rapeseed meal tannins and egg taint. *Animal Feed Science & Technology,* **6,** 421–31.

Ferti, C., Curti, B., Simonetta, M. P., Ronchi, S., Galliano, M. & Minchiotti, L. (1981). Reactivity of D-amino acid oxidase with 1,2-cyclohexanedione: evidence for one arginine in the substrate binding site. *European Journal of Biochemistry,* **119,** 553–7.

Finazzi-Agro, A. (1989). Copper-containing amine oxidases. In *PQQ and Quinoproteins,* ed. J. A. Jongejan & J. A. Duine. Kluwer Academic Publishers, Dordrecht/Boston/London, pp. 279–82.

Fitzpatrick, P. F. & Massey, V. (1983). The reaction of 8-mercaptoflavins and flavoproteins with sulfite. *Journal of Biological Chemistry,* **258**(16), 9700–5.

Floris, G., Giartosio, A. & Rinaldi, A. (1983). Diamine oxidase from *Lens esculenta* seedlings: purification and properties. *Phytochemistry,* **22,** 1871–4.

Fonda, M. L. (1988). Pyridoxamine (pyridoxine) phosphate oxidase activity in mammalian tissues. *Comparative Biochemistry and Physiology,* **90B**(4), 731–7.

Gahl, W. A., Vale, A. M. & Pitot, H. C. (1982). Spermidine oxidase in human pregnancy serum. *Biochemical Journal,* **201,** 161–6.

Goh, Y. K., Robblee, A. R. & Claudium, D. R. (1985). Influence of glucosinolates and free oxazolidinethione in a laying diet containing a constant amount of sinapine on the thyroid size and hepatic trimethylamine oxidase activity of brown-egg layers. *Canadian Journal of Animal Science,* **65,** 921–7.

Gouygou, J. P., Sinquin, C., Etienne, M., Landrein, A. & Durand, P. (In press). Quantitative and qualitative determination of biogenic amines in fish. In *Pelagic Fish: the Resource and its Exploitation,* ed. J. R. Burt, R. Hardy & K. J. Whittle. Fishing News Books, Oxford.

Greenawalt, J. W. (1974). The isolation of outer and inner mitochondrial membranes. In *Methods of Enzymology,* Vol. XXXI, ed. S. Fleischer & L. Packer. Academic Press, London, pp. 310–23.

Greenawalt, J. W. (1979). Survey and update of outer and inner mitochondrial membrane separation. In *Methods in Enzymology,* Vol. LV, ed. S. Fleischer & L. Packer. Academic Press, London, pp. 88–98.

Guilbault, G. C. (1976). Enzyme electrodes and solid surface fluorescence methods. In *Methods in Enzymology,* Vol. XLIV, ed. K. Mosbach. Academic Press, London, pp. 579–633.

Hall, T. R. & Urueña, G. (1983). Pharmacology and physiology of monoamine oxidase activity in vertebrates—a comparative study. *Comparative Biochemistry and Physiology,* **76B**(3), 393–7.

Hare, M. L. C. (1928). Tyramine oxidase: a new enzyme system in liver. *Biochemical Journal,* **22,** 968–79.

Harel, S. & Kanner, J. (1985). Hydrogen peroxide generation in ground muscle tissues. *Journal of Agricultural and Food Chemistry,* **33,** 1186–8.

Hill, J. M. (1971). Diamine oxidase (pea seedling). In *Methods in*

Enzymology, Vol. XVIIB, ed. H. Tabor & C. W. Tabor. Academic Press, London, pp. 730–5.

Hölttä, E. (1977). Oxidation of spermidine and spermine in rat liver. Purification and properties of polyamine oxidase. *Biochemistry*, **16**(1), 91–100.

Hölttä, E. (1983). Polyamine oxidase (rat liver). In *Methods in Enzymology*, Vol. 94, ed. H. Tabor & C. W. Tabor. Academic Press, London, pp. 306–11.

Husain, M., Edmondson, D. E. & Singer, T. P. (1982). Kinetic studies on the catalytic mechanism of liver monoamine oxidase. *Biochemistry*, **21**, 595–600.

Huss, H. H. (1988). *Fresh Fish—Quality and Quality Changes*. A training manual prepared for the FAO/DANIDA Training Programme on Fish Technology and Quality Control. FAO Fisheries Series No 29. FAO, Rome, p. 40.

Ijomah, P., Clifford, M. N., Walker, R., Wright, J., Hardy, R. & Murray, C. K. (In press). Further volunteer studies on scombrotoxicosis. In *Pelagic Fish: the Resource and its Exploitation*, ed. J. R. Burt, R. Hardy & K. J. Whittle. Fishing News Books, Oxford.

Ikeda, S. (1980). Other organic components and inorganic components. In *Advances in Fish Science and Technology*, ed. J. J. Connell. Fishing News Books, Farnham, Surrey, UK, pp. 111–24.

International Union of Biochemistry (IUB) (1984). *Enzyme Nomenclature*. Academic Press, London.

Jahns, F. D. & Rand, A. G. Jr (1977). Enzyme methods to assess marine food quality. In *ACS Symposium Series*, No. 47, pp. 266–78.

Jahns, F. D. & Rand, A. G. (1978). Immobilized enzyme method to assess fish quality. US Patent 4,105,800.

Johnston, J. P. (1968). Some observations upon a new inhibitor of monoamine oxidase in brain tissue. *Biochemical Pharmacology*, **17**, 1285–97.

Kagan, H. M., Gachern, S. N., Trackman, P. C., Calaman, S. D. & Greenaway, F. T. (1989). Enzymology of lysyl oxidase. In *PQQ and Quinoproteins*, ed. J. A. Jongejan & J. A. Duine. Kluwer Academic Publishers, Dordrecht/Boston/London, pp. 317–26.

Kappeller-Adler, R. (1970). *Amine Oxidases and Methods for their Study*. Wiley–Interscience, New York.

Kappeller-Adler, R. (1971). Determination of amine oxidases. In *Methods of Biochemical Analysis*, Supplemental Volume, ed. D. Glick. Interscience Publishers, John Wiley & Sons, New York, London, Sydney, Toronto, pp. 35–87.

Karube, I. & Suzuki, S. (1982). Biosensors for Food Process Control. In *Use of Enzymes in Food Technology*, ed. P. Dupuy. Technique and Documentation, Lavoisier, Paris, France.

Karube, I., Satoh, I., Araki, Y., Suzuki, S. & Yamada, H. (1980). Monoamine oxidase electrode in freshness testing of meat. *Enzyme Microbiology & Technology*, **2**, 117–20.

Kazarinoff, M. N. & McCormick, D. B. (1975). Rabbit liver pyridoxamine (pyridoxine) 5′-phosphate oxidase: purification and properties. *Journal of Biological Chemistry*, **250**(9), 3436–42.

Keevil, T. & Mason, H. S. (1978). Molecular oxygen in biological oxidations. In *Methods in Enzymology,* Vol. LII, ed. S. Fleischer & L. Packer. Academic Press, London, pp. 3–40.

Kenney, M. C. (1980). Flavin coenzymes covalently linked to proteins. In *Flavins and Flavoproteins,* ed. K. Yagi & T. Yamano. Japan Scientific Societies Press, Tokyo; University Park Press, Baltimore, pp. 237–43.

Kinemuchi, H., Sudo, M., Yoshino, M., Kawaguchi, T., Sunami, Y. & Kamijo, K. (1983). A new type of mitochondrial monoamine oxidase distinct from Type A and Type B. *Life Sciences,* **32,** 517–24.

Klinman, J. P., Hartmann, C. & Janes, S. M. (1989). Mechanistic probes of copper amine oxidases. In *PQQ and Quinoproteins,* ed. J. A. Jongejan & J. A. Duine. Kluwer Academic Publishers, Dordrecht/Boston/London, pp. 297–305.

Knoop, F. (1910). Über den physiologischen Abbau der Säuren und die Synthese einer Aminosäure im Tierkörper. *Zeitschrift für Physiologische Chemie,* **67,** 489–502 (cited by Meister, 1965).

Krebs, H. A. (1935). Metabolism of amino acids, III. Deamination of amino acids, *Biochem. J.,* **29,** 1620–44.

Kumazawa, T. & Suzuki, O. (1987). Occurrence of polyamine oxidase in tissues of the catfish. *Comparative Biochemistry and Physiology,* **88B**(3), 727–8.

Kung, H.-F. & Wagner, C. (1970). Oxidation of C_1 Compounds by *Pseudomonas* sp. MS. *Biochemical Journal,* **116,** 357–65.

Kusakabe, H., Kodama, K., Kuninaka, A., Yoshino, H., Misono, H. & Soda, K. (1980). A new antitumor enzyme, L-lysine oxidase from *Trichoderma viride. Journal of Biological Chemistry,* **255,** 976–81.

Kusakabe, H., Midorikawa, Y., Kuninaka, A. & Yoshino, H. (1983). Purification and properties of a new enzyme, L-glutamate oxidase, from *Streptomyces* sp. X-119-6 grown on wheat bran. *Agricultural and Biological Chemistry,* **47,** 1323–8.

Kusakabe, H., Midorikawa, Y. & Fujishima, T. (1984). Methods for determining L-glutamate in soy sauce with L-glutamate oxidase. *Agricultural and Biological Chemistry,* **48**(1), 181–4.

Langham, L., Garber, B. M., Roe, D. A. & Kazarinoff, M. N. (1982). A radiometric assay of pyridoxamine (pyridoxine) 5′-phosphate oxidase. *Analytical Biochemistry,* **125,** 329–34.

Leegwater, D. C. & Tas, A. C. (1980). Identification of an enzymic oxidation product of lysinoalanine. *Lebensmittel-Wissenschaft & Technologie,* **14,** 87–91.

Lerke, P. A., Porcuna, M. N. & Chin, H. B. (1983). Screening test for histamine in fish. *Journal of Food Science,* **48,** 155–7.

Levene, C. I. (1989). Chick embryo lysyl oxidase: evidence for a pyridoxal derivative as an essential cofactor. In *PQQ and Quinoproteins,* ed. J. A. Jongejan & J. A. Duine. Kluwer Academic Publishers, Dordrecht/Boston/London, pp. 335–44.

LiMuti, C. M. & Paulson, T. C. (1989). Colorimetric detection of microbially excreted lysine directly on agar plates. *Journal of Microbiological Methods,* **9,** 129–37.

Lin, J.-K., Chang, H.-W. & Lin-Shiau, S.-Y. (1984). Abundance of dimethylamine in seafoods: possible implications in the incidence of human cancer. *Nutrition and Cancer,* **6**(3), 148–59.

Lobenstein-Verbeek, C. L., Jongejan, J. A., Frank, J. & Duine, J. A. (1984). Bovine serum amine oxidase: a mammalian enzyme having covalently bound PQQ as prosthetic group. *FEBS Letters,* **170,** 305–9.

McCormick, D. B. & Merrill, A. H. Jr (1980). Pyridoxamine (pyridoxine) 5′-phosphate oxidase. In *Vitamin B6*: *Metabolism and Role in Growth,* ed. G. P. Tryfiates. Food & Nutrition Press, Trumbull, CT, USA, pp. 1–26.

Matheis, G. & Whitaker, J. R. (1987). A review: enzymatic cross-linking of proteins applicable to foods. *Journal of Food Biochemistry,* **11,** 309–27.

Matsui, M., Ishibashi, T. & Kawabata, T. (1984*a*). Effect of broiling temperatures on the formation of *N*-nitrosodimethylamine from dried squid products. *Bulletin of the Japanese Society of Scientific Fisheries,* **50**(1), 151–4.

Matsui, M., Ishibashi, T. & Kawabata, T. (1984*b*). Precursors of *N*-nitrosodimethylamine formed in dried squid upon broiling. *Bulletin of the Japanese Society of Scientific Fisheries,* **50**(1), 155–9.

Matsumiya, M. & Otake, S. (1981). The investigation of diamine oxidase-like enzymes in the various organs of fishes. *Bulletin of the Japanese Society of Scientific Fisheries,* **47**(12), 1599–604.

Matsumiya, M. & Otake, S. (1983). On the diamine oxidase in the pyloric caeca and the intestine of common mackerel *Scomber japonicus. Bulletin of the Japanese Society of Scientific Fisheries,* **49**(11), 1695–9.

Matsumiya, M. & Otake, S. (1984). Comparison of the characteristics of the diamine oxidase in the digestive organs of several teleostan fishes. *Bulletin of the Japanese Society of Scientific Fisheries,* **50**(7), 1261–5.

Matsumiya, M. & Otake, S. (1985). Effect of feeding condition on the diamine oxidase activity of several teleostan fishes. *Bulletin of the Japanese Society of Scientific Fisheries,* **51**(2), 273–8.

Matsumiya, M. & Otake, S. (1986). Purification and properties of diamine oxidase in the pyloric caeca of yellowtail *Seriola quinqueradiata. Bulletin of the Japanese Society of Scientific Fisheries,* **52**(9), 1617–23.

Meer, R. A. van der, Wassenaar, P. D. van, Brourshaven, J. H. van & Duine, J. A. (1989). Primary structure of a PQQ containing peptide from porcine kidney diamine oxidase. In *PQQ and Quinoproteins,* ed. J. A. Jongejan & J. A. Duine. Kluwer Academic Publishers, Dordrecht/Boston/London, pp. 348–50.

Meister, A. (1965). *Biochemistry of the Amino Acids,* Vol. I. Academic Press, New York, London, pp. 269–437.

Merrill, A. H., Kazarinoff, M. N., Tsuge, H., Horiike, K. & McCormick, D. B. (1979). Pyridoxamine (pyridoxine) 5′-phosphate oxidase from rabbit liver. In *Methods in Enzymology,* Vol. 62, ed. D. B. McCormick & L. D. Wright. Academic Press, London, pp. 568–74.

Metzler, D. E. (1977). *Biochemistry*: *the Chemical Reactions of Living Cells.* Academic Press, New York/San Francisco/London, p. 664.

Mizushima, S. (1957). Studies on the metabolism of D-amino acid in microorganisms. 4. On the nature of the purified D-glutamic-aspartic oxidase of *Aspergillus ustus. Journal of General and Applied Microbiology,* **3,** 233–9.

Mondovì, B. & Riccio, P. (1984). Copper amine oxidases: structure and function. In *Advances in Inorganic Biochemistry*, Vol. 6, ed. G. L. Eichorn & L. G. Marzilli. Elsevier, Amsterdam, pp. 225–44.

Mondovì, B. & Riccio, P. (1989). On the role of copper in PQQ amine oxidases. In *PQQ and Quinoproteins,* ed. J. A. Jongejan & J. A. Duine. Kluwer Academic Publishers, Dordrecht/Boston/London, pp. 289–95.

Mondovì, B., Rotilio, G., Costa, M. T. & Agro, A. F. (1971). Diamine oxidase (pig kidney). In *Methods in Enzymology,* Vol. XVIIB, ed. H. Tabor & C. W. Tabor. Academic Press, London, pp. 735–46.

Mondovì, B., Turini, P., Befani, O. & Sabatini, S. (1983). Purification of bovine plasma amine oxidase. In *Methods in Enzymology,* Vol. 94, ed. H. Tabor & C. W. Tabor. Academic Press, London, pp. 314–18.

Murray, J. & Burt, J. R. (1974). Transaminase activity in extracts of cod muscle studied by the use of a ^{14}C-labelled amino acid mixture. *Journal of the Science of Food and Agriculture,* **25,** 11–20.

Nakano, M. & Danowski, T. S. (1971). L-Amino acid oxidase (rat kidney). In *Methods in Enzymology,* Vol. XVIIB, ed. H. Tabor & C. W. Tabor. Academic Press, London, pp. 601–5.

Nasu, S., Wicks, F. D. & Gholson, R. K. (1982). L-Aspartate oxidase, a newly discovered enzyme of *Escherichia coli,* is the B protein of quinolate synthetase. *Journal of Biological Chemistry,* **257,** 626–32.

Neubauer, O. (1909). Uber den Abbau der Aminosäuren im gesunden und kranken Organisms. *Deutsche Archiv fur Klinische Medizin,* **95,** 211–56 (cited by Meister, 1965).

Nicotra, A. & Senatori, O. (1989). Some characteristics of mitochondrial monoamine oxidase activity in eggs of carp (*Cyprinus carpio*) and rainbow trout (*Salmo gairdneri*). *Comparative Biochemistry and Physiology,* **92C,** 401–4.

Nicotra, A., Serafino, A. L. & Senatori, O. (1986). Monoamine oxidase activity in tissues of the starfish *Marthasterias glacialis. Comparative Biochemistry and Biophysics,* **84C**(1), 195–7.

Nishino, T., Massey, V. & Williams, C. H. Jr (1980). Chemical modifications of D-amino acid oxidase. Evidence for active site histidine, tyrosine and arginine residues. *Journal of Biological Chemistry,* **255**(8), 3610–16.

Nutter, L. M., Meisler, N. T. & Thanassi, J. W. (1983). Absence of pyridoxine-(pyridoxamine-) 5′-phosphate oxidase in Morris hepatoma 7777. *Biochemistry,* **22,** 1599–604.

Oi, S., Shimada, K., Inamasu, M. & Yasunoba, K. T. (1970). Mechanistic studies of beef liver mitochondrial amine oxidase. XVIII. Amine oxidase. *Archives of Biochemistry and Biophysics,* **139,** 28–37.

Padmanabhan, R. & Kim, K.-H. (1965). Oxidation of spermidine by a *Pseudomonas. Biochemical and Biophysical Research Communications,* **19**(1), 1–5.

Pavelka, J. (1982). Physical and chemical effects underlying the reduction of histamine levels in the meat of mackerels and related fishes and possibilities of using these factors in the production of food from fish. *Veterinarni Medicina, Praha,* **27**(4), 237–46.

Pearson, A. W., Butler, E. J. & Fenwick, G. R. (1980). Rapeseed meal and egg taint. The role of sinapine. *Journal of the Science of Food and Agriculture,* **31,** 898–904.

Pflug, W. & Lingens, F. (1983). Occurrence of pyridoxamine-phosphate oxidase and pyridoxal kinase in Gram-negative and Gram-positive bacteria (in German). *Hoppe-Seyler's Zeitschrift fur Physiologische Chemie,* **364,** 1627–30.

Pogell, B. M. (1958). Enzymatic oxidation of pyridoxamine phosphate to pyridoxal phosphate in rabbit liver. *Journal of Biological Chemistry,* **232,** 761–76.

Pogell, B. M. (1963). Pyridoxamine 5-phosphate oxidase. In *Methods in Enzymology,* Vol. 6, ed. S. P. Colowick & N. O. Kaplan. Academic Press, London, pp. 331–4.

Rehbein, H. (1988). Relevance of trimethylamine oxide demethylase activity and haemoglobin content to formaldehyde production and texture deterioration in frozen stored minced fish muscle. *Journal of the Science of Food and Agriculture,* **43**(3), 261–7.

Robinson, D. S. (1987). *Food-Biochemistry and Nutritional Value.* Longman Scientific and Technical, Harlow, UK, p. 451.

Ronchi, S., Minchiotti, L., Curti, B. & Bridgen, J. (1976). D-amino acid oxidase primary structure: cyonogen bromide fragments and sequences of the cysteinyl peptides. in *Flavins and Flavoproteins,* ed. T. P. Singer. Elsevier. Amsterdam, pp. 702–7.

Salach, J. I. Jr (1978). Preparation of monoamine oxidase from *beef liver* mitochondria. In *Methods in Enzymology,* Vol. LIII, ed. S. Fleischer & L. Packer. Academic Press, London, pp. 495–501.

Salach, J. I. & Weyler, W. (1987). Preparation of the flavin-containing aromatic amine oxidases of human placenta and beef liver. In *Methods in Enzymology,* Vol. 142, ed. S. Kaufman. Academic Press, London, pp. 627–37.

Salach, J. I., Singer, T. P., Yasunobu, K. T., Minamiura, N. & Youdim, M. B. H. (1976). In *Monoamine Oxidase and its Inhibition,* ed. J. Knight. Elsevier, Amsterdam.

Sikorski, Z. E. (1988). Smoking of fish and carcinogens. In *Fish Smoking and Drying,* ed. J. R. Burt. Elsevier Applied Science, London, and New York, pp. 73–83.

Simonetta, M. P., Galliani, S., Vanoni, M. A., Ronchi, S. & Curti, B. (1984). The role of arginines in D-amino acid oxidase. In *Flavins and Flavoproteins,* ed. R. C. Bray, P. C. Engel & S. G. Mayhew. Walter de Gruyter & Co., Berlin, pp. 295–9.

Smith, T. A. (1975). Recent advances in the biochemistry of plant amines. *Phytochemistry,* **14,** 865–90.

Smith, T. A. (1981). Amines in food. *Food Chemistry,* **6**(3), 169–200.

Smith, T. A. (1983). Polyamine oxidase (oat seedlings). In *Methods in Enzymology,* Vol. 94, ed. H. Tabor & C. W. Tabor. Academic Press, London, pp. 311–14.

Stewart, L. C. & Klinman, J. P. (1988). Dopamine beta hydroxylase. *Annual Review of Biochemistry,* **57,** 551–92.

Stubbe, J. A. (1989). Protein radical involvement in biological catalysis? *Annual Review of Biochemistry,* **58,** 257–85.

Sumner, S. S. (1987). Histamine production in Swiss cheese. PhD thesis, University of Wisconsin, Madison. Dissertation Abstracts DA 8800388.

Sumner, S. S. & Taylor, S. L. (1989). Detection method for histamine-producing, dairy-related bacteria using diamine oxidase and leucocrystal violet. *Journal of Food Protection,* **52**(2), 105–8.

Suzuki, O., Matsumoto, T. & Katsumata, Y. (1984). Determination of polyamine oxidase activities in human tissues. *Experientia,* **40,** 838–9.

Swenson, R. P., Williams, C. H. & Massey, V. (1984). The effect of the methylation of histidine-217 in pig kidney D-amino acid oxidase on ligand binding and on catalysis. In *Flavins and Flavoproteins,* ed. R. C. Bray, P. C. Engel & S. G. Mayhew. Walter de Gruyter & Co., Berlin, pp. 301–4.

Syed, M. A. & Coombs, T. L. (1982). Copper metabolism in the plaice *Pleuronectes platessa* (L.). *Journal of Experimental Marine Biology and Ecology,* **63,** 281–96.

Tabor, C. W. & Tabor, H. (1984). *Annual Review of Biochemistry,* **53,** 749–90.

Tanaka, F. & Yagi, K. (1980). Cooperative binding of flavin adenine dinucleotide accompanied by change in quaternary structure of D-amino acid oxidase. In *Flavins and Flavoproteins,* ed. K. Yagi & T. Yamano. Japan Scientific Societies Press, Tokyo, University Park Press, Baltimore, pp. 387–94.

Tipton, K. F. (1971). Monoamine oxidase (pig brain mitochondrial). In *Methods in Enzymology,* Vol. XVIIB, ed. H. Tabor & C. W. Tabor. Academic Press, London, pp. 717–22.

Todoriki, S., Tajima, M. & Senda, M. (1988). Enzymatic oxygen electrode method for determination of polyamines in foods. *Analytical Sciences,* **4,** 583–5.

Tojo, H., Horiike, K., Shiga, K., Nishina, Y., Miura, R., Watari, H. & Yamano, T. (1982). Thermodynamic characterization of hog kidney D-amino acid oxidase apoenzyme in concentrated guanidine hydrochloride solution. Preferential interaction with the solvent components and the molecular weight of the monomeric unit. *Journal of Biochemistry,* **92,** 1741–52.

Tojo, H., Horiike, K., Shiga, K., Nishina, Y., Watari, H. & Yamano, T. (1984). Stoichiometry of the self-association of D-amino acid oxidase. In *Flavins and Flavoproteins,* ed. R. C. Bray, P. C. Engel & S. G. Mayhew. Walter de Gruyter, Berlin, New York, pp. 305–8.

Tokieda, T., Niimura, T., Takamura, F. & Yamaha, T. (1977). Purification and some properties of cyclohexylamine oxidase from a *Pseudomonas* sp. *Journal of Biochemistry,* **81,** 851–8.

Tolbert, N. E. (1978). Peroxisomal redox enzymes. In *Methods in Enzymology,* Vol. LII, ed. S. Fleischer & L. Packer. Academic Press, London, pp. 493–505.

Tsuge, H. & Okada, T. (1984). Immobilization of yeast pyridoxamine phosphate oxidase to halogenoacetyl polysaccharides. *Biotechnology and Bioengineering,* **26,** 412–18.

Tsuge, H., Sen-Maru, K. & Ohashi, K. (1978). Immobilization of yeast pyridoxamine (pyridoxine) 5′-phosphate oxidase by organomercurial agarose. *FEBS Letters,* **93**(2), 331–4.

Tsuge, H., Ozeki, K., Sen-Maru, K. & Ohashi, K. (1979). Purification and properties of pyridoxamine (pyridoxine) 5′-phosphate oxidase from baker's yeast. *Agricultural and Biological Chemistry,* **43**(9), 1801–7.

Tsuge, H., Ozeki, K. & Ohashi, K. (1980). Molecular and enzymatic properties of pyridoxamine (pyridoxine) 5′-phosphate oxidase from baker's yeast. *Agricultural and Biological Chemistry,* **44**(10), 2329–35.

Tsuge, H., Kuroda, Y., Iwamoto, A. & Ohashi, K. (1982*a*). Partial purification and property of pyridoxine (pyridoxamine)-5′-phosphate oxidase isozymes from wheat seedlings. *Archives of Biochemistry and Biophysics,* **217**(2), 479–84.

Tsuge, H., Itoh, K., Ozeki, K. & Ohashi, K. (1982*b*). Separation and characterization of two pyridoxamine (pyridoxine) 5′-phosphate oxidases from baker's yeast. *Agricultural and Biological Chemistry,* **46**(4), 1075–7.

Tsuge, H., Iwamoto, A., Yamamoto, I., Miyata, H., Hattori, K. & Ohashi, K. (1986). Distribution of pyridoxaminephosphate oxidase activity in various plants and calluses. *Agricultural and Biological Chemistry,* **50**(2), 289–96.

Tsushida, T. & Takeo, T. (1985). Purification and some properties of tea leaf amine oxidase. *Agricultural and Biological Chemistry,* **49**(2), 319–26.

Tu, S.-C., Edelstein, S. J. & McCormick, D. B. (1973). A modified purification method and properties of pure porcine D-amino acid oxidase. *Archives of Biochemistry and Biophysics,* **159,** 889–96.

Voigt, M. N. & Eitenmiller R. R. (1978). Role of histidine and tyrosine decarboxylases and mono- and diamine oxidases in amine build up in cheese. *Journal of Food Protection,* **41**(3), 182–6.

Wada, H. & Snell, E. E. (1961). The enzymatic oxidation of pyridoxine and pyridoxamine phosphates. *Journal of Biological Chemistry,* **236,** 2089–95.

Wollenberger, U., Scheller, F. W., Böhmer, A., Passarge, M. & Muller, H.-G. (1989). A specific enzyme electrode for L-glutamate—development and application. *Biosensors,* **4,** 381–91.

Yagi, K. (1971). Reaction mechanism of D-amino acid oxidase. In *Advances in Enzymology,* Vol. 34, ed. F. F. Nord. Interscience Publishers, New York, London, Sydney, Toronto, pp. 41–78.

Yamada, H., Kumagai, H., Kawasaki, H., Matsui, H. & Ogata, K. (1967). *Biochemical and Biophysical Research Communications,* **29,** 723–7.

Yamada, H., Kumagai, H. & Uwajima, T. (1971). Tyramine oxidase (*Sarcina lutea*). In *Methods in Enzymology,* Vol. XVIIB, ed. H. Tabor & C. W. Tabor. Academic Press, London, pp. 722–6.

Yamamoto, S. & Nagasaki, S. (1969). *Kochi Daigaku Gakujutsu Kenkyu Hokoku,* **18,** 195. (In Japanese, cited by Tsuge *et al.*, 1980.)

Yanigasawa, H., Hirasawa, E. & Suzuki, Y. (1981). Purification and properties of diamine oxidase from pea epicotyls. *Phytochemistry,* **20,** 2105–8.

Yasunobu, K. T. & Gomes, B. (1971). Mitochondrial amine oxidase (monoamine oxidase) (beef liver). In *Methods in Enzymology,* Vol. XVIIB, ed. H. Tabor & C. W. Tabor. Academic Press, London, pp. 709–17.

Youdim, M. B. H. (1976). Rat liver mitochondrial monoamine oxidase—an iron-requiring flavoprotein. In *Flavins and Flavoproteins,* ed. T. P. Singer. Elsevier, Amsterdam, pp. 593–604.

Youdim, M. B. H. & Tenne, M. (1987). Assay and purification of liver monoamine oxidase. In *Methods in Enzymology,* Vol. 142, ed. S. Kaufman. Academic Press, London, pp. 617–27.

Zeller, E. A. (1963). Diamine oxidases. In *The Enzymes,* Vol. 2, Part 1, 1st edn, ed. J. B. Sumner & K. Myrbäck. Academic Press, New York, pp. 313–35.

4

The Lactoperoxidase Systems of Bovine and Human Milk

Kenneth M. Pruitt & David N. Kamau
Department of Biochemistry, The University of Alabama at Birmingham, Alabama USA

HISTORICAL BACKGROUND

The first suggestion that antibacterial properties of milk might be related to peroxidase activity was made in a landmark paper by Hanssen (1924):

> We venture to suggest the theory that the bactericidal properties of milk bear some relation to the oxydizing enzymes of the milk, and we believe that the variability of the bactericidal properties of milk is caused by variations of the quantities of oxydases and peroxydases contained in different milk samples.

This prescient study anticipated many of the significant developments in peroxidase chemistry which were to follow in succeeding decades. Hanssen showed that the antibacterial properties of bovine milk survived heating at 70°C for 15 min but were destroyed when heated at 75°C. He based his peroxidase theory, in part, on his observations that the heat stability of milk peroxidase activity was identical to the heat stability of milk antibacterial properties. He showed that the antibacterial properties of milk correlated with the cow's diet but he wrongly assumed that the variation was due to variation in milk peroxidase content. He could not have known then that the enzyme is consistently present in high amounts in bovine milk and that the availability of hydrogen peroxide (H_2O_2) and the concentration of the thiocyanate ion (SCN^-) were the limiting factors for lactoperoxidase (LPO) antibacterial activity in bovine milk.

Wright & Tramer (1958) attributed the antistreptococcal activity of milk to the presence of LPO. The role of LPO in the antimicrobial activity of milk was confirmed by Portmann & Auclair (1959) when they demonstrated that partially purified LPO could restore the antimicrobial activity of heated milk. The involvement of H_2O_2 in the LPO system was established by Jago & Morrison (1962) when they observed that high concentrations of catalase reversed the antimicrobial activity of milk.

The presence of the peroxidase enzyme and hydrogen peroxide alone were not enough to explain the antimicrobial properties of the bovine LPO system. The last piece of the puzzle fell into place with the first published report, by Reiter *et al.* (1963), that SCN was required to activate the bovine LPO system.

PURIFICATION

The first attempts to purify milk peroxidase were made by Jones & Simms (1930). They isolated a relatively crude fraction from bovine milk which was 200 times higher in bactericidal activity than the original sample. They named this substance 'lactenin' (*lac*, milk: *teneo*, hold).

The early history of the isolation and identification of peroxidase enzymes in milk has been reviewed by Paul (1987), Paul & Ohlsson (1985) and Reiter (1985). At the present time, the most effective methods of purifying LPO from bovine milk begin by coagulation of proteins in unpasteurized, skimmed milk with addition of rennin. LPO is separated from the whey fraction and purified by a combination of ion exchange and hydrophobic chromatography (Paul *et al.*, 1980). Commercial preparations of LPO (now available from several suppliers) may be further purified by hydrophobic chromatography or isoelectric focusing (Rule *et al.*, 1976).

COMPOSITION

Immunochemical Studies

The first immunochemical studies of bovine LPO were made by Morrison & Allen (1963). Immunochemically similar peroxidases were soon identified in goat and sheep milk (Allen & Morrison, 1966) and in human saliva (Morrison *et al.*, 1965). Because of its strong cross-reaction with antibodies raised against bovine LPO and similar kinetic properties, human salivary peroxidase became known as 'lactoperoxidase'. However, subsequent studies have shown that the structural (Månsson-Rahemtulla *et al.*, 1988) and kinetic (Pruitt *et al.*, 1988) properties of human salivary peroxidase (SPO) are significantly different from those of bovine LPO.

Amino Acid Composition

The various reports (Morrison & Steele, 1968; Carlström, 1969; Månsson-Rahemtulla *et al.*, 1988) of the amino acid composition of LPO are reasonably consistent. The data of Carlström (1969) were

Table 1. Amino Acid Composition of Bovine Lactoperoxidase, Adapted from Carlström (1969), Expressed as Nearest Integral Number of Residues per 78 500 g of Protein[a,b]

Non-polar side chains		*Polar side chains*	
Alanine	37	Asparagine + Glutamine[a]	62
Cystine	8	Serine	33
Cysteine	0	Threonine	32
Glycine	40	Tyrosine	15
Isoleucine	27		Total = 142
Leucine	68		
Methionine	8		
Phenylalanine	30		
Proline	42		
Tryptophan	15		
Valine	28		
	Total = 303		
Basic side chains		*Acidic side chains*	
Arginine	37	Asparagine + Aspartic acid	71
Lysine	34	Glutamine + Glutamic acid	61
Histidine	14		Total = 132
	Total = 85		
		Aspartic acid + Glutamic acid = 70[b]	

[a] Estimated from total amide nitrogen (62).
[b] Calculated from total acidic side chains (132) – total amide nitrogen (62).

reported in units of amino acid residues/78 500 g protein. These units are convenient for relating composition to structural properties. The data have been regrouped and are reproduced in Table 1.

The amino acid analysis does not give separate reports of the number of residues of asparagine and aspartic acid nor of glutamine and glutamic acid. The sum of the asparagine and glutamine residues may be estimated from the total amide nitrogen. The sum of the aspartic acid and glutamic acid residues may be calculated by subtracting the total amide nitrogen from the total for the acidic side chains. The total number of amino acid residues in the polypeptide backbone is given by the sum of the numbers of residues listed in Table 1 excluding the number listed for asparagine + glutamine. These residues are counted under the 'Acidic side chain' heading. The total for the LPO molecule is 600 amino acid residues.

Carbohydrate Composition

Several reports (Rombauts *et al.*, 1967; Morrison & Steele, 1968; Carlström, 1969 and Månsson-Rahemtulla *et al.*, 1988) of carbohydrate composition are in good agreement. One of the earlier reports (Carlström, 1969) and a recent report (Månsson-Rahemtulla *et al.*, 1988) based on high-pressure liquid chromatography (HPLC) are compared in Table 2. The composition data indicate that the majority of the oligosaccharides are of 'high-mannose' type. The side chains are *N*-linked to the polypeptide (Månsson-Rahemtulla *et al.*, 1988). The small amounts of glucose reported in the later study may be contaminants detected by the very sensitive HPLC method.

Table 2. Carbohydrate Composition of Bovine Lactoperoxidase[a,b]

	Residues/78 500 g	
	a	*b*
Galactose	2	NR
Mannose	26	26
Fucose	1	NR
Glucose	2	NR
Glucosamine	12	14
Galactoseamine	4	4

[a] Calculated by multiplying the reported (Månsson-Rahemtulla, *et al.*, 1988) value of micrograms carbohydrate per milligram of dry weight of sample (which is equivalent to grams of carbohydrate/1000 g sample) by 78·5, dividing by the gram molecular weight of the carbohydrate, and rounding to the nearest integer.
[b] Carlström, 1969.
NR = not reported.

Heterogeneity

As isolated from bovine milk, LPO is heterogeneous (reviewed in Paul & Ohlsson, 1985). Two major fractions, designated 'A' and 'B' are found. Depending upon the method of isolation and the method of analysis, four or more subfractions of each of the major fractions may be detected. All of these subfractions have the same iron content and possess full enzymatic activity (Carlström, 1965). Carlström (1969) presented evidence suggesting that the major fraction A is derived from fraction B by loss of carbohydrate groups and that subfractions within each of the major fractions are derived by successive deamidation of asparagine and/or glutamine residues. He reported molecular weights of 78 500 (fraction B) and 76 500 (fraction A).

Functional Significance of the Carbohydrate Component

Deglycosylation of plant peroxidases is claimed to be accompanied by loss of enzymatic activity (Tadros *et al.*, 1983; Hu & van Huystee, 1989). This is apparently not the case for LPO since fractions A and B have the same specific enzyme activities (Carlström, 1965) in spite of significant differences in carbohydrate content. The functional significance of the carbohydrate moieties of LPO is not known.

MOLECULAR STRUCTURE

Primary Structure

The amino acid sequence of bovine LPO has not been reported. The enzyme exists as a single chain with leucine as the N-terminal amino acid (Sievers, 1981). The structure is stabilized by eight disulfide bonds (Carlström, 1969). Reduction of the disulfide bonds destroys enzymatic activity (Månsson-Rahemtulla *et al.*, 1988).

Secondary and Tertiary Structure

Studies of the far-ultraviolet circular dichroism spectrum of LPO indicate that the molecule contains 65% β-structure, 23% α-helix and

12% unordered structure (Sievers, 1980). Sedimentation analyses suggest that the molecule has an ellipsoidal form in solution (Paul & Ohlsson, 1985). Its thermodynamic properties in solution are similar to those of other small, globular proteins (Pfeil & Ohlsson, 1986).

Inferences about the secondary and tertiary structure of LPO may be made based on its amino acid composition (Table 1) and its properties. The enzyme contains 50% non-polar side chains. The fact that hydrophobic chromatography is an effective means for purification indicates that these non-polar side chains exist in part as exposed clusters on the surface of the molecule. The molecule also has exposed positive charges on its surface as indicated by its affinity for negatively charged ion-exchange resins. This affinity is consistent with the amino acid composition which shows that the number of basic side chains exceeds the number of acidic side chains by 15.

These hydrophobic and charged clusters on the surface of LPO provide a mosaic which gives the molecule strong affinity for many different kinds of surfaces. It is strongly adsorbed to glass (Honka *et al.*, 1982), enamel (Pruitt & Adamson, 1977), bacterial cell surfaces (Pruitt *et al.*, 1979), and to other proteins (Tenovuo *et al.*, 1982).

Adsorbed LPO retains its enzymatic activity which means that attachment to surfaces does not block access to the heme group. Thus, the surface binding sites and the heme group must be in separate locations on the LPO surface. This separation is consistent with the domain structure proposed by Pfeil & Ohlsson (1986). These authors studied the thermal unfolding of LPO using differential scanning calorimetry and optical methods. Their results indicated that LPO has at least two structural domains of different thermal stabilities and that the heme group is in the less stable domain.

Heme Structure

Sievers (1979) isolated and identified protoheme IX (1,3,5,8-tetramethyl-2,4-divinylporphyrine-6,7-dipropionic acid) from a pronase digest of bovine LPO. From this work, she concluded that protoheme IX was the prosthetic group in LPO. She found no evidence that the heme was covalently bound to the apoenzyme.

However, recent reports (Nichol *et al.*, 1987; Thanabal & La Mar, 1989; Modi *et al.*, 1990) do provide evidence that the prosthetic group in bovine LPO is covalently attached. The suggested structure is

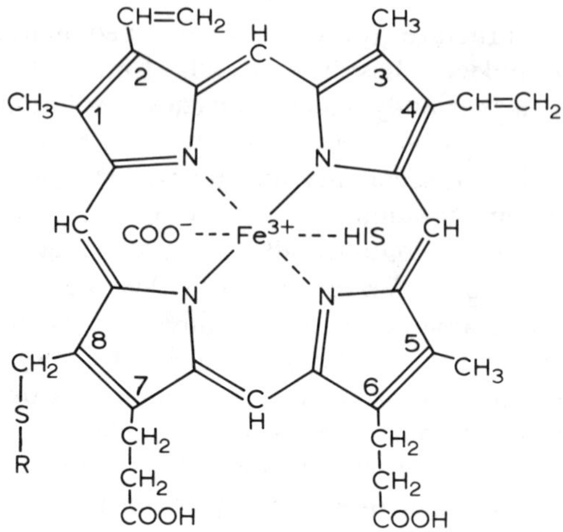

Fig. 1. Structure of the lactoperoxidase heme. The heme group is proposed to be bound to the apoenzyme by a disulfide bond linking the 8—CH_2 group to a cysteinyl residue. The proximal and distyl axial ligands are proposed to be a histidyl residue and a carboxyl group.

shown in Fig. 1. The modification of protoheme IX is that the heme group is covalently bound to the peptide backbone via a disulfide bond at the 8—CH_2 group. The Fe^{3+} is held in the porphyrin ring by four bonds to nitrogen. The fifth and sixth axial ligands are thought to be a histidyl residue and a carboxylate ion, respectively (Sievers, 1980; Sievers *et al.*, respectively 1983).

A Non-heme Lactoperoxidase in Bovine Milk

Allen & Morrison (1963) reported that crude bovine LPO preparations contained hemin-free polypeptides which reacted with anti-LPO immune serum. No further studies of this non-heme LPO were published until the work of Dumontet & Rousset (1983). These authors reported isolation by cation exchange chromatography of comparable amounts of LPO and non-heme LPO from bovine whey. The two forms gave a single line in immunodiffusion using anti-LPO antibodies. They had the same apparent molecular weight in velocity sedimentation on sucrose gradients, similar amounts of carbohydrate, and similar peptide maps after limited proteolysis by subtilisin or trypsin. The non-heme LPO gave no spectrum in the Soret region and was devoid of enzymatic activity. Although no experimental details were provided, these authors reported that the non-heme LPO was also found in goat, sheep, and human milk. The origin of the non-heme form and its biological significance are unknown.

REACTION MECHANISMS

Reaction with Hydrogen Peroxide

Our basic understanding of the physical chemistry of the reactions of peroxidases with hydrogen peroxide began 50 years ago with the pioneering work of Hugo Theorell, Karl Gustav Paul, Britton Chance and their colleagues (reviewed by Chance, 1951; Chance *et al.*, 1984). The reactions are complex and follow different pathways depending upon the concentration of hydrogen peroxide and whether or not exogenous electron donors are present (Kohler & Jenzer, 1989). The major intermediates for LPO are ferric peroxidase (the native enzyme in its ground oxidation state), Compound I, Compound II, Compound III, and ferrous peroxidase. In this section we will discuss the reactions which generate these intermediates and the resulting state of the heme group in the absence of exogenous electron donors.

The system may be described by the following reactions:
[ferric enzyme] $\geq$ [H_2O_2]:

$$\text{ferric enzyme} + H_2O_2 \rightarrow \text{Compound I} + H_2O \tag{1}$$

$$H^+ + \text{Compound I} + \bar{e} \rightarrow \text{Compound II} \tag{2}$$

$$\text{Compound II} + \bar{e} \rightarrow \text{ferric enzyme} \tag{3}$$

[ferric enzyme] $<$ [H_2O_2]:

$$\text{Compound II} + H_2O_2 \rightarrow \text{Compound III} + H_2O \tag{4}$$

$$\text{Compound III} \rightarrow \text{ferrous enzyme} + O_2 \tag{5}$$

$$\text{ferrous enzyme} + H_2O_2 \rightarrow \text{Compound II} + H_2O \tag{6}$$

In the above scheme, $\bar{e}$ represents an electron. In the absence of exogenous electron donors, this electron may be supplied by impurities in the enzyme preparation and by oxidizable groups within the enzyme. In reaction (1) both oxidizing equivalents of the peroxide are transferred to the heme. One oxidation equivalent is stored as a porphyrin-centered, π cation radical and the second as a low-spin oxyferryl center, $Fe^{4+}{=}O$ (Kohler & Jenzer, 1989). The π cation radical is reduced by the addition of an electron to Compound I to form Compound II in reaction (2) (Makino *et al.*, 1986; Sitter *et al.*, 1985; Terner *et al.*, 1985). The oxyferryl center retains a single bond between iron and oxygen, $Fe^{4+}{-}O^-$ (Chance *et al.*, 1984). In this reaction, the enzyme also takes up a single proton. The actual site occupied by this proton is a matter of controversy (Yamada & Yamazaki, 1974; La Mar *et al.*, 1983). Addition of an electron to the $Fe^{4+}{-}O^-$ [reaction (3)] regenerates the ferric enzyme.

At high concentrations of peroxide and in the absence of exogenous electron donors, Compound II may react with peroxide to generate Compound III, reaction (4). In Compound III, the iron has been reduced to the ferrous state and an adduct with dioxygen is formed, $Fe^{2+}{-}O_2$ (Dunford & Stillman, 1976). Dissociation of dioxygen from Compound III, reaction (5), generates the ferrous enzyme in which the heme group has the same structure as the ferric enzyme but with iron in the lower oxidation state. The ferrous enzyme can be oxidized by peroxide to regenerate Compound II, reaction (6). Reactions (4), (5), and (6) may be accompanied by irreversible enzyme inactivation by cleavage of the heme group (Jenzer *et al.*, 1986, 1987). The net result of reactions (4), (5), and (6) is the catalytic degradation of 2 molecules of peroxide to 2 molecules of water and 1 molecule of dioxygen (Kohler *et al.*, 1988).

Reactions in the Presence of One-Electron Donors

The following reaction mechanism describes the one-electron reductions:

$$\text{Compound I} + \text{AH} \rightarrow \text{Compound II} + \text{A}^{\bullet} \tag{7}$$

$$\text{Compound II} + \text{AH} \rightarrow \text{Ferric enzyme} + \text{A}^{\bullet} \tag{8}$$

$$2\text{A}^{\bullet} \rightarrow \text{A}_2 \tag{9}$$

Where AH represents a broad range of donors and $A^{\bullet}$ is a free radical. The experimental evidence supporting this mechanism is reviewed in Kohler & Jenzer (1989). Reaction (7) is very fast. The half-life of Compound I is 0·2 s (Kimura & Yamazaki, 1979). Thus, in the presence of one-electron donors and low concentrations of peroxide, Compound II is the intermediate which will be observed under steady-state conditions.

A comprehensive discussion of the vast literature describing the many different types of donors represented by AH is beyond the scope of this review. We will consider only a few selected examples. Many spectrophotometric peroxidase assays are based on the stepwise, one-electron oxidations of aromatic electron donors which generate colored products. Some typical donors are pyrogallol (Chance & Maehly, 1955), guaiacol (Chance & Maehly, 1955), and ABTS(2,2′-azino-di-(3-ethylbenzythiazoline-6-sulfonic acid) (Shindler & Bardsley, 1975; Schindler *et al.*, 1976). The one-electron peroxidations of phenol, catechol, and hydroquinone generate free radicals which bind to proteins and to DNA (reviewed by Smith *et al.*, 1989). These donors are in-vivo metabolites of benzene. Their subsequent peroxidation has been suggested to play a role in benzene toxicity and carcinogenicity. Free radicals are also generated by one-electron peroxidations of Dopa methyl ester(dihydroxyphenylalanine methyl ester) and 6-hydroxy-Dopa(trihydroxyphenylalanine) (Metodiewa *et al.*, 1989*b*), norepinephrine (Metodiewa *et al.*, 1989*a*), and benzidine (Josephy, 1985).

Reactions in the Presence of Two-electron Donors

Glutathione (G–SH) is an important component of oxidative defense mechanisms. Hydrogen peroxide is consumed by a two-electron transfer reaction catalyzed by glutathione peroxidase during which G–SH is oxidized to the disulfide G–S–S–G. However, lactoperoxidase *will not* catalyze directly the oxidation of G–SH (Nakamura *et al.*, 1986).

Halide ions (Cl^-, Br^-, I^-) and the thiocyanate ion (SCN^-) are oxidized by Compound I by a direct two-electron transfer of oxidizing equivalents (reviews: Morrison & Schonbaum, 1976; Thomas, 1985). The net reaction is:

$$\text{Compound I} + X \rightarrow \text{Ferric enzyme} + XO, \qquad (10)$$

where X represents the halide or the thiocyanate ion and XO is the oxidized product. Equation (10) describes the net reaction for these complex, two-electron transfers. The detailed mechanisms vary depending upon the particular peroxidase and the particular donor. Peroxidase specificities and reaction products are given in Table 3. The products of the peroxidation of two-electron donors kill or inhibit the growth and metabolism of many species of micro-organisms (Pruitt & Reiter, 1985).

Table 3. Peroxidase Specificities and Reaction Products[a]

Two-electron donor	*May be oxidized by*[b]	*Some reaction products*[c]
F^-	No known peroxidase	—
Cl^-	MPO, EPO, CPO	Cl_2, OCl^-
Br^-	MPO, EPO, CPO, LPO, TPO	Br_2, OBr^-
SCN^-	MPO, EPO, CPO, LPO, TPO	$(SCN)_2$, $OSCN^-$
I^-	MPO, EPO, CPO, LPO, HPO, TPO	I_2, I_3^-, OI^-

[a] Adapted from Thomas, 1985.
[b] The abbreviations are: MPO (myeloperoxidase), EPO (eosinophil peroxidase), CPO (choloroperoxidase), LPO (lactoperoxidase), TPO (thyroid peroxidase), HPO (horse-radish peroxidase).
[c] The products generated, and the relative amounts vary with the enzyme, relative and absolute concentrations of donors and H_2O_2, with pH, and with other reaction conditions. Other products in addition to those listed may also be formed. The hypohallide ions are in equilibrium with the conjugate hypohalous acids and the pK_as are different for the various hypohalous acids and for HOSCN.

The concentrations of the halides and SCN^- in human and bovine milk (Tables 4 and 5) are derived from the diet and are highly variable from place to place, from season to season, and from subject to subject. The data listed in Tables 4 and 5 were converted from the units (usually ppm, μg/ml, or mg/litre) in the original reports to mM or μM. A report from the UK (Phillips *et al.*, 1988) points out that the iodide content of bovine milk has increased significantly because of addition of iodide to cattle feed. These authors suggest that iodide concentrations in milk and dairy products in the UK may now have reached harmful levels based on the similarities which they observed between the monthly variations in milk iodide and the seasonality in the onset of thyrotoxicosis in their subjects. There has been one report of chloride deficiency in one individual (Hill & Bowie, 1983). Chloride was undetectable in this mother's milk and her fully breast-fed infant suffered from dietary chloride deficiency syndrome which was resolved when the child was placed on a cows' milk formula. The data reported by Macy (1949) show that chloride concentrations in colostrum tend to be higher than those in mature milk.

Table 4. Halide and Thiocyanate Concentrations in Human Milk

Halide	*Concentrations*		*No. obs*	*Location*	*Ref.*
	Mean	*Range*			
Chloride (mM)					
days post-partum					
0–5	16·5	12·2–28·4	8	USA	1
6–10	12·9	0·86–20·3	21	USA	1
>15	10·6	2·48–20·7	216	USA	1
	0	[a]	3	South Africa	2
Bromide (μM)	20	3·8–55	278	Netherlands	3
Iodide (μM)	0·4	0·007–1·2	167	Italy	4
	[b]	0·24–0·32	[b]	Germany	5
	2·1	[b]	[c]	Finland	6
	1·4	0·2–3·9	61	USA (NC)	7
Thiocyanate (μM)	122	[b]	[c]	Finland	6
	[b]	<80	[b]	Sweden	8
	21	3–58	8	USA (TN)	9

[a] Chloride was undetectable in milk samples taken from a single individual on three different occasions.
[b] Not given.
[c] Multiple samples taken from nine donors on two different days.

References

1. Macy, 1949. Data compiled from several sources.
2. Hill & Bowie, 1983.
3. Goewie & Hogendoorn, 1985.
4. Etling *et al.*, 1986.
5. Wiechen & Kock, 1984.
6. Mäkinen & Tenovuo, 1980.
7. Gushurst *et al.*, 1984. Comparable results were reported from California by Bruhn & Franke, 1983.
8. Gotherfors & Marklund, 1975.
9. Funderburk & van Middlesworth, 1967.

For LPO in human and bovine milk, the significant two-electron donor is SCN^-. LPO will not catalyze the oxidation of Cl^-, and the concentrations of the other halides in milk are too low to enable them to compete effectively with SCN^- for the available oxidizing equivalents of Compound I.

Table 5. Hallide and Thiocyanate Concentrations in Bovine Milk

Halide	*Concentrations*		*No. obs*	*Location*	*Ref.*
	Mean	*Range*			
Chloride (μM)	22	18–32	18	Japan	5
	32	26–43	10	South Africa	7
	40	30–110	[a]	Scotland	8
Bromide (μM)	93	70–133	18	Japan	5
Iodide (μM)	0·2–0·8[b]	0·08–2·0	848	France	1
	3·3	2·4–4·2[c]	[d]	Saskatchewan	9
	1·0–1·6[b]	0·7–2·3	84	UK	4
	2·8	0·4–28	[d]	USA (MI)	2
	0·4	0·3–0·5	[d]	FRG	10
Thiocyanate (μM)	135	130–140[c]	[c]	Saskatchewan	9
	[d]	25–250	[d]	UK, France	6
	18	12–31	6	USA (TN)	3

[a] Number of samples not given. Multiple samples were taken from six cows over a 30-day period before and after parturition. The data were presented graphically. The numbers shown here were estimated visually from the graphs.
[b] Mean values varied from town to town, season to season and year to year.
[c] The range of the mean ± standard error. The actual number of samples was not given, but multiple samples were taken from six different cows over a period of several weeks.
[d] Number not given.

References

1. Aumont *et al.,* 1987.
2. Curtis & Hamming, 1982.
3. Funderburk & van Middlesworth, 1967.
4. Nelson *et al.,* 1987.
5. Ohno *et al.,* 1988.
6. Reiter, 1978.
7. van Staden, 1986.
8. Thompson, 1988.
9. Laarveld, *et al.,* 1981.
10. Wiechen & Kock, 1984.

The Peroxidation of SCN^-

The first detailed studies of the major product of LP-catalyzed oxidation of SCN^- were reported by Oram & Reiter (1966*a*). That work confirmed earlier reports that the product was a short-lived reactive oxidant which increased extinction at 235 nm, and the authors suggested that the product might be sulfur dicyanide ($S(CN)_2$) because the two substances had similar biochemical properties. However, subsequent investigations have shown that the major product of the LPO catalyzed oxidation of SCN^- at pH 7 is the hypothiocyanite ion, $OSCN^-$ (Aune & Thomas, 1977; Hoogendorn *et al.*, 1977; Pruitt *et al.*, 1982), which exists in a dynamic equilibrium with hypothiocyanous acid ($pK_a = 5{\cdot}3$, Thomas, 1981):

$$HOSCN \leftrightarrow H^+ + OSCN^- \qquad (11)$$

The increased extinction at 235 nm at pH 7 is a consequence of the contributions of HOSCN ($\lambda_{max} = 240$ nm) and $OSCN^-$ ($\lambda_{max} = 220$ nm) to the observed spectrum (Tenovuo, *et al.*, 1986).

However, the reaction system is complex and $OSCN^-$ may not be the first product released from the enzyme's active site. Other short-lived intermediates which may be formed in varying amounts depending upon reaction conditions (reviewed by Thomas, 1985) include thiocyanogen ($(SCN)_2$), cyanogen thiocyanate (NC–SCN), cyanosulfuric acid (HO_2SCN) and cyanosulfurous acid (HO_3SCN).

The LPO-catalyzed reactions shows pH optima which are $[SCN^-]$ dependent (Wever *et al.*, 1982; Pruitt *et al.*, 1988). At high concentrations of SCN^-, LPO activity is reduced by a pH-dependent binding of SCN^- to the enzyme. Using ^1H and ^{15}N nuclear magnetic resonance studies at pH 6·1, Modi *et al.* (1989) estimated a $K_D = 90$ mM for the dissociation constant of the LPO—SCN^- complex. Based on kinetic studies, Pruitt *et al.* (1988) reported a value of 46 M^{-1} for the binding constant at pH 6·0. The reciprocal of this number is $K_D = 22$ mM. Sievers (1985) used spectrophotometric titrations to study the binding. From her data a $K_D = 32$ mM may be estimated at pH 6·0. The agreement of these estimates of the K_D based on very different methods of analysis is remarkable. Also there is some discrepancy in the reported values of the pK_a of the ionizable group involved in the SCN^- binding. Modi *et al.* (1989) found $pK_a = 6{\cdot}4$, Pruitt *et al.* (1988), estimated a value of 6·0, and Sievers (1985), 5·1.

The complex dependence of the initial thiocyanate peroxidation rates as a function of pH and of $[SCN^-]$ is consistent with the following mechanism:

$$E + H_2O_2 \xrightarrow{k_1} EO + H_2O$$

$$EO + H^+ \overset{K_1}{\longleftrightarrow} EOH^+$$

$$EOH^+ + SCN^- \xrightarrow{k_4} E + OSCN^- + H^+$$

$$E + SCN^- \overset{K_i}{\longleftrightarrow} ESCN^-$$

$$EOH^+ + H^+ \overset{K_2}{\longleftrightarrow} EOH_2{}^+$$

$$ESCN^- + H^+ \overset{K_3}{\longleftrightarrow} ESCNH$$

Compound I is represented by EO. The mechanism assumes that protonated Compound I (EOH^+) is the enzymatically active species and that $ESCN^-$, $EOH_2{}^+$, and ESCNH are inactive. This mechanism accurately predicts the pH optima and the pH dependence of the SCN^- inhibition. Values of the rate constants (k_1, k_4) and the equilibrium constants (K_1, K_i, K_2, K_3) fro LPO and for human SPO have been estimated on the basis of steady-state kinetic analyses (Pruitt *et al.*, 1988).

The products HOSCN and $OSCN^-$ react rapidly with sterically accessible protein sulfhydryl groups (Aune & Thomas, 1977; Aune *et al.*, 1987) to yield sulfenyl thiocyanates (R–S–SCN). This reaction may be reversed by high concentrations of reducing agents. At low concentrations of sulfhydryls (R′–SH), the R–S–SCN may react to form mixed disulfides (R–S–S–R′). At higher concentrations of R′–SH, the R–S–SCN may be reduced back to R–SH. The R–S–SCN may also be hydrolyzed to sulfinic acids (R–S–OH). The reactions are:

$$OSCN^- + R\text{–}SH \rightarrow R\text{–}S\text{–}SCN + OH^- \quad (12)$$

$$R\text{–}S\text{–}SCN + H_2O \rightarrow R\text{–}S\text{–}OH + H^+ + SCN^- \quad (13)$$

$$R\text{–}S\text{–}SCN + R'\text{–}SH \rightarrow R\text{–}S\text{–}S\text{–}R' + H^+ + SCN^- \quad (14)$$

$$R\text{–}S\text{–}S\text{–}R' + R'\text{–}SH \rightarrow R\text{–}SH + R'\text{–}S\text{–}S\text{–}R' \quad (15)$$

The effects of the LPO/SCN^-/H_2O_2 system on micro-organisms (Pruitt & Reiter, 1985; Carlsson, 1987; Pruitt, 1987) can be understood in terms of these reactions. The active product of the system is generated by reaction (10). Inhibition of micro-organisms is a consequence of reaction (12) when the oxidized R–SH group is a critical factor in microbial metabolism. For example, oxidation of the sulfhydryl groups of glyceraldehyde–phosphate dehydrogenase will stop microbial glycolysis. Inhibited micro-organisms may recover from the effects of reaction (12) by regeneration of the critical R–SH groups through reactions (14) and (15). Failure to reverse the consequences of reaction (12) may lead to cell death.

A threshold ratio of [$OSCN^- + HOSCN$] to number of susceptible sites in each bacterial cell is required for the bactericidal effect of the LPO system (Thomas *et al.*, 1983; Pruitt *et al.*, 1984; Månsson-Rahemtulla *et al.*, 1987). The ratio is pH-dependent and is lower at low pH. Near pH 7 the value is about 10^8 molecules of $HOSCN + OSCN^-$ per cell. With low concentrations of $OSCN^- + HOSCN$ or a high inoculum, the threshold ratio is not attained. Hoogendoorn *et al.* (1977) reported complete inhibition of the metabolic activity of *S. mutans* by $0{\cdot}3 \times 10^8$ molecules of $HOSCN + OSCN^-$ per cell. A higher ratio of $3{\cdot}1 \times 10^8$ molecules/cell was reported by Månsson-Rahemtulla *et al.* (1987) to be required for complete inhibition of this micro-organism at pH 7·1. From the data of Thomas & Aune (1978) the number of $HOSCN + OSCN^-$ molecules/bacterial cell required for complete inhibition of the respiration of *E. coli* can be estimated as 5×10^8.

Some bacteria are resistant to the LP system. Oram & Reiter (1966*b*) isolated from a resistant strain of *Streptococcus cremoris* a 'reversal factor' which, in the presence of $NADH_2$, catalyzed the reduction of the thiocyanate oxidation product of the LP system back to SCN^-. Carlsson *et al.* (1983) found that a cell-free extract from strains of *Streptococcus sanguis* and *Streptococcus mitis* contained an oxidoreductase enzyme which catalyzed the reduction of $OSCN^-$ to SCN^- by $NADH_2$. They attributed the ability of these bacteria to recover rapidly from the effects of the LP system to the activity of this enzyme.

POTENTIAL HUMAN TOXICITY OF THE LACTOPEROXIDASE/SCN^-/H_2O_2 SYSTEM

Data reviewed elsewhere in this report show that activation of the bovine lactoperoxidase system by addition of hydrogen peroxide and thiocyanate is an effective strategy for preservation. Is bovine milk treated in this way safe for human consumption? Here we review the potential toxicity of the components and products of the lactoperoxidase system.

The Lactoperoxidase Enzyme

This protein is a normal component of bovine and human milk and is present in all mammalian milks which have been tested so far. It is secreted by the mammary glands and is present at concentrations of several $\mu g/ml^{-1}$ (Kiermeier & Kayser, 1960; Korhonen, 1973). LPO is the most abundant enzyme in bovine milk. It accounts for up to 1% of whey proteins (Korhonen, 1980). Therefore, preservation methods do not require further addition of the enzyme. Obviously its presence in milk poses no threat to human health.

The Thiocyanate Ion

The SCN^- ion is also a normal milk component (Table 4). It is derived from diet (Wood, 1975) primarily by hydrolysis of glucosinolates which are widely distributed in plants. The glucosinolates have the following general structure:

$$\begin{array}{c} \text{S-glucose} \\ | \\ R\text{—}C\text{=}N\text{—}OSO_3^- \end{array}$$

Following ingestion, glucosinolates are hydrolyzed by a complex sequence of enzyme-catalyzed reactions to yield glucose, HSO_4^-, SCN^- and other products. Another source of SCN^- is the cyanogenic glycosides found in cassava, sweet potatoes, millet, and sugar-cane. Upon ingestion these compounds are hydrolyzed and release cyanide which is detoxified by a reaction which yields SCN^-. The cyanide which is in tobacco smoke is detoxified by the same reaction and this accounts for the higher concentrations of SCN^- found in the serum, saliva, and milk of smokers (van Middlesworth, 1986).

Because of variations in diet, the SCN^- concentrations in milk show seasonal variations. The normal concentrations (Table 4) must be supplemented in order to generate inhibitory levels of HOSCN +

$OSCN^-$ *in vitro*. Supplementation up to levels as high as 1–2 mM SCN^- pose no threat to human health because SCN^- is present as a normal component of the saliva of healthy human beings at concentrations within this range (Pruitt *et al.*, 1986). Ingested SCN^- is excreted with a half life of 2–5 days (Wood, 1975). Since high serum concentrations of SCN^- interfere with the iodine uptake of the thyroid gland and the iodination of thyroxine, it is important to establish that a prolonged intake of thiocyanate-supplemented milk does not result in any disturbance of thyroid function. Dahlberg *et al.* (1984) fed 37 healthy subjects 200 ml of milk supplemented with SCN^- and H_2O_2 (initial concentrations of $[SCN^-] = 344$ μM, $[H_2O_2] = 250$ μM) twice daily for a 12-week period. They concluded that this regimen had no effect on thyroid function because the measured serum levels of thyroxine, triiodothyronine, and thyrotropic hormone were normal before and after the period of supplemented-milk ingestion. This same group studied the effect of ingestion of supplemented milk on thyroid function in a group of 55 goitrous subjects living in an endemic goiter region of western Sudan. They found no evidence of a negative effect on the thyroid function of these iodine deficient subjects after a 4-week period of daily ingestion of 250 ml of SCN^--supplemented ($[SCN^-] = 327$ μM) milk.

The suggestion of adding SCN^- to milk is often met with strong negative reaction on the part of the general public. This response may be based in part on the fact that the word 'thiocyanate' sounds similar to 'cyanide' which is a widely known 'poison'.

Hydrogen Peroxide

The only approved additive for the preservation of milk in the absence of refrigeration is hydrogen peroxide. It may be added at concentrations of 100–800 ppm (FAO, 1957; CFR, 1984) which corresponds to 2·9–23·5 mM H_2O_2. At the low $[SCN^-]$ found in bovine milk, LPO is destroyed by such high concentrations of peroxide. The preservative effects are therefore due to H_2O_2 toxicity. Excess hydrogen peroxide can react with $HOSCN/OSCN^-$ to generate products which are more cytotoxic than hydrogen peroxide itself (Carlsson *et al.*, 1984). The added peroxide is decomposed by addition of catalase prior to pasteurization.

Peroxide is highly toxic for mammalian cells. However, at low $[H_2O_2]$ (100 μM or less) and in the presence of LPO and SCN^- mammalian cells are protected from this toxicity (Hänström *et al.*, 1983; Tenovuo & Larjava, 1984). Since the mammary gland is a secretory tissue with very high metabolic activity, H_2O_2 would be an expected by-product of normal aerobic metabolism. However, any H_2O_2 produced by the mammary gland would be rapidly consumed by reactions catalyzed by the LPO and catalase enzymes in the milk (Reiter, 1985). These same reactions would prevent the long-term persistence of any low concentrations of H_2O_2 added to milk to activate the LPO system.

The fate of H_2O_2 added to milk is illustrated by the following calculations. Suppose H_2O_2 and SCN^- sufficient to give initial concentrations of 1 mM each were added to a milk sample. The net result of reactions (1)–(10) would be the consumption of H_2O_2 by first-order kinetics (Pruitt *et al.*, 1986):

$$[H_2O_2] = [H_2O_2]_0 \exp(-k_{1app}t), \qquad (16)$$

where $[H_2O_2]$ = concentration of peroxide at time t, $[H_2O_2]_0$ = initial concentration of peroxide, and k_{1app} = the apparent first-order rate constant = 1·43 min^{-1}.

We can use eqn (16) to calculate the amount of peroxide which would remain in the milk sample at any time after addition of the SCN^- and H_2O_2. Substituting $t = 10$ min and $[H_2O_2]_0 = 1$ mM, gives $[H_2O_2] = 1 \times 10^{-10}$ M, which is an insignificant amount. In any event, such small amounts of peroxide consumed orally by human beings would be rapidly converted to non-toxic products by the SPO system (Carlsson, 1987).

Reaction Products

In the absence of sufficient concentrations of exogenous electron donors and at high (relative to [LPO]) concentrations of peroxide, major reaction products would be water and dioxygen generated as a result of catalatic reactions [eqns (4), (5), and (6)]. In milk supplemented with SCN^- the major reaction products would be HOSCN and $OSCN^-$. Since the pK_a of HOSCN is 5·3 (Thomas, 1981), the amount of HOSCN present at the normal pH of milk (6·6) would not exceed 5% of the total HOSCN + $OSCN^-$. The HOSCN + $OSCN^-$ are normal constituents of human saliva. Concentrations in resting human whole saliva are near 50 μM but are highly variable (Pruitt *et al.*, 1986). Obviously, HOSCN + $OSCN^-$ are not toxic at these concentrations.

In-vitro cell culture studies (Tenovuo & Larjava, 1984) have shown that concentrations of up to 300 μM HOSCN + $OSCN^-$ are not toxic for mammalian cells. Other in vitro studies have shown that HOSCN + $OSCN^-$ are not mutagenic and do not oxidize mammalian DNA (White *et al.*, 1983). Dentifrices which elevate HOSCN + $OSCN^-$ in the mouth have been in widespread use for many years with no reports of harmful effects on human beings (Hoogendorn & Piessens, 1987).

When human erythrocytes were incubated at pH 6·0 for 90 min at 37°C with a concentration of HOSCN + $OSCN^-$ which was initially 250 μM, 5% hemolysis was reported (Grisham & Ryan, 1990). These authors found no hemolysis under these conditions at pH 6·5 or higher even after 2 h of incubation. These in-vitro observations suggest that mammalian cells may be slightly affected by exposure to high concentrations of HOSCN + $OSCN^-$ for extended periods of time at pH 6. However, these conditions would not be produced *in vivo* by addition of low concentrations (1–2 mM) of SCN^- and H_2O_2 *in vitro* to bovine milk and the subsequent human consumption of that milk many hours later. The HOSCN + $OSCN^-$ generated at the pH of milk (6·6) by initial concentrations of SCN^- and H_2O_2 of the order of 1–2 mM would decompose spontaneously to non-toxic products (Thomas, 1981; Tenovuo *et al.*, 1986) by the time the milk was consumed.

When H_2O_2 is added in excess over the amount of SCN^- present, oxidation products which are toxic for mammalian cells may be produced transiently (Carlsson *et al.*, 1984). However, these transient products are rapidly consumed by reactions with $OSCN^-$ and other components (Pruitt *et al.*, 1982).

HUMAN MILK PEROXIDASES

Human milk was first analyzed for peroxidase activity by Wüthrich *et al.* (1963). These authors reported that peroxidase activity was higher in cows' milk than it was in human milk. Similar differences were reported by Kiermeier & Kuhlmann (1972). The latter authors also found that peroxidase activity was highest in human colostrum and slowly fell during lactation. They found a very strong positive correlation between cell content of human milk and peroxidase activity during the 2nd to 9th day post-partum, and that a higher cell count was maintained longer in human milk than in bovine milk. These observations suggest significant contribution of myeloperoxidase (MPO) to the observed peroxidase activity (see below). Gotherfors & Marklund (1975) also reported that cows' milk had higher peroxidase activity than human colostrum and that peroxidase activity in human milk declined significantly during lactation. However, they did not report a correlation between cell content of human colostrum and peroxidase activity.

After the above studies were carried out, an improved and very sensitive assay for peroxidases based on the oxidation of 2,2′-azino-di-(3-ethylbenzythiazoline-6-sulphonic acid) or ABTS was developed (Shindler & Bardsley, 1975; Shindler *et al.*, 1976). Since the publication of the ABTS method, there have been many reports of peroxidase activities in human colostrum and milk (reviewed by Reiter, 1981; Reiter, 1985). The major conclusions of this body of work are: in contrast to bovine milk, peroxidase activity is highest in human colostrum and declines rapidly during the post-partum period; total colostral peroxidase activity is correlated with the presence of PMN leukocytes; there is great variation among the different reports; some human milk samples contain very high peroxidase activity while other samples show little or none.

Does human milk contain a true LPO or is the observed activity due entirely to MPO derived from milk PMN leukocytes? Gotherfors & Marklund (1975) concluded that human milk peroxidase activity was due primarily to LPO and not MPO. However, Moldoveanu *et al.* (1982) found that human milk peroxidase gave an elution profile in molecular exclusion chromatography which was similar to that of MPO derived from blood PMN leukocytes. They also reported that human SPO reacted with an antibody directed against bovine LPO, but they detected no reaction of that same antibody preparation with either human milk peroxidase or MPO. Hashinaka & Yamada (1986) prepared an antiserum against human MPO and found that the peroxidase in human colostrum gave a single precipitin line against this antiserum. The peroxidase activity in human colostrum was precipitated completely by anit-MPO IgG. They purified human colostral whey peroxidase to homogeneity and found that this preparation was composed of subunits whose size corresponded to the subunits of

MPO. The conclusions of the authors of these latter two studies were that the peroxidase enzyme in human milk was MPO.

In contrast to the above reports, Langbakk & Flatmark (1984) found that all of the peroxidase activity in pooled whey from human colostrum was absorbed to phenyl sepharose and eluted with a retention time characteristic of bovine LPO. MPO cannot be eluted from phenyl sepharose. Total peroxidase activity was also removed from the whey by an immuno-affinity column utilizing antibovine LPO antibodies. The colostral peroxidase also had an elution pattern from ConA-Sepharose which was similar to that obtained with bovine LPO. They found that the colostral peroxidase peak was bound by Sepharose beads coated with antibovine LPO antibodies but that human MPO and eosinophil peroxidase were not. In subsequent studies (Langbakk & Flatmark, 1989) they obtained an apparent 1450-fold purification of the colostral enzyme. This purified preparation had spectral properties (Soret peak at 412 nm) and size (M_r 80 000) similar to those of bovine LPO.

In hindsight, it would seem apparent that *both* a true secretory LPO and an MPO derived from leukocytes may be found in human colostrum in variable absolute and relative amounts. The methods used by Moldoveanu *et al.* (1982) and by Hashinaka & Yamada (1986) may not have been sufficiently sensitive or selective to detect LPO. On the other hand the methods used by Langbakk & Flatmark (1984, 1989) would not have detected MPO in their colostral samples because the MPO would *not* have appeared in the elution profile from the phenyl sepharose columns. Furthermore, the MPO may *not* have been bound to their anti-LPO affinity columns and might have been passed through during the washing procedures carried out prior to the step of eluting the bound LPO with glycine–HCl. Apparently they did not analyze these column washings for peroxidase activity.

Work from our laboratory (Pruitt *et al.*, 1991) has confirmed the presence of enzyme activities in human milk attributable to both human LPO and MPO. Using a method (Månsson-Rahemtulla *et al.*, 1986) which permits quantitative evaluation of the separate contributions of LPO and MPO to the total peroxidase activity, we analyzed 27 milk samples from 21 different subjects. MPO accounted for 38% (SD = 18) of the mean total peroxidase activity measured for these samples, and LPO for 62%.

BIOLOGICAL SIGNIFICANCE OF MILK PEROXIDASE SYSTEMS

Human Milk

Since human milk contains MPO, LPO, halides and SCN^-, all of the reaction products listed in Table 3 may potentially be generated. There have been no reports of measurement of these substances in human milk. However, the oxidized forms of SCN^- are probably the major oxidation products. Both MPO and LPO will catalyze the peroxidation of this ion. The SCN^- competes very effectively with Cl^- for available oxidizing equivalents of MPO Compound I, and the concentrations of Br^- and I^- are far less than that of SCN^-.

The oxidized forms of SCN^- kill or inhibit the growth and metabolism of many different species of micro-organisms (Pruitt & Reiter, 1985) so that the LPO system in human milk could function as an antibody-independent defense mechanism. Human milk will kill *Escherichia coli* when it is supplemented with SCN^- and H_2O_2 to concentrations of 0·22 mM (Reiter, 1981). Work from our laboratory (Pruitt *et al.*, 1991) has shown that human milk with a high intrinsic peroxidase activity will kill *E. coli* and *Salmonella typhimurium* when it is supplemented with SCN^- and H_2O_2 to concentrations of 3 mM. The biological significance of these experiments is not clear. They demonstrate the antibacterial properties of the human milk peroxidase system, but these effects require addition of SCN^- and H_2O_2 to concentrations not normally found *in vivo*. It may be that the most important function of the human milk peroxidase system is to protect the mammary gland from the toxic products of aerobic metabolism (Carlsson, 1987).

Bovine Milk

Although LPO is the most abundant enzyme in bovine milk, the concentrations of SCN^- and H_2O_2 in fresh bovine milk from healthy cows are usually too low to generate inhibitory levels of $HOSCN + OSCN^-$ *in vitro*. No measurements of hydrogen peroxide in fresh, uncontaminated milk have been reported. Hydrogen peroxide could be supplied by leukocytes and some strains of streptococci (Reiter, 1985). Exposure to light has been reported to generate H_2O_2 in milk, but the mechanism was not identified (Toyosaki *et al.*, 1984). The defensive role which the LPO system plays in the bovine udder is complex (Reiter & Perraudin, 1990). Major difficulties include the identification of an in-vivo source for hydrogen peroxide, the variable and often low levels of SCN^-, and the survival of streptococci infused into lactating udders. However, the LPO system may protect the nonlactating gland from infection by *Staphylococcus aureus* and *Streptococcus uberis*.

When H_2O_2 and SCN^- are added *in vitro* to bovine milk, whey or growth media containing LPO, the systems inhibit various mastitis-causing bacteria (Mickelson & Brown, 1985; Marshall *et al.*, 1986; Sandholm *et al.*, 1988). Some of these micro-organisms are human pathogens and can be shed into milk from udders of infected cows. The LPO system is also inhibitory against enterococci and enterobacteriaceae which cause gastroenteritis and are associated with high infant morbidity and mortality, especially in developing countries (Banks & Board, 1985). Thus, the LPO system may provide an in-vivo defense mechanism against udder infections in the cow, and may protect the newborn from gastroenteritis. There is strong evidence for such protection of calves, which is found in an important series of experiments described by Reiter (1985). An inoculum of 1×10^7 cfu ml^{-1} *E. coli* in 200 ml of heated (to inactivate LPO) milk was fed orally to calves. They were then fed 2 liters of heated milk which contained no added bacteria. After 1 h, the viable count in the calves' abomasal fluid was 1×10^6 cfu ml^{-1}. In a second set of experiments, calves were fed 2 liters of raw, unheated milk and the viable count was reduced to 1×10^4 cfu ml^{-1}. When the raw milk was supplemented with MgO_2 (which hydrolyzes to liberate H_2O_2) in a third set of experiments, the 1-h count of the calves' abomasal fluid was reduced to 100 cfu ml^{-1}.

The results of the above experiments are striking. They raise an interesting question. In many countries, pasteurized bovine milk is routinely fed to human infants. As much as 50% of bovine LPO survives normal pasteurisation. Does the consumption by humans of bovine milk provide augmentation of natural defense mechanisms as well as nutrition? Calf saliva contains very little peroxidase activity but the saliva of human infants has high concentrations of SPO which retains its activity in the gastric juice (Gotherfors & Marklund, 1975). Therefore, it would seem that the protective effects of bovine milk consumption would be less significant for human infants than they would be for newborn calves.

PRACTICAL APPLICATIONS

Control of Spoilage Bacteria

The widespread use of refrigeration during storage and transportation of raw milk suppresses growth of mesophilic and thermoduric bacteria. However, this leads to selective growth of psychrotrophic and cold-tolerant bacteria in milk (Cousin, 1982). Most psychrotrophs in raw milk pose no health hazard, but can produce heat-stable extracellular lipases (Stead, 1986) and/or proteinases (Cousin, 1982; Fairbarn & Law, 1986). These enzymes can cause undesirable changes in milk and milk products (Law *et al.*, 1976, 1977, 1979). Thus, additional means of controlling growth of these bacteria should be viewed as complementary to refrigeration.

The antibacterial activity of the LPO system in milk against psychrotrophic spoilage organisms has been widely investigated (Björck *et al.*, 1975; Björck, 1978; Reiter *et al.*, 1976). Using a glucose/glucose oxidase system to generate H_2O_2, and supplementing milk with 0·17–0·26 mM SCN^-, Björck *et al.* (1975) demonstrated that the LPO system was bactericidal against *Pseudomonads* and *Escherichia coli*. When various Gram-negative rods isolated from milk were exposed to the LP system, at least 91% of the inoculum was killed within 4 h at 30°C (Björck *et al.*, 1975). Reiter *et al.* (1976) observed that the bactericidal effect against *E. coli* was increased when SCN^- concentration was raised from 0·015 to 0·15 mM, and was reduced when the initial inoculum was increased. Björck (1978) observed a similar increase in the bactericidal effect of the LPO system against *P. fluorescens* when both the SCN^- and H_2O_2 concentrations were increased up to an equimolar concentration of 0·3 mM.

Apart from controlling psychrotrophic spoilage bacteria, the LPO system has been widely investigated for its potential to control mesophilic spoilage bacteria, especially in bovine milk (Björck *et al.*, 1979; Reiter & Harnuv, 1982; Zajac *et al.*, 1983; Kamau & Kroger, 1984). Preservation of buffalo milk by the LPO system has also been investigated both at ambient and refrigeration temperatures (Thakar & Dave, 1986; Chakraborty *et al.*, 1986). For bacteria that survive the initial bactericidal activity of the LPO system, there is an extended lag phase or recovery period. The length of this lag period is highly temperature dependent, being much longer at cold storage than at high temperatures (Björck *et al.*, 1979; Zajac *et al.*, 1983; Kamau &

Kroger, 1984). On average, this lag period is about 8 h at ambient temperatures (22–37°C). After recovery, most bacteria resume normal growth. Thus, the LPO system can prevent spoilage of milk during short storage or transportation periods at ambient temperatures.

Control of Pathogenic Bacteria

In the mid-1950s, refrigeration was considered adequate for controlling growth and/or toxin production by food-borne pathogens. However, certain food-borne pathogens are capable of growth and/or toxin production under cold storage as well as at normal growth temperatures. These pathogens include *Listeria monocytogenes, Staphylococcus aureus, Campylobacter jejuni, Bacillus cereus* and *Salmonella typhimurium* (Palumbo, 1986). Although most of these pathogens are destroyed by proper pasteurisation, their presence in milk poses a health hazard. Furthermore, some of these pathogens, such as *S. aureus,* can produce heat-stable toxins that can withstand pasteurisation and cause food poisoning (reviewed by Smith *et al.*, 1983).

The LPO system is both bactericidal and bacteriostatic against *S. typhimurium,* a well-recognized food-borne pathogen (Reiter *et al.*, 1976; Wray & McLaren, 1987). The rough strains of *Salmonella* are more susceptible to the LPO system than the smooth strains (Purdy *et al.*, 1983; Wray & McLaren, 1987). Purdy *et al.* (1983) suggested that cell wall permeability was responsible for the observed difference in susceptibility to the LPO system. *Campylobacter jejuni* is a major cause of acute enteritis in humans (Skirrow, 1982). Milk has been associated with several outbreaks of *C. jejuni* enteritis (Potter *et al.*, 1984; Sharp *et al.*, 1985; Barret, 1986). The occurrence of this pathogen in dairy cows and in milk has been demonstrated (Humphrey & Beckett, 1987). The bactericidal effect of the LPO system against *C. jejuni* in milk has been demonstrated (Beumer *et al.*, 1985). Poultry meat is also a major source of *C. jejuni* infections (Shanker *et al.*, 1982; Engavall *et al.*, 1986). Borch *et al.* (1989) showed that the LPO system has a strong bactericidal activity against strains of *C. jejuni* isolated from poultry. The bactericidal effect of the LPO system against this pathogen is more pronounced at higher temperatures (Beumer *et al.*, 1985; Borch *et al.*, 1989).

Staphylococcus aureus is a major causative agent of bovine mastitis and poses a human health problem since this pathogen can be shed into milk from mastitic udders. Although *S. aureus* is readily destroyed by proper pasteurisation, enterotoxins produced by strains of this pathogen can withstand pasteurization and cause food poisoning in the absence of viable cells (reviewed by Smith *et al.*, 1983). The LPO system is both bactericidal and bacteriostatic against *S. aureus* in milk (Kamau *et al.*, 1990*a*). As with most other bacteria, the bactericidal effects are more pronounced at high than at low temperatures, while the bacteriostatic effects are short-lived at high temperatures.

Listeria monocytogenes is an intracellular pathogen capable of survival and growth within phagocytes, a characteristic associated with virulence (Wilder & Edberg, 1973). Listeric infections can cause septicemia, meningitis and abortion, and the main groups at risk include neonates, pregnant women and other immunocompromised individuals (Gray & Killinger, 1966; Marth, 1988). Since 1981, three food-borne outbreaks of human listeriosis have been reported, in North America, one involving coleslaw in Canada (Schlech *et al.*, 1983), two in the USA associated with milk (Fleming *et al.*, 1985; Ho *et al.*, 1986) and cheese (James *et al.*, 1985; Linnan *et al.*, 1988). A fourth outbreak associated with soft cheese was reported in Switzerland in 1987 (Anon., 1987*a*). The mortality rate for the three outbreaks in North America was about 30%. In addition, numerous dairy products (milks, buttermilks, soft cheeses) have been recalled because of contamination with this pathogen (Anon., 1985, 1986*a,b,c*) costing the dairy industry over $80 million dollars. The risk of listeriosis is further amplified by the ability of *L. monocytogenes* to grow at refrigeration temperatures (Wilkins *et al.*, 1972) and its relatively higher heat resistance compared to other non-spore-forming bacteria (Bearns & Girard, 1958; Doyle *et al.*, 1985, 1987; Fernandez-Garayzabal *et al.*, 1986).

The LPO system is both bactericidal and bacteriostatic against *L. monocytogenes* (Dennis & Ramet, 1989; Earnshaw & Banks, 1989; Kamau *et al.*, 1990*a*; Siragusa & Johnson, 1989). Dennis & Ramet observed that the initial bactericidal effect against *L. monocytogenes* was more pronounced at 15 than at 4°C, and complete inactivation was achieved faster when the initial inoculum was low. Thus, the LPO system in conjunction with refrigeration holds promise for controlling the growth of this pathogen in milk.

Enhanced Thermal Destruction of Bacteria

Studies carried out with *L. Monocytogenes* and *S. aureus* in milk revealed that exposure to the LPO system prior to thermal processing dramatically enhances thermal inactivation of these bacteria (Kamau *et al.*, 1990*b*). Since there are no previous reports on this aspect of the LPO system these studies (Kamau *et al.*, 1990*b*) will be summarized in this section. Milk samples for *L. monocytogenes* studies were supplemented with initial concentrations of 2·4 mM SCN^- and 0·6 mM H_2O_2, while for *S. aureus* the respective concentrations were 1·2 mM SCN^- and 0·3 mM H_2O_2. The lower concentrations for the *S. aureus* studies were used in order to obtain observable results since higher concentrations resulted in rapid killing of this micro-organism by the LPO system as well as H_2O_2 alone prior to and during heating.

After exposure to the LPO system, the D values for *L. monocytogenes* in milk were on average reduced by 75% at 52·2 and 57·8°C. The respective maximum specific killing rates for *L. monocytogenes* at these temperatures were increased by 176, 398 and 388% in the presence of the LPO system, compared to untreated milk. For *S. aureus,* the D values were reduced by 90% on average, both at 52·2 and 55·2°C. The maximum specific killing rate for *S. aureus* was increased by 143 and 667%, respectively, at 52·2 and 55·2°C. In the presence of the LPO system, no viable cells of *S. aureus* were recovered after heating for less than 1 min at 57·8°C. The LPO system also reduced the initial lag in killing even at the lower heating temperatures. These observations suggest a synergistic nonreversible cell injury by the LPO system in conjunction with heat. Damage to bacterial cell membranes caused by the LPO system may account, in part, for the enhanced thermal destruction of bacteria after exposure to the LPO system (Kamau *et al.*, 1990*b*).

Because of the ability of bacteria to recover from the LPO system, it is critical that heat processing follows immediately after activation of the LPO system for the synergistic effects to be realized. For example, when *L. monocytogenes* was exposed to the LPO system in milk and samples held at 35°C for different times prior to heating at 55·2°C, there was gradual increase in the D values, approaching normal values after 8 h (Kamau *et al.*, 1990*b*). Since the LPO system enhances thermal destruction of bacteria, it is logical that activation of this system prior to heat processing should improve the shelf-life of pasteurised milk or other thermally processed foods. Based on an unacceptable total plate count of 2×10^4 CFU ml^{-1} for grade A pasteurised milk (USPHS, 1977), the predicted shelf-life of milk samples with the activated LPO system is more than doubled, compared to untreated samples (Kamau *et al.*, 1991).

Inactivation of Microbial Toxins?

There is evidence that the MPO system can inactivate diphtheria and tetanus toxins (Agner, 1974, 1950). Oxidative inactivation of aflatoxin B_1 by the MPO system has also been reported (Odajima, 1981). These toxins belong to a class of cytolytic bacterial toxins which are active in the reduced form (free SH-groups) and are inactivated by oxidation reactions (Aouf & Geoffrey, 1984). Bacteria producing this class of toxins include *L. monocytogenes, B. cereus, Streptococcus pneumoniae, S. pyogenes, C. botulinum, C. perfringens* and *C. tetani.* The $HOSCN + OSCN^-$ produced by the LPO system react rapidly with –SH groups. Therefore, it seems likely that the LPO system could be used effectively to neutralize –SH–dependent bacterial toxins. Furthermore, the intracellular pathogenesis of certain bacteria could be due to production of enough–SH toxins by these bacteria to reduce the oxidation products of leukocyte peroxidase systems to levels which the bacteria could survive.

Effect on Food Processing and Quality

If milk is preserved by the LPO system, starter culture inhibition resulting in delayed fermentation can occur during the manufacture of cultured dairy products (Devaldez *et al.*, 1988). It is therefore critical to screen starter cultures for their sensitivity to the LPO system. However, heating of milk or addition of small amounts of cysteine can be used to inactivate the LPO system prior to inoculation with starter culture.

It is conceivable that food components containing oxidizable groups may be altered by activation of the LPO system. Destruction of β-carotene by the LPO system has been reported (Kanner & Kinsella, 1983). However, as discussed above, the oxidation products $HOSCN + OSCN^-$ are normal components of human saliva, and they rapidly decay under in-vitro conditions. Therefore, it seems unlikely that these products generated by activating the milk LPO system could adversely affect food components.

SUMMARY OF POTENTIAL INDUSTRIAL APPLICATIONS

The most widely recommended industrial application of the LPO system in food production is in the dairy industry for the preservation of raw milk during storage and/or transportation to processing plants. There are already detailed guidelines for use of the LPO system in preservation of milk, especially in the absence of refrigeration (IDF, 1983, 1986; Codex Alimentarius Commission, 1988). The proper use of LPO system in milk preservation poses no known human health hazard and can help improve milk production, expecially in developing countries where refrigeration is not available. If the LPO system is activated immediately prior to application of approved thermal processes, the shelf-life of dairy products may be extended significantly and high-temperature processes may be replaced with more economical lower temperature treatments. In addition to energy savings, LPO-low temperature thermal processes may provide better nutrient and/or quality retention for highly heat-sensitive foods such as salad dressings, spreads, beverages, dips, desserts and soft cheeses.

REFERENCES

Agner, K. (1947). Detoxicating effect of verdoperoxidase on toxins. *Nature,* **159,** 271–2.

Agner, K. (1950). Studies on peroxidative detoxification of purified diphtheria toxin. *J. Exp. Med.,* **92,** 337–47.

Allen, P. Z. & Morrison, M. (1963). Lactoperoxidase. IV. Immunological analysis of bovine lactoperoxidase preparations obtained by a simplified fractionation procedure. *Arch. Biochem. Biophys.,* **102,** 106–13.

Allen, P. Z. & Morrison, M. (1966). Lactoperoxidase VI. Immunochemical studies on lactoperoxidase from the milk of several species. *Arch. Biochem. Biophys.,* **113,** 540–47.

Anon. (1985). FDA finds *Listeria* in three brands of General Foods soft cheeses. *Food Chem. News,* **27**(25), 33.

Anon. (1986*a*). Class I recall: milk products. FDA Enforcement Report 25 June 1986.

Anon. (1986*b*). FDA finds *Listeria* in Brie cheese from French-certified plant. *Food Chem. News,* 17 Feb. 1986.

Anon. (1986*c*). Brie cheese recalls extended by General Foods, US Importer. *Food Chem. News,* 3 March 1986.

Anon. (1987*a*). *N. Y. Times* reports *Listeria* outbreak in Switzerland from soft cheese. *Food Chem. News,* 7 Dec. 1987.

Aouf, J. E. & Geoffrey, C. (1984). Structure and activity relationships in sulfhydryl-activated toxins. In *Bacterial Protein Toxins,* ed. J. E. Aouf, F. J. Fehrebach, J. H. Freer & J. Jeljaszwicz. FEMS Symposium No. 24, pp. 165–71.

Aumont, G., Le Querrec, F., Lamand, M. & Tressol, J. C. (1987). Iodine content of dairy milk in France in 1983 and 1984. *J. Food Prot.,* **50,** 490–93.

Aune, T. M. & Thomas, E. L. (1977). Accumulation of hypothiocyanite ion during peroxidase-catalyzed oxidation of thiocyanate ion. *Eur. J. Biochem.*, **80,** 209–14.

Aune, T. M., Thomas, E. L. & Morrison, M. (1987). Lactoperoxidase-catalyzed incorporation of thiocyanate ion into a protein substrate. *Biochem.*, **16,** 4611–15.

Banks, J. G. & Board, R. G. (1985). Preservation by the lactoperoxidase system (LPO) of a contaminated infant milk formula. *Lett. Appl. Microbiol.*, **1,** 81–85.

Barret, N. J. (1986). Communicable disease associated with milk and dairy products in England and Wales: 1983–1984. *J. Infection,* **12,** 265–72.

Bearns, R. E. & Girard, K. F. (1958). The effect of pasteurization on *Listeria monocytogenes. Can. J. Microbiol.*, **4,** 55–61.

Beumer, R. R., Noomen, A., Marijs, J. A. & Kampelmacher, E. H. (1985). Antibacterial action of the lactoperoxidase system on *Campylobacter jejuni* in cow's milk. *Neth. Milk Dairy J.*, **39,** 107–14.

Björck, L. (1978). Antibacterial effect of the lactoperoxidase system on psychrotrophic bacteria in milk. *J. Dairy Res.*, **30,** 199–204.

Björck, L., Rosén, C.-G., Marshall, V. E. & Reiter, B. (1975). Antibacterial activity of the lactoperoxidase system in milk against pseudomonads and other Gram-negative bacteria. *Appl. Microbiol.*, **30,** 199–204.

Björck, L., Claesson, O. & Schuthes, W. (1979). The lactoperoxidase-thiocyanate-hydrogen peroxide system as a temporary preservative for raw milk in developing countries. *Milchwissenschaft,* **34,** 726–9.

Borch, E., Wallentin, C., Rosén, M. & Björck, L. (1989). Antibacterial effect of the lactoperoxidase/thicyanate/hydrogen peroxide system against strains of *Campylobacter* isolated from poultry. *J. Food Prot.*, **52,** 638–41.

Bruhn, J. C. & Franke, A. A. (1983). Iodine in human milk. *J. Dairy Sci.*, **66,** 1396–8.

Carlsson, J. (1987). Salivary peroxidase: An important part of our defense against oxygen toxicity. *J. Oral. Pathol.*, **16,** 412–16.

Carlsson, J., Iwami, Y. & Yamada, T. (1983). Hydrogen peroxide excretion by oral streptococci and effect of lactoperoxidase–thiocyanate–hydrogen peroxide. *Infect. Immun.*, **40,** 70–80.

Carlsson, J., Edlund, M.-B. K. & Hänström, L. (1984). Bactericidal and cytotoxic effects of hypothiocyanite–hydrogen peroxide mixtures. *Infect. Immun.*, **44,** 581–6.

Carlström, A. (1965). The heterogeneity of lactoperoxidase. *Acta Chemica Scand.*, **19,** 2387–94.

Carlström, A. (1969). Physical and compositional investigations of the subfractions of lactoperoxidase. *Acta Chemica Scand.*, **23,** 185–202.

CFR (1984). Code of Federal Regulations. Foods and Drugs, 21 CFR 170.3. US Government Printing Office, Washington, DC.

Chakraborty, B. K., Chaudry, S. S., Alex, K. A., Jacob, G. & Soni, G. J. (1986). Application of the Lactoperoxidase System for preserving buffalo milk produced in Indian villages. *Milchwissenschaft,* **41,** 16–19.

Chance, B. (1950). The properties of the enzyme–substrate compounds of lactoperoxidase. *J. Am. Chem. Soc.*, **72,** 1577–83.

Chance, B. (1951). The enzyme–substrate compounds and mechanism of action of the hydroperoxidases. In *The Enzymes,* Vol. II, ed. J. B. Sumner & K. Myrback. Academic Press, New York, pp. 428–53.
Chance, B. & Maehly, A. C. (1955). Assay of catalases and peroxidases. In *Methods in Enzymology,* Vol. II, ed. S. P. Colowick & N. O. Kaplan. Academic Press, New York, pp. 764–75.
Chance, B., Powers, L., Ching, Y., Poulos, T., Schonbaum, G. R., Yamazaki, I. & Paul, K. G. (1984). X-ray absorption studies of intermediates in peroxidase activity. *Arch. Biochem. Biophys.,* **235,** 596–611.
Codex Alimentarius Commission (1988). Draft guidelines for the preservation of raw milk by use of the lactoperoxidase system where refrigeration is virtually impossible. Doc CX/FH/88/12.
Cousin, M. A. (1982). Presence and activity of psychrotrophic microorganisms in milk and dairy products: a review. *J. Food Prot.,* **45,** 209–19.
Curtis, A. R. & Hamming, P. (1982). Differential pulse polarographic determination of total iodine in milk. *J. Assoc. Off. Anal. Chem.,* **65,** 20–23.
Dahlberg, P.-A., Bergmark, A., Björck, L., Bruce, A., Hambraeus, L. & Claesson, O. (1984). Intake of thiocyanate by way of milk and its possible effect on thyroid function. *Am. J. Clin. Nutr.,* **39,** 416–20.
Dahlberg, P.-A., Bergmark, A., Eltom, M., Björck, L. & Claesson, O. (1985). Effect of thiocyanate levels in milk on thyroid function in iodine deficient subjects. *Am J. Clin. Nutr.,* **41,**1010–14.
Dennis, F. & Ramet, J.-P. (1989). Antibacterial activity of the lactoperoxidase system on *Listeria monocytogenes* in trypticase soy broth, UHT milk and French soft cheese. *J. Food Prot.,* **52,** 706–11.
Devaldez, G. F., Bibi, W. & Bachmann, M. R. (1988). Antibacterial effect on the lactoperoxidase/thiocyanate/hydrogen peroxide [LP] system on the activity of thermophilic starter culture. *Milchwissenschaft,* **43,** 350–52.
Doyle, M. P., Meske, L. M. & Marth, E. H. (1985). Survival of *Listeria monocytogenes* during the manufacture and storage of nonfat dry milk. *J. Food Prot.,* **48,** 740–42.
Doyle, M. P., Glass, K. A., Beery, J. T., Garcia, G. A., Pollard, D. J. & Schultz, R. D. (1987). Survival of *Listeria monocytogenes* in milk during high-temperature, short-time pasteurization. *Appl. Environ. Microbiol.,* **53,** 1433–8.
Dumontet, C. & Rousset, B. (1983). Identification, purification, and characterization of a non-heme lactoperoxidase in bovine milk. *J. Biol. Chem.,* **258,** 14166–72.
Dunford, H. B. & Stillman, J. S. (1976). On the function and mechanism of action of peroxidases. *Coord. Chem. Reviews,* **19,** 187–251.
Earnshaw, R. G. & Banks, J. G. (1989). A note on the inhibition of *Listeria monocytogenes* NCTC 11994 in milk by an activated lactoperoxidase system. *Lett. Appl. Microbiol.,* **8,** 203–5.
Engavall, A., Sandstedt, B. K. & Danielsson-Tham, M.-L. (1986). Colonization of broilers with *Campylobacter* in conventional Broiler-Chicken Flocks. *Acta Vet. Scand.,* **27,** 540–47.
Etling, N., Padovani, E., Fouque, F. & Tato, L. (1986). First-month

variations in total iodine content of human breast milks. *Early Human Dev.*, **13,** 81–5.

Fairbairn, D. J. & Law, B. A. (1986). Proteinases of psychrotrophic bacteria: Their production, properties, effects and control. *J. Dairy Res.*, **53,** 139–77.

FAO (1957). Report on the meeting of experts on the use of hydrogen peroxide and other preservatives in milk, Rome, Doc. 57/11/8655.

Fernandez-Garayzabal, J. F., Dominguez-Rodriguez, L., Vazquez-Boland, J. A., Blanco-Cancelo, J. L. & Suarez-Fernandez, G. (1986). *Listeria monocytogenes* dans le Lait Pasteurise. *Can. J. Microbiol.*, **32,** 149–50.

Fleming, D. W., Cochi, S. L., MacDonald, K. L., Brondum, J., Hayes, P. S., Plikaytis, B. D., Holmes, M. B., Audurier, A., Broome, C. V. & Reingold, A. L. (1985). Pasteurized milk as a vehicle of infection in an outbreak of Listeriosis. *New Engl. J. Med.*, **312,** 404–7.

Funderburk, C. F. & Middlesworth, L. van (1967). Effect of lactation and perchlorate on thiocyanate metabolism. *Am. J. Physiol.*, **213,** 1371–7.

Goewie, C. E. & Hogendoorn, E. A. (1985). Liquid chromatographic determination of bromide in human milk and plasma. *J. Chrom.*, **344,** 157–65.

Gotherfors, L. & Marklund, S. (1975). Lactoperoxidase activity in human milk and in saliva of newborn infants. *Inf. Immun.*, **11,** 1210–15.

Gray, M. L. & Killinger, A. H. (1966). *Listeria monocytogenes* and listeric infections. *Bacteriol. Rev.*, **30,** 309–82.

Grisham, M. B. & Ryan, R. (1990). Cytotoxic properties of salivary oxidants. *Am. J. Physiol.*, **258,** C115–21.

Gushurst, C. A., Mueller, J. A., Green, J. A. & Sedor, F. (1984). Breast milk iodide: reassessment in the 1980s. *Pediatrics,* **73,** 354–7.

Hanssen, F. S. (1924). The bactericidal property of milk. *Brit. J. Exp. Path.*, **5,** 271–80.

Hänstrom, L., Johansson, L. & Carlsson, J. (1983). Lactoperoxidase and thiocyanate protect cultured mammalian cells against hydrogen peroxide toxicity. *Med. Biol.*, **61,** 268–74.

Hashinaka, K. & Yamada, M. (1986). Indentification of myeloperoxidase in human colostrum. *Arch. Biochem. Biophys.*, **247,** 91–6.

Hill, I. D. & Bowie, M. D. (1983). Chloride deficiency syndrome due to chloride-deficient breast milk. *Arch. Dis. Child.*, **58,** 224–6.

Ho, J. L., Shands, K. N., Friedland, G., Eckind, P. & Fraser, D. W. (1986). An outbreak of type 4b *Listeria monocytogenes* infection involving patients from eight Boston hospitals. *Arch. Intern. Med.*, **146,** 520–24.

Honka, E., Ohlsson, P. I. & Paul, K. G. (1982). The adsorption of lactperoxidase to glass. *Acta Chem. Scand.*, **B36,** 273–4.

Hoogendoorn, H. & Piessens, J. P. (1987). Treatment of aphthous patients by enhancement of the salivary peroxidase system. *J. Oral Pathol,* **16,** 425–7.

Hoogendorn, H., Piessens, J. P., Scholtes, W. & Stoddard, L. A. (1977). Hypothiocyanite ion: the inhibitor formed by the system lactoperoxidase–thiocyanate–hydrogen peroxide. *Caries Res.*, **11,** 77–84.

Hu, C. & Huystee, R. B. van (1989). Role of carbohydrate moieties in peanut (*Arachis hypogaea*) peroxidases. *Biochem. J.*, **263,** 129–35.

Humphrey, T. J. & Beckett, P. (1987). *Campylobacter jejuni* in dairy cows and raw milk. *Epidem. Infect.*, **98,** 263–9.
Jago, G. R. & Morrison, M. (1962). Anti-streptococcal activity of lactoperoxidase. *Proc. Soc. Exp. Biol. Med.*, **111,** 585–8.
IDF (1983). Temporary preservation of raw milk by activation of the lactoperoxidase system. International Dairy Federation. F-Doc 96.
IDF (1986). Code of practice for the preservation of raw milk by the lactoperoxidase system. International Dairy Federation. D-Doc 150.
Jago, G. R. & Morrison, M. (1962). Anti-streptococcal activity of lactoperoxidase. *Proc. Soc. Exp. Biol. Med.*, **111,** 585–8.
James, S. M., Fannin, S. L., Agee, B. A., Hall, B., Parker, E., Vogt, J., Run, G., Williams, J., Lieb, L., Salminen, C., Pendergast, J., Werner, S. B. & Chin, J. (1985). Listeriosis outbreak associated with Mexican-style cheese in California. *Morbid. Mortal. Weekly Report,* **34,** 357–9.
Jenzer, H., Jones, W. & Kohler, H. (1986). On the molecular mechanism of lactoperoxidase-catalyzed H_2O_2 metabolism and irreversible enzyme inactivation. *J. Biol. Chem.*, **261,** 15550–6.
Jenzer, H., Kohler, H. & Broger, C. (1987). The Role of hydroxyl radicals in irreversible inactivation of lactoperoxidase by excess H_2O_2. *Arch. Biochem. Biophys.*, **258,** 381–90.
Jones, F. S. & Simms, H. S. (1930). The bacterial growth inhibitor (Lactenin) of milk. The preparation in concentrated form. *J. Exp. Med.*, **51,** 327–39.
Josephy, P. D. (1985). Oxidative activation of benzidine and its derivatives by peroxidases. *Environmental Health Perspectives,* **64,** 171–8.
Kamau, D. N. (1989). The potential use of the naturally occurring bovine milk lactoperoxidase antimicrobial system to inhibit growth and enhance thermal destruction of milkborne pathogens: *L. monocytogenes* and *S. aureus.* PhD Dissertation in Food Science, The Pennsylvania State University, USA.
Kamau, D. N., Doores, S. & Pruitt, K. M. (1990*a*). Antibacterial activity of the lactoperoxidase system against *Listeria monocytogenes* and *Staphylococcus aureus* in milk. *J. Food Prot.*, **53,** 1010–4.
Kamau, D. N., Doores, S. & Pruitt, K. M. (1990*b*). Enhanced thermal destruction of *Listeria monocytogenes* and *Staphylococcus aureus* by the lactoperoxidase system (1990). *Appl. Environ. Microbio.*, **56,** 2711–16.
Kamau, D. N., Doores, S. & Pruitt, K. M. (1991). Activation of the lactoperoxidase system prior to pasteurization for shelf-life extension of milk. *Milchwissenschaft,* **46,** 213–14.
Kanner, J. & Kinsella, J. E. (1983). Lipid deterioration: β-Carotene destruction and oxygen evolution in a system containing lactoperoxidase, hydrogen peroxide and halides. *Lipids,* **18,** 198–203.
Kiermeier, F. & Kayser, C. (1960). Zur kenntnis der lactoperoxidase activität in kuhmilch und abhängigkeit von biologischen einflüssen. *Zeitschfrift für Lebensmittel Untersuchung und Forschung,* **112,** 481–98.
Kiemeier, F. & Kuhlmann , F. (1972). Laktoperoxidase—activität in human und kuhmilch. *Müch. med. Wschr.*, **114,** 2144–6.
Kimura, S. & Yamazaki, I. (1979). Comparison between hog intestinal

peroxidase and bovine lactoperoxidase-compound I formation and inhibition by benzhydroxamine acid. *Arch. Biochem. Biophys.*, **198,** 580–88.

Kohler, H. & Jenzer, H. (1989). Interaction of lactoperoxidase with hydrogen peroxide. *Free Radical Biology & Medicine,* **6,** 323–39.

Kohler, H., Taurog, A. & Dunford, H. B. (1988). Spectral studies with lactoperoxidase and thyroid peroxidase: interconversions between native enzyme, compound II, and compound III. *Arch. Biochem. Biophys.*, **264,** 438–49.

Korhonen, H. (1973). Untersuchungen zur Bakterizidie der Milch und Immunisierung der Bovimen Milchdrüse. PhD Dissertation, Helsinki, Finland, p. 88.

Korhonen, H. (1980). A new method for preserving raw milk: Lactoperoxidase antibacterial system. *Animal World,* **35,** 24–9.

Laarveld, B., Brockman, R. P. & Christensen, D. A. (1981). The Effects of the level of iodine in canola meal concentrate on milk iodine and thiocyanate content and thyroid function in dairy cows. *Can. J. Anim. Sci.*, **61,** 625–32.

La Mar, G. N., Ropp, J. S. de, Latos-Grazynski, L., Balch, A. L., Smith, K. M., Parish, D. W. & Cheng, R. (1983). Proton NMR characterization of the ferryl group in model complexes and hemoproteins: evidence for the $Fe_{iv}{=}O$ group in ferryl myoglobin and compound II of horseradish peroxidase. *J. Am. Chem. Soc.*, **105,** 782–7.

Langbakk, B. & Flatmark, T. (1984). Demonstration and partial purification of lactoperoxidase from human colostrum. *Federation of European Biochemical Socieites (FEBS), Letters* **174,** 300–303.

Langbakk, B. & Flatmark, T. (1989). Lactoperoxidase from human colostrum. *Biochem. J.*, **259,** 627–31.

Laven, C. T. & Pruitt, K. M. (1985). Bactericidal properties of the human milk peroxidase system. *Fed. Proc.*, **42,** 1328 abstr. No. 6075.

Law, B. A., Sharpe, M. E. & Chapman, H. R. (1976). The effect of lipolytic Gram-negative psychrotrophs in stored milk on the development of rancidity in cheddar cheese. *J. Dairy Res.*, **43,** 459–68.

Law, B. A., Andrews, A. T. & Sharpe, M. E. (1977). Gelation of ultra-high temperature sterilized milk by proteases from a strain of *Pseudomonas fluorescens* isolated from raw milk. *J. Dairy Res.*, **44,** 145–8.

Law, B. A., Andrews, A. T., Cliffe, A. J., Sharpe, M. E. & Chapman, H. R. (1979). Effect of proteolytic raw milk psychrotrophs on Cheddar cheesemaking with stored milk. *J. Dairy Res.*, **46,** 497–509.

Linnan, M. J., Mascola, L., Lou, X. D., Goulet, V., May, S., Salminen, C., Hird, D., Yonekura, M. L., Hayes, P., Weaver, R., Audurier, A., Plikaytis, B. D., Fannin, S. L., Kleks, A. & Broome, C. V. (1988). Epidemic listeriosis associated with Mexican-style cheese. *New Engl. J. Med.*, **319,** 823–8.

Macy, I. G. (1949). Composition of human colostrum and milk. *Am. J. Dis. Childhood,* **78,** 589–603.

Mäkinen, K. K. & Tenovuo, J. (1980). Effect of peroral administration of xylitol on human milk. *Nutr. Rep. Int.*, **22,** 793–800.

Makino, R., Uno, T., Nishimura, Y., Iizuka, T., Tsuboi, M. & Ishimura, Y.

(1986). Coordination structures and reactivities of compound II in iron and manganese horseradish peroxidases. *J. Biol. Chem.*, **261**, 8376–82.
Månsson-Rahemtulla, B., Baldone, D. C., Pruitt, K. M. & Rahemtulla, F. (1987). Effects of variations in pH and hypothiocyanite concentrations on *S. mutans* glucose metabolism. *J. Dent. Res.*, **66**, 486–91.
Månsson-Rahemtulla, B., Rahemtulla, F., Baldone, D. C., Pruitt, K. M. & Hjerpe, A. (1988). Purification and characterization of human salivary peroxidase. *Biochemistry*, **27**, 233–9.
Marshall, V. M. E., Cole, W. M. & Bramley, A. J. (1986). Influence of the lactoperoxidase system on susceptibility of the udder to *Streptococcus uberis* infection. *J. Dairy Res.*, **53**, 507–14.
Marth, E. H. (1988). Disease characteristics of *Listeria monocytogenes*. *Food Technol.*, **42**, 165–8.
Metodiewa, D., Reszka, K. & Dunford, H. B. (1989*a*). Evidence for a peroxidatic oxidation of norepinephrine, a catecholamine, by lactoperoxidase. *Biochem. Biophys. Res. Commun.*, **160**, 1183–8.
Metodiewa, D., Reszka, K. & Dunford, H. B. (1989*b*). Oxidation of the substituted catechols dihydroxyphenylalanine methyl ester and trihydroxyphenylalanine by lactoperoxidase and its compounds. *Arch. Biochem. Biophys.*, **274**, 601–8.
Mickelson, M. N. & Brown, R. W. (1985). Physiological characteristics of *Streptococcus dysgalactiae* and *Streptococcus uberis* on the effect of the lactoperoxidase complex on their growth in a chemically defined medium and milk. *J. Dairy Sci.*, **68**, 1095–102.
Middlesworth, L. van (1986). Potential metabolic significance of blood thiocyanate. *Endocrin. Exp.*, **20**, 17–22.
Modi, S., Behere, D. V. & Mitra, S. (1989). Binding of thiocyanate to lactoperoxidase: ^{1}H and ^{15}nuclear magnetic resonance studies. *Biochemistry*, **28**, 4689–94.
Modi, S., Behere, D. V. & Mitra, S. (1990). Coordination geometry of heme in lactoperoxidase: pH-dependent ^{1}H relaxivity and optical spectral studies. *J. Inorg. Biochem.*, **38**, 17–25.
Moldoveanu, Z., Tenovuo, J., Mestecky, J. & Pruitt, K. M. (1982). Human milk peroxidase is derived from milk leukocytes. *Biochim. Biophys. Acta*, **718**, 103–8.
Morrison, M. & Allen, P. Z. (1963). The identification and isolation of lactoperoxidase from salivary gland. *Biochem. Biophys. Res. Commun.*, **13**, 490–94.
Morrison, M., Allen, P. Z., Bright, J. & Jayasinghe, W. (1965). Lactoperoxidase V. Identification and isolation of lactoperoxidase from salivary gland. *Arch. Biochem. Biophys.*, **111**, 126–33.
Morrison, M. & Schonbaum, G. R. (1976). Peroxidase-catalyzed halogenation. *Ann. Rev. Biochem.*, **45**, 861–88.
Morrison, M. & Steele, W. F. (1968). In *Biology of the Mouth*, ed. P. Person. American Association for the Advancement of Science, Washington, DC, pp. 89–110.
Nakamura, M., Yamazaki, I., Ohtaki, S. & Nakamura, S. (1986). Charac-

terization of one- and two-electron oxidations of glutathione coupled with lactoperoxidase and thyroid peroxidase reactions. *J. Biol. Chem.*, **261,** 13923–7.

Nelson, M., Phillips, D. I. W., Morris, J. A. & Wood, T. J. (1987). Urinary iodine excretion correlates with milk iodine content in seven British towns. *J. Epidemio. Comm. Health,* **42,** 72–5.

Nichol, A. W., Angel, L. A., Moon, T. & Clezy, P. S. Lactoperoxidase haem, an iron-porphyrin thiol (1987). *Biochem. J.,* **247,** 147–50.

Odajima, T. (1981). Oxidative destruction of the microbial aflatoxin by the myeloperoxidase-chloride system. *Arch. Oral Biol.,* **26,** 339–40.

Ohno, S., Hond, Y., Teramoto, H. & Horiguchi, S. (1988). Simultaneous determination of bromide and chlorine in cow's milk by radioactivation analysis. *Analyst,* **113,** 515–17.

Oram, J. D. & Reiter, B. (1966*a*). The inhibition of streptococci by lactoperoxidase, thiocyanate and hydrogen peroxide. *Biochem. J.,* **100,** 382–8.

Oram, J. D. & Reiter, B. (1966*b*). The inhibition of streptococci by lactoperoxidase, thiocyanate and hydrogen peroxide. The effect of the inhibitory system on susceptible and resistant strains of Group N streptococci. *Biochem. J.,* **100,** 373–81.

Palumbo, S. A. (1986). Is refrigeration enough to restrain foodborne pathogens? *J. Food Prot.,* **49,** 1003–9.

Paul, K. G. (1987). Peroxidases: past and present. *J. Oral. Pathol.,* **16,** 409–11.

Paul, K.-G. & Ohlsson, P.-I (1985). The chemical structure of lactoperoxidase. In *The Lactoperoxidase System,* ed. K. M. Pruitt and J. O. Tenovuo. Marcel Dekker, New York, pp. 15–29.

Paul, K. G., Ohlsson, P. J. & Henriksson, A. (1980). The isolation and some liganding properties of lactoperoxidase. *FEBS Letters,* **110,** 200–204.

Pfeil, W. & Ohlsson, P. I. (1986). Lactoperoxidase consists of domains: A scanning calorimetric study. *Biochim. Biophys. Acta,* **872,** 72–5.

Phillips, D. I. W., Nelson, M., Barker, D. J. P., Moris, J. A. & Wood, T. J. (1988). Iodine in milk and the incidence of thyrotoxicosis in England. *Clin. Endocrin.,* **28,** 61–6.

Portmann, A. & Auclair, J. E. (1959). Relation entre la lactenine L_2 et la lactoperoxydase. *Lait,* **39,** 147–58.

Potter, M. E., Kaufmann, A. F., Blake, P. A. & Feldman, R. A. (1984). Unpasteurized milk: The hazard of a health fettish. *J. Am. Med. Assoc.,* **252,** 2048–52.

Pruitt, K. M. (1987). The salivary peroxidase system: Thermodynamic, kinetic, and antibacterial properties. *J. Oral. Pathol.,* **16,** 417–20.

Pruitt, K. M. & Adamson, M. (1977). Enzyme activity of salivary lactoperoxidase adsorbed to human enamel. *Inf. Immun.,* **17,** 112–16.

Pruitt, K. M. & Reiter, B. (1985). Biochemistry of peroxidase system: antimicrobial effects. In *The Lactoperoxidase System,* ed. K. M. Pruitt & J. O. Tenovuo. Marcel Dekker, New York, pp. 143–78.

Pruitt, K. M., Adamson, M. & Arnold, R. (1979). Lactoperoxidase binding to streptococci. *Inf. Immun.,* **25,** 304–9.

Pruitt, M., Tenovuo, J., Andrews, R. W. & McKane, T. (1982). Lactoperoxidase-catalyzed oxidation of thiocyanate: Polarographic study of the oxidation products. *Biochemistry,* **21,** 562–7.

Pruitt, K. M., Smith, J., Hazelrig, J. & Månsson-Rahemtulla, B. (1984). A model for analysis of protein defense mechanisms in secretions. *Protides of the Biological Fluids,* **32,** 17–20.

Pruitt, K., Månsson-Rahemtulla, B. & Baldone, D. C. (1985). Does human milk contain a secretory peroxidase. *Fed. Proc.,* **44,** Abstr. 2557.

Pruitt, K. M., Tenovuo, J., Månsson-Rahemtulla, B., Harrington, P. & Baldone, D. C. (1986). Is thiocyanate peroxidation at equilibrium *in vivo*? *Biochim. Biophys. Acta,* **870,** 385–91.

Pruitt, K. M., Månsson-Rahemtulla, B., Baldone, D. C. & Rahemtulla, F. (1988). Steady-state kinetics of thiocyanate oxidation catalyzed by human salivary peroxidase. *Biochemistry,* **27,** 240–45.

Pruitt, K. M., Rahemtulla, F., Månsson-Rahemtulla, B., Baldone, D. C. & Laver, G. T. (In press). Peroxidases in human milk. In *Immunology of Milk and* the Neonate, eds J. Mestecky, C. Blair & P. L Ogra, Plenum Publishing Corporation, New York.

Purdy, M. A., Tenovuo, J., Pruitt, K. M. & White, W. E. (1983). Effect of growth phase and cell envelope on the susceptibility of *Salmonella typhimurium* to the lactoperoxidase–thiocyanate–hydrogen peroxide system. *Infect. Immun.,* **39,** 1187–95.

Reiter, B. (1978). Review of nonspecific antimicrobial factors in colostrum. *Ann. Rech. Vét.,* **9,** 205–24.

Reiter, B. (1981). The contribution of milk to resistance to intestinal infection in the new-born. In *Immunological Aspects of Infection in the Fetus and New-Born,* ed. H. P. Lambert & C. B. S. Wood. Academic Press, London, pp. 155–95.

Reiter, B. (1985). The lactoperoxidase system of bovine milk. In *The Lactoperoxidase System,* ed. K. M. Pruitt & J. O. Tenovuo. Marcel Dekker, New York, pp. 123–41.

Reiter, B. & Harnuv, G. (1982). The preservation of refrigerated and uncooled milk by its natural lactoperoxidase system. *Dairy Ind. Intern.,* **47,** 13–19.

Reiter, B., Pickering, A., Oram, J. D. & Pope, G. S. (1963). Peroxidase-thiocyanate inhibition of streptoccci in raw milk. *J. Gen. Microbiol.,* **33,** xii.

Reiter, B., Pickering, A., Oram, J. D. & Pope, G. S. (1963). Peroxidase-thiocyanate inhibition of streptoccci in raw milk. *J. Gen. Microbiol.,* **33,** xii.

Reiter, B., Marshall, V. M. E., Björck, L. & Rosén, C.-G. (1976). Nonspecific bactericidal activity of the lactoperoxidase-thiocyanate-hydrogen peroxide system of milk against *Escherichia coli* and some gram-negative pathogens. *Infect. Immun.,* **13,** 800–807.

Rombauts, W. A., Schroeder, W. A. & Morrison, M. (1967). Bovine lactoperoxidase. Partial characterization of the further purified protein. *Biochemistry,* **6,** 2965–77.

Rule, A., H., Schaumburg-Lever, G., Patel, R. P., Schmidt-Ullrich, B. &

Okun, M. R. (1976). Purification of peroxidase by isoelectric focusing. *Immunochemistry,* **13,** 819–21.

Sandholm, M., Ali-Vehmas, T., Kaartinen, L. & Junnila, M. (1988). Glucose oxidase (GOD) as a source of hydrogen peroxide for the lactoperoxidase (LPO) system in milk: Antibacterial effect of the GOD-LPO system against mastitis pathogens. *J. Vet. Med.,* B **35,** 346–52.

Shindler, J. S. & Bardsley, W. G. (1975). Steady state kinetics of lactoperoxidase with ABTS as a chromogen. *Biochem. Biophys. Res. Comm.,* **67,** 1307–12.

Shindler, J. S., Childs, R. E. & Bardsley, W. G. (1976). Peroxidase of human cervical mucus. Isolation and characterization. *Eur. J. Biochem.,* **65,** 325–31.

Schlech, W. F., Lavigne, P. M., Bortolussi, R. A., Allen, A., Haldane, E. V., Wort, A. J., Hightower, A. W., Johnson, S. E., King, S. H., Nicholls, E. S. & Broome, C. V. (1983). Epidemic listeriosis-evidence for transmission by food. *New Engl. J. Med.,* **308,** 203–6.

Shanker, S., Rosenfield, J. A., Davey, G. R. & Sorell, T. C. (1982). *Campylobacter jejuni*: Incidence in processed broilers and biotype distribution in human and broiler isolates. *Appl. Environ. Microbiol.,* **43,** 1219–20.

Sharp, J. C. M., Patterson, G. M. & Barrett, N. J. (1985). Pasteurization and the control of milkborne infections in Britain. *Br. Med. J.,* **291,** 463–4.

Sievers, G. (1979). The prosthetic group of milk lactoperoxidase is protoheme IX. *Biochim. Biophys. Acta,* **579,** 181–190.

Sievers, G. (1980). Structure of milk lactoperoxidase. A study using circular dichroism and difference absorption spectroscopy. *Biochim. Biophys. Acta,* **624,** 249–59.

Sievers, G. (1981). Structure of milk lactoperoxidase. Evidence for a single polypeptide chain. *FEBS Letters,* **127,** 253–6.

Sievers, G. (1985). The thiocyanate binding to lactoperoxidase. *Protides Biol. Fluids,* **32,** 129–32.

Sievers, G., Gadsby, P. M. A., Peterson, J. & Thomson, A. J. (1983). Assignment of the axial ligands of the haem in milk lactoperoxidase using magnetic circular dichroism spectroscopy. *Biochim. Biophys. Acta,* **742,** 659–68.

Siragusa, G. R. & Johnson, M. G. (1989). Inhibition of *Listeria monocytogenes* growth by the lactoperoxidase–thiocyanate–H_2O_2 antimicrobial system. *Appl. Environ. Microbiol.,* **55,** 2802–5.

Sitter, A. J., Reczek, C. M. & Terner, J. (1985). Heme-linked ionization of horseradish peroxidase compound II monitored by the resonance Raman Fe(IV)═O stretching vibration. *J. Biol. Chem.,* **260,** 7515–22.

Skirrow, M. B. (1982). *Campylobacter* enteritis—the first five years. *J. Hyg.,* **89,** 175–84.

Smith, J. L., Buchanan, R. L. & Palumbo, S. A. (1983). Effect of food environment on staphylococcal enterotoxin synthesis: a review. *J. Food Prot.,* **46,** 545–55.

Smith, M. T., Yager, J. W., Steinmetz, K. L. & Eastmond, D. A. (1989). Peroxidase-dependent metabolism of benzene's phenolic metabolites and its potential role in benzene toxicity and carcinogenicity. *Environmental Health Perspectives,* **82,** 23–9.

Staden, J. F. van (1986). Flow-injection analysis of chloride in milk with a dialyzer and a coated tubular inorganic chloride-selective electrode. *Anal. Let.*, **19,** 1407–19.

Stead, D. (1986). Microbial lipases: their characheristics, role in food spoilage and industrial uses. *J. Dairy Res.*, **53,** 481–505.

Tadros, L. K., Andrei, D. & Motas, C. (1983). The role of carbohydrate moiety in the intracellular degradation of glycoproteins. XIX. The involvement of specific sugar residues in the active site of peroxidase isoenzyme A. *Rev. roum. Biochim.*, **20,** 293–300.

Tenovuo, J. & Larjava, H. (1984). The protective effect of peroxidase and thiocyanate against hydrogen peroxide toxicity assessed by the uptake of [^{3}H]-thymidine by human gingival fibroblasts cultured *in vitro*. *Arch. Oral Biol.*, **29,** 445–451.

Tenovuo, J., Moldoveanu, Z., Mestecky, J., Pruitt, K. M. & Månsson-Rahemtulla, B. (1982). Interaction of specific and innate factors of immunity: IgA enhances the antimicrobial effect of the lactoperoxidase system against *Streptococcus mutans*. *J. Immunol.*, **128,** 726–31.

Tenovuo, J., Pruitt, K., Månsson-Rahemtulla, B., Harrington, P. & Baldone, D. C. (1986). Products of thiocyanate peroxidation: properties and reaction mechanisms. *Biochim. Biophys. Acta*, **870,** 377–84.

Terner, J., Sitter, A. J. & Reczek, C. M. (1985). Resonance raman spectroscopic characterization of horseradish peroxidase. Observation of the $Fe^{IV}{=}O$ stretching vibration of Compound II. *Biochim. Biophys. Acta*, **828,** 73–80.

Thakar, R. P. & Dave, J. M. (1986). Application of the activated lactoperoxidase-thiocyanate-hydrogen peroxide system in enhancing the keeping quality of milk at higher temperatures. *Milchwissenschaft*, **41,** 20–3.

Thanabal, V. & La Mar, G. N. (1989). A nuclear overhauser effect investigation of the molecular and electronic structure of the heme crevice in lactoperoxidase. *Biochemistry*, **28,** 7038–44.

Thomas, E. L. (1981). Lactoperoxidase-catalyzed oxidation of thiocyanate: equilibria between oxidized forms of thiocyanate. *Biochemistry*, **20,** 3273–80.

Thomas, E. L. (1985). Products of lactoperoxidase-catalyzed oxidation of thiocyanate and halides. In *The Lactoperoxidase System*, ed. K. M. Pruitt & J. O. Tenovuo. Marcel Dekker, New York, pp. 31–53.

Thomas, E. L. & Aune, T. M. (1978). Oxidation of *Escherichia coli* sulfhydryl components by the peroxidase–hydrogen peroxide–iodide antimicrobial system. *Antimicrob. Agents Chemother.*, **13,** 1006–10.

Thomas, E. L., Pera, K. A., Smith, K. W. & Chwang, A. K. (1983). Inhibition of *Streptococcus mutans* by the lactoperoxidase antimicrobial system. *Infect. Immun.*, **39,** 767–78.

Thompson, G. E. (1988). Electrolytes and free fatty acids in milk during mammary involution in the cow. *Res. Vet. Sci.*, **44,** 175–7.

Toyosaki, T., Yamamoto, A. & Mineshita, T. (1984). The photolysis mechanism of riboflavin in milk serum: The correlation between the superoxide and riboflavin decomposition. *Agric. Biol. Chem.*, **48,** 2919–22.

USPHS (1977). Grade A Pasteurized Milk Ordinance—Recommendations of

US Food and Drug Administration. US Department of Health, Education and Welfare. USPHS-FDA, Washington, DC.
Wever, R., Kast, W. M., Kasinoedin, J. H. & Boelens, R. (1982). The peroxidation of thiocyanate catalysed by myeloperoxidase and lactoperoxidase. *Biochim. Biophys. Acta,* **709,** 212–19.
White, W. E. Jr, Pruitt, K. M. & Mânsson-Rahemtulla, B. (1983). peroxidase–thiocyanate–peroxide antibacterial system does not damage DNA. *Antimicrob. Agents Chemother.,* **23,** 267–72.
Weichen, A. & Kock, B. (1984). Zur routinemässigen bestimmung niedriger iodkonzentrationen in milch. *Fresenius Z. Anal. Chem.,* **319,** 569–72.
Wilder, M. S. & Edberg, J. C. (1973). Interaction of virulent and avirulent *Listeria monocytogenes* with cultured mouse peritoneal macrophages. *Infect. Imm.,* **7,** 409–15.
Wilkins, P. O., Bourgeois, R. & Murray, R. G. E. (1972). Psychrotrophic properties of *Listeria monocytogenes. Can. J. Microbiol.,* **18,** 543–51.
Wood, J. L. (1975). Biochemistry. In *Chemistry and Biochemistry of Thiocyanic Acid and its Derivatives,* ed. A. A. Newman. Academic Press, New York, pp. 156–221.
Wray, C. & McLaren, I. (1987). A note on the effect of the lactoperoxidase system on Salmonellas *in vitro* and *in vivo. J. Appl. Bacteriol.,* **62,** 115–18.
Wright, R. C. & Tramer, J. (1958). Factors influencing the activity of cheese starters. The role of milk peroxidase. *J. Dairy Res.,* **25,** 104–18.
Wüthrich, S., Richterich, R. & Hostettler, H. (1963). Untersuchengun über Milchenzyme. *Zeitschrift für Lenensmittle-Untersuchung und-Forschung,* **124,** 336–44.
Yamada, H. & Yamazaki, I. (1974). Proton balance in conversions between five oxidation–reduction states of horseradish peroxidase. *Arch. Biochem. Biophys.,* **165,** 728–38.
Zajac, M., Gladys, J., Skarzynska, M., Harnuv, G. & Björck, L. (1983). Changes in bacteriological quality of raw milk stabilized by activation of its lactoperoxidase system and stored at different temperatures. *J. Food Prot.,* **46,** 1065–8.

5

Lipoxygenases

John R. Whitaker

Department of Food Science and Technology, University of California, California, USA

INTRODUCTION

Lipid oxidation is of major importance for the quality of our foods and for our health. Lipid oxidation changes the aroma, flavor, taste, texture, color and nutritional quality of foods. Peroxidation of lipids in tissues of animals as a result of insufficient amounts of Vitamin E, selenium, or glutathione peroxidase or the result of certain diseases leads to major damage to tissues, especially to membranes.

Changes in aroma, flavor and taste of some foods frequently arise from products formed directly from oxidation of polyunsaturated lipids. Of even more concern are the free radical intermediates formed that can interact with and alter other constituents, such as proteins, amino acids, nucleic acids, carbohydrates, vitamins and pigments. Some of the products formed may be toxic, such as the potential carcinogen, malonaldehyde (Shamberger, 1977, 1980). Malonaldehyde may act as a catalyst in formation of *N*-nitrosamines in foods containing secondary amines and nitrite (Kikugawa *et al.*, 1980; Kurechi *et al.*, 1980). Cholesterol oxidation products may be angiotoxic (Peng *et al.*, 1978; Taylor *et al.*, 1979) or even carcinogenic (Smith & Kulig, 1975; Chan & Black, 1976). Fatty acid hydroperoxides may be catalysts in the formation of *N*-nitrosopyrrolidine, for example in grilled or fried bacon (Coleman, 1978).

Oxidation of unsaturated lipids includes free radical chain reactions. The reactions can be divided in the three separate processes: initiation, propagation and termination (eqns (1)–(6); Chan, 1987).

Initiation: $$X^{\cdot} + RH \xrightarrow{k_i} R^{\cdot} + XH \tag{1}$$

Propagation: $$R^{\cdot} + O_2 \xrightarrow{k_p} RO_2^{\cdot} \tag{2}$$

$$RO_2 + RH \xrightarrow{k_p} ROOH + R^{\cdot} \tag{3}$$

Termination: $$R^{\cdot} + R^{\cdot} \xrightarrow{k_t} RR \tag{4}$$

$$RO_2^{\cdot} + R \xrightarrow{k_t} ROOR \tag{5}$$

$$RO_2^{\cdot} \xrightarrow{k_t} R^{\cdot} + O_2 \quad \text{etc.} \tag{6}$$

Initiation can be the result of metal ion, light or enzymic catalysis. Initiation is furthered by ROOH (produced in eqn (3)), especially as catalyzed by metal ions (eqns (7) and (8)).

$$ROOH + M^{2+} \rightarrow RO^{\cdot} + OH^{-} + M^{3+} \tag{7}$$

$$ROOH + M^{3+} \rightarrow ROO^{\cdot} + H^{+} + M^{2+} \tag{8}$$

The enzyme lipoxygenase, responsible for a substantial part of lipid oxidation in raw food sources, is an initiator of peroxidation, but also a propagator via eqn (2).

Gardner (1983) schematically illustrated the time course of the three processes involved in oxidation of polyunsaturated lipids as shown in Fig. 1. Also included is a relative indication of the concentration of reactants and products.

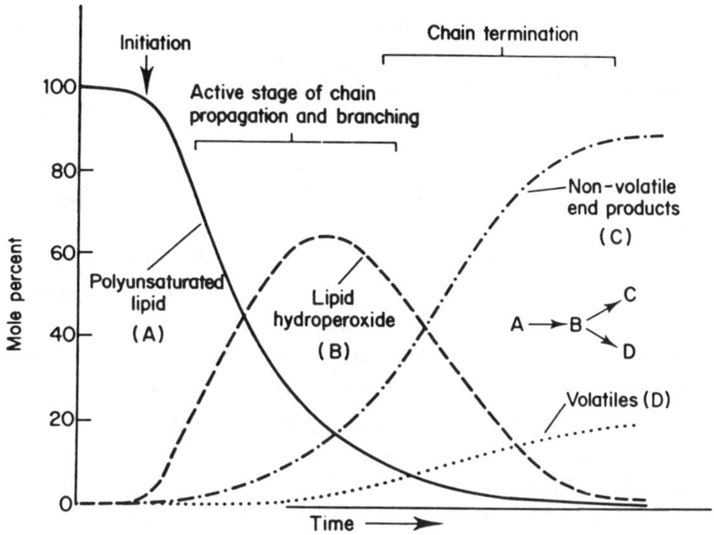

Fig. 1. Typical autoxidation of a polyunsaturated lipid as a function of time showing the various stages in the reaction. From Gardner (1983) with permission.

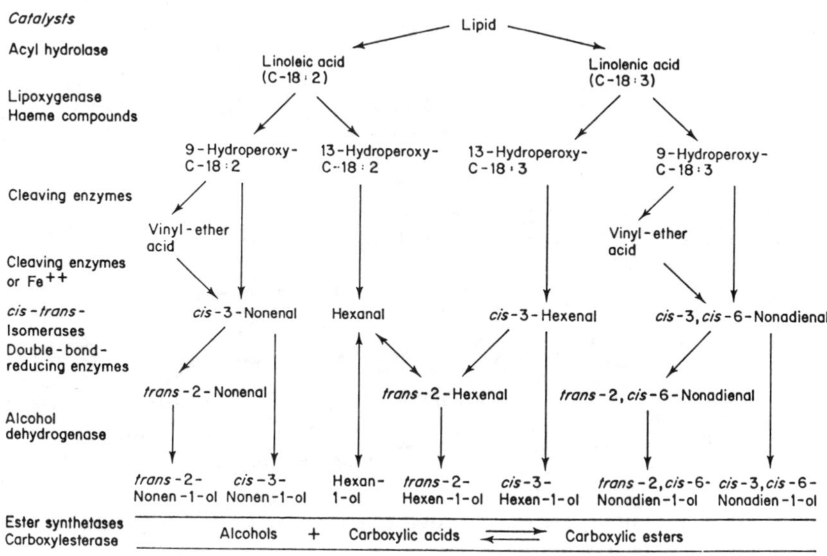

Fig. 2. Formation of C-6 and C-9 aldehydes and alcohols from plant lipids. From Eriksson (1979) with permission.

Peroxidation of unsaturated lipids gives rise to a large spectrum of products. Lipoxygenase is not responsible for all of these products. The formation of lipid oxidation products from linoleic and linolenic acids occurs by a series of reactions (Fig. 2; Eriksson, 1979, 1987):

(a) The lipids, phospholipids and glycolipids are hydrolyzed by acylhydrolases to give free fatty acids, including linoleic and linolenic acids.
(b) These two unsaturated fatty acids (and others) are oxidized by lipoxygenase, or metal ions including heme compounds, in the presence of O_2 to give the hydroperoxy compounds.
(c) The hydroperoxy compounds are cleaved by hydroperoxide lyase (cleaving enzymes) or Fe^{II} to give aldehydes.
(d) The aldehydes may undergo *cis, trans* isomerization (*cis-trans* isomerases) or double bond saturation (double bond reducing enzymes) to give rise to other aldehydes.
(e) The aldehydes may be reduced to alcohols by alcohol dehydrogenase.
(f) The alcohols may react with carboxylic acids (catalyzed by ester synthetases and carboxylesterases) to give carboxylic esters.

The hydroperoxides may also form other products (Fig. 3; Gardner, 1988). Hydroperoxide lyase produces aldehydes and aldehyde acids

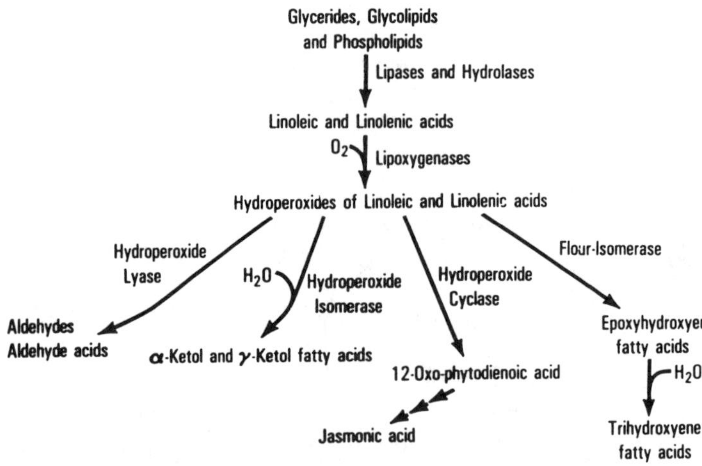

Fig. 3. Products formed enzymically from linoleic and linolenic acids in plants. From Gardner (1988) with permission.

from hydroperoxides; hydroperoxide isomerase produces α-ketol and γ-ketol fatty acids; hydroperoxide cyclase produces 12-oxo-phytodienoic acid, a precursor of jasmonic acid; and flour-isomerase produces epoxyhydroxyene fatty acids that are hydrolyzed to trihydroxyene fatty acids. The non-enzymatic oxidation of linolenic acid has been shown to give as many as 74 compounds (3 hydrocarbons, 25 aldehydes, 2 ketones, 6 alcohols, 12 acids, 19 esters and 7 miscellaneous compounds; Grosch, 1987).

The remainder of this chapter will deal only with lipoxygenases. Therefore, the chemistry of oxidation of unsaturated fatty acids will be limited to those reactions catalyzed directly by lipoxygenases.

HISTORICAL ASPECTS

Lipid oxidation was studied at least 160 years ago by Berzelius (Lillard, 1978). He showed that lipid oxidation of linseed oil involved an induction period, oxygen uptake, carbon dioxide formation and polymerization. However, it was not until the 1920s that evidence began to accumulate, suggesting the presence of enzymes that could perform a similar reaction. In 1928, Haas & Bohn found that additions of small amounts of soybean flour to wheat flour dough decreased the normal yellow color of wheat (Bohn & Haas, 1928; Haas & Bohn, 1934). The reaction required molecular oxygen. About the same time, Andre & Hou (1932*a*) in France reported that the fat in soybean curd differed depending on whether it was obtained from unboiled or heated soybean extract. They concluded that the difference was due to a heat-labile enzyme, which they named 'lipoxydase' ('lipoxidase' in English). They also found this 'lipoxidase' in the seeds of the common bean, *Phaseolus vulgaris* and the seeds of another legume, *Dolichos lablab* (Andre & Hou, 1932*b*). In 1929, Navez reported that aqueous suspensions of ground seeds of *Pisum, Phaseolus* and *Lupinus* consumed oxygen (Navez, 1929). This observation was studied further by Craig, who showed that the oxygen consumed far exceeded the carbon dioxide produced. Further, a crude ammonium sulfate fraction of the extract, when added to a lipid, caused the same reaction (Craig, 1936). He called this enzyme 'unsaturated fat oxidase'.

In completely unrelated studies Hauge and co-workers, on determining that carotene was lost during curing of alfalfa, suggested that alfalfa contained a carotene-destroying enzyme (Hauge & Aitkenhead,

1931; Hauge, 1935). In 1943, Sumner demonstrated the presence of lipoxygenase in alfalfa (Sumner, 1943).

As early as 1940 Sumner & Sumner and Tauber reported that 'lipoxydase', 'unsaturated fat oxidase' and 'carotene oxidase' were identical (Sumner & Sumner, 1940; Tauber, 1940). They showed that the primary reaction catalyzed by the enzyme was the peroxidation of unsaturated fats and that carotene degradation occurred only as a result of the lipid oxidation. Conventional protein purification methods led to substantial purification of lipoxygenase (Balls *et al.*, 1943) and in 1947 the enzyme was crystallized (Theorell *et al.*, 1947).

The soybean lipoxygenase crystallized by Theorell *et al.* (1947) was shown to be homogeneous by electrophoresis, diffusion and sedimentation (ultracentrifuge). It had a molecular weight of 102 000, isoelectric point of 5·4 and consisted only of amino acids. No prosthetic group or metal ions were found (Holman, 1947, 1949). It had a pH optimum of 9 (Holman, 1947; Fukuba, 1952).

Evidence for other lipoxygenases (isoenzymes) in soybeans soon arose. Kies (1947) reported that the carotene bleaching activity of soy extracts was much more heat labile than was the peroxide-forming activity. This was confirmed by Schroeder (1968). In 1958, Koch and colleagues (Koch *et al.*, 1958) reported that soybeans contained two lipoxygenases, one with a preference for triglycerides as substrates and the other with a preference for free polyunsaturated acids as substrates. In 1969, a second lipoxygenase was purified from soybean extract (Schormüller *et al.*, 1969). In 1970, Christopher and co-workers (Christopher *et al.*, 1970) also purified a second lipoxygenase isoenzyme to homogeneity. This isoenzyme, designated lipoxygenase-2, differed from the Theorell enzyme, lipoxygenase-1, in pH optimum and in ester/acid preference. Later, the same workers purified a third isoenzyme, lipoxygenase-3, that differed from the other two isoenzymes in elution from a DEAE-Sephadex column, isoelectric point, pH-activity profile and stimulation by Ca^{2+} (Christopher *et al.*, 1972*a*). Isoenzyme-3 could be further resolved into two fractions on columns of hydroxyapatite or CM-Sephadex (Christopher *et al.*, 1972*a*). The more rapidly eluting enzyme was labeled lipoxygenase-3 and the slower eluting component lipoxygenase-4. Lipoxygenases-2, -3 and -4 had pH optima of 6·5–7·0, in contrast to lipoxygenase-1 with a pH optimum of 9. Supporting evidence for lipoxygenase isoenzymes came from gel electrophoresis on polyacrylamide gels containing starch (Guss *et al.*, 1967). The isoenzymes were located by immersing the gels in linoleic acid and then in acidic potassium iodide to give a blue band wherever there was lipoxygenase activity. The same technique gave evidence for lipoxygenase isoenzymes in wheat (Guss *et al.*, 1967) and peas and beans (Hale *et al.*, 1969).

Axelrod and co-workers and others have determined the amino acid sequences for three of the four soybean lipoxygenase isoenzymes (Shibata *et al.*, 1987, 1988; Yenofsky *et al.*, 1988). They have very similar amino acid sequences. Lipoxygenase-1 and -3 are 81 and 74% identical to lipoxygenase-2. A region of 40 amino acid residues containing a cluster of 6 histidine and 2 tyrosine residues, highly conserved in all three isoenzymes, is discussed as a possible Fe-binding region (Shibata *et al.*, 1988). Lipoxygenase-1 contains 838 amino acid residues (mol. wt 94 038); lipoxygenase-2 contains 865 amino acids (mol. wt 97 036) and lipoxygenase-3 contains 859 amino acid residues (mol. wt 96 380). The relative amounts of the isoenzymes differ among cultivars and lipoxygenase-1 can be largely eliminated without apparent effect on the plant.

Lipoxygenase isoenzymes have been reported in other legume seeds. Hale *et al.* (1969) found two to three isoenzymes in fresh green peas. Eriksson & Svensson (1970) separated two major isoenzymes with pIs of 5·80 and 5·82 and a minor isoenzyme from dry peas. Arens *et al.* (1973) separated two active fractions on DEAE-cellulose chromatography from pea seed extracts. Isoelectric focusing gave two components with pIs at 6·0 and 6·5. Four isoenzymes were separated from pea seed extract by CM-cellulose chromatography (Antis & Friend, 1974). Reynolds (1982) also reported four isoenzymes in pea seeds. Disk gel electrophoresis and DEAE-cellulose chromatography showed at least three lipoxygenase isoenzymes in extracts of immature English peas (*Pisum sativum* L. var. Dark Skinned Perfection; Chen & Whitaker, 1986).

Lipoxygenase isoenzymes have been reported in cereal seeds (Table 1). Wheat has been reported to contain up to four isoenzymes (Guss *et al.*, 1967), barley two after germination (Yabuuchi, 1976), maize two after germination (Vick & Zimmerman, 1982), rice three (Shastry & Raghavendra Rao, 1975), rye three (Heimann & Klaiber, 1977) and oats two (Heimann *et al.*, 1973*a,b*).

The mechanism of lipoxygenase action evolved slowly and with a degree of difficulty. While workers in the field expected that some sort of metal prosthetic group should be involved, none was found despite repeated attempts. Lipoxygenase appeared to be an exception to the

Table 1. Some Properties of Plant Lipoxygenases

Source	*Mol. wt*	*pH opt*	*pI*	*Substrate spec.*		*Product(s)*		*Ref.*
				Lino-leate	*Methyl-linoleate*	*9-OOH*	*13-OOH*	
Soybean								
L-1	102 000	9·0–9·5	5·68	1	0·1		+	Chisom (1985)
L-2	100 000	6–7	6·25	1	0·52	+	+	Chisom (1985)
L-3	100 000	6–7	6·15	1	2·0	+	+	Chisom (1985)
Wheat (3 isoenzymes)	90 000–95 000[a]	6–6·5[a]	5·1; 5·4; 6·25[a]			+[b]		[a]Nicolas *et al.* (1982) [b]Kühn *et al.* (1985)
Barley								
L-1	64 000	5·9–6·0	4·75	+		+		Führling (1975)
L-2 (germinating barley)		7·0–7·5	6·6	+	+		+	Yabuuchi (1976)
Maize		5[a]				+[b]		[a]Fritz & Beevers (1955); [b]Gardner (1975)
Rice								
L-1		4·5[a]				+[a]	+[a]	[a]Ida *et al.* (1983)
L-2		5·5[a]				+[a]	+[a]	[b]Yamamoto & Fujii (1980)
L-3	95 000[b]	7·0[a]		+[a]		+[a]		
Rye (2 isoenzymes)	102 000	7·3–7·5					+	Heimann & Klaiber (1977)
Oat								
L-1	250 000							Heimann *et al.* (1973*a,b*)
L-2	100 000	6·75	5·7			+		

[a] Refers to specific reference.

rule that all oxygenases must possess a prosthetic group, cofactor or metal ion. Even disulfide bonds as the transient recipient of the unpaired electron were considered (Walker, 1963). In June 1972 at the 11th World Congress of the International Society for Fat Research held in Gothenburg, Sweden, H. W.-S. Chan reported finding iron in stoichiometric amounts in purified commercial lipoxygenase and that extensive dialysis of the enzyme against a variety of metal chelators resulted in loss of enzymatic activity (Chan, 1973). These results were confirmed by Roza & Francke (1973) and Pistorius & Axelrod (1973, 1974). Evidence for involvement of iron included ESR, spectrophotometry and stopped-flow experiments. It is now accepted that iron plays a key role in the mechanism of lipoxygenase.

The history of lipoxygenases of animal tissues is relatively short. Until the 1970s, it was believed that lipid peroxidation in animal tissues resulted exclusively from heme compounds ubiquitously present (Tappel, 1953; Boyd & Adams, 1955). Extensive prostaglandin research led to discovery of lipoxygenase activity in platelets (Hamberg & Samuelsson, 1974; Nugteren, 1975; Ho *et al.*, 1977) and leucocytes (Borgeat *et al.*, 1976). Also, a lipoxygenase from rabbit reticulocytes was discovered during this period (Schewe *et al.*, 1975).

Within the last 10 years, much information has accumulated on the lipoxygenase metabolites in animal cells (Hansson *et al.*, 1983). However, few data exist on the physical and chemical structure, kinetics and mechanism of animal lipoxygenases. In part, this has been a result of the difficulty of purifying the enzymes. The first animal lipoxygenase purified to homogeneity and partly characterized is the one from rabbit reticulocytes (Rapoport *et al.*, 1979; Schewe *et al.*, 1981). More recently, a homogeneous lipoxygenase has been obtained from RBL cells (Goetze *et al.*, 1985). Several other animal lipoxygenases with different specificities are known to exist. Rapid progress on purifying and characterizing additional animal lipoxygenases is expected.

The complete sequences of rabbit reticulocyte (RBC) 15-lipoxygenase (Fleming *et al.*, 1989) and of human leukocyte 5-lipoxygenase (Dixon *et al.*, 1988; Matsumoto *et al.*, 1988) have been determined. There is a cluster of five conserved histidine residues that may be part of the active site responsible for chelating the iron moiety. There was relatively little amino acid sequence homology between rabbit reticulocyte 15-lipoxygenase and human leukocyte 5-lipoxygenase or soybean lipoxygenase-3 (Yenofsky *et al.*, 1988).

DISTRIBUTION OF LIPOXYGENASES

Lipoxygenases could be ubiquitous in all plants and animals and in most micro-organisms (Table 2). The amounts of lipoxygenases in tissues vary widely and can differ among cultivars/strains of the same organisms, with age, tissue and with environmental factors. Soybean seeds contain more lipoxygenase than do English green peas and green beans. Cereal seeds generally have less lipoxygenase than legumes and the leafy vegetables have even less. Lipoxygenase of the seeds of soybeans, English peas, green beans and corn are responsible for aroma and flavor deterioration (Williams *et al.*, 1986; Velasco *et al.*, 1989), unless properly blanched, while lipoxygenases in cauliflower and broccoli appear to be of little or no consequence (Lim *et al.*, 1989).

Lipoxygenase is found primarily in the cytosol of the storage

Table 2. Plant and Animal Sources of Lipoxygenase

Plants	Peanut (seed)
Apple (fruit)	Potatoes (tuber)
Alfalfa (leaf)	Pumpkin (seed)
Articum lappa (leaf)	Rape (seed)
Barley (seed)	*Robinia pseudoacacia* (seed)
Beans (*Phaseolus,* various; seed)	*Sinapis alba* (seedlings)
	Solanum dulcamara (leaf)
Cauliflower (florets)	Squash (seedling)
Centaurea spp. (leaf)	Soybean (seed)
Chlorella (whole cell)	Sunflower (seed)
Cirsium arvense (leaf)	Tomato (fruit)
Digitalis purpurea (leaf)	Trefoil (seed)
Dimorphotheca sinuata (seed)	Urd (seed)
Dornicium plantagineum (leaf)	*Urtica dioica* (leaf)
Egg plant (fruit)	Vetch, purple
Fava bean (seed)	Wheat (seed, leaf)
Flax (seed)	
Garbanzo bean (seed)	**Animals**
Locust, honey (leaf, pod, seed)	Cow (platelets, lung, renal medulla)
Lupine (seed)	Human (platelets, leukocytes)
Maize (germ)	Pig (leukocytes, lung, pancreas, ileum, thymus, lymph nodes)
Marijuana (seed)	
Olive (seed)	Rabbit (reticulocytes)
Peas (seed)	Rat (basophiles)

Adapted in part from Axelrod (1974).

parenchyma of mature soybean seeds (Vernooy-Gerritsen *et al.*, 1983, 1984). On germination, however, lipoxygenase decreases in the storage parenchyma and increases in the abaxial hypodermis, epidermis and vascular bundle sheaths (Vernooy-Gerritsen *et al.*, 1983). Lipoxygenase in cereal seeds appears to be associated primarily with the germ, although the subcellular location is not known. The germ and bran of wheat were found to contain 17- and 4-fold more activity, respectively, than the endosperm (Averman *et al.*, 1971). Lipoxygenase activity varied more than 10-fold in wheat samples of different varieties and conditions of cultivation (Colas & Chargelegue, 1974). Lipoxygenase activity increased substantially after 3·5 days germination (Guss *et al.*, 1968) and was located primarily in the coleoptile plus the first leaf and in the roots. The amount of lipoxygenase decreased throughout the seedling within 7 days of germination. Lipoxygenase was not detected in mitochondria of wheat seedlings (Goldstein *et al.*, 1980). It has been found in the chloroplasts of a number of plants, including soybean leaves (Vernooy-Gerritsen *et al.*, 1983), pea leaves (Haydar & Hadziyev, 1973), wheat (Douillard & Bergeron, 1979, 1981) and possibly cauliflower and calabrese florets (Wardale & Galliard, 1975, 1977).

ISOLATION OF LIPOXYGENASES

Numerous methods have been used to purify lipoxygenases from various sources. A method for purifying a lipoxygenase from immature English green peas developed in the author's laboratory will be presented. Most of the methods in the literature have resulted in a five- to 40-fold purification while Chen & Whitaker (1986) obtained a 199-fold purification.

Isolation of lipoxygenase from immature English green peas proved particularly difficult because of instability of the enzyme. Siddiqi & Tappel (1956) partially purified lipoxygenase from immature Thomas Laxton English peas by water extraction of an acetone powder and treatment with barium acetate, acetone and basic lead acetate. Lee & Wagenknecht (1958) obtained a 35-fold increase in lipoxygenase specific activity over that of lyophilized immature peas by ammonium sulfate fractionation and heat treatment, with a recovery of 49% of the original activity. Klein (1976) purified lipoxygenase from the acetone powder of immature English green peas by extraction with buffer containing 2 mM $CaCl_2$, 30–60% ammonium sulfate fractionation, Sephadex G-150 gel filtration chromatography, and DEAE-Sephadex chromatography; she obtained a nine-fold purification with 3% recovery of activity.

Chen & Whitaker (1986) purified two of the lipoxygenase isoenzymes from immature dark-skinned Perfection variety of English peas (*Pisum sativum* L.) to homogeneity. They started with an acetone powder. The acetone powder was prepared by blending frozen peas with cold acetone (−25°C) in a ratio of 1:2·5 (w/v). The powder was collected on a Buchner funnel and washed with another two volumes of cold acetone. The acetone powder was extracted with four volumes (1:4, w/v) of cold 0·2 M sodium phosphate buffer, pH 7·0, containing 0·85 M sodium chloride for 1–2 h. The extract was centrifuged at 17 300*g* for 20 min, the pellet discarded and the supernatant was adjusted to pH 7·0. Solid ammonium sulfate was added slowly to the supernatant at 4°C with continuous stirring to give 25% saturation. After stirring for 1 h, the suspension was centrifuged at 17 300*g* for 20 min, the precipitate discarded, and the supernatant was adjusted to 50% saturated ammonium sulfate with continuous stirring for 1 h. The yellowish white precipitate was collected by centrifugation as above. The pellet was dissolved in 5 mM sodium phosphate buffer, pH 7·4, and dialyzed against several changes of the same buffer. The preparation was centrifuged as above and the pellet discarded.

The supernatant liquid was applied to the top of a DEAE-cellulose column (2·6 × 40 cm) previously equilibrated with 5 mM sodium phosphate buffer, pH 7·4. The column was washed with the same buffer until the absorbance returned to base line. The column was eluted with a linear salt gradient of 0–0·2 M sodium chloride in 5 mM sodium phosphate buffer, pH 7·4. The active fractions from the DEAE-cellulose column (Fig. 4) were pooled and the enzyme was precipitated from solution between 50 and 90% saturated ammonium sulfate. The supernatant, with very low lipoxygenase activity, was discarded and the pellet was dissolved in 5 mM sodium phosphate buffer, pH 7·2, containing 0·03 M NaCl sodium chloride.

The dissolved enzyme above was added directly to the top of a Sepharose GB-100 column (1 × 98 cm) pre-equilibrated with the same buffer and the column washed with the same buffer. Only one peak with lipoxygenase activity was observed (Fig. 5a). The fractions with lipoxygenase activity were pooled and concentrated to 2 ml by ultrafiltration on a millipore membrane. The concentrated solution

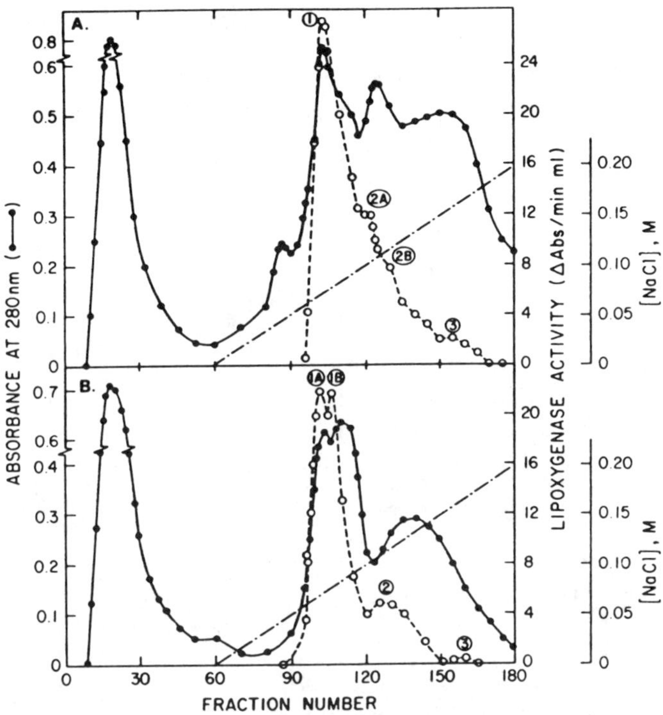

Fig. 4. DEAE-cellulose chromatographic purification of immature English pea lipoxygenase. ●, Absorbance at 280 nm; ○, lipoxygenase assay; ·–·, sodium chlodride gradient. (a) Ammonium sulfate cut applied was 25–50% fraction (see Table 3); (b), ammonium sulfate cut applied was 20–50% fraction. From Chen & Whitaker (1986) with permission.

was added to the top of a polybuffer 94 column (0·9 × 48 cm) pre-equilibrated with 0·025 M imidazole–HCl buffer, pH 7·4. The enzymes were eluted with a polybuffer 74 solution, pH 4·0, to form a pH 7·4–4·0 gradient in the column. One major peak and one minor peak, both with lipoxygenase activity, were obtained (Fig. 5b). The fractions containing most of the lipoxygenase activity (peak 1) were pooled and the enzyme precipitated by 90% saturated $(NH_4)_2SO_4$ to remove the polybuffer 74, desalted, lyophilized and the sample stored in the freezer.

Progress of the purification was followed by disk gel electrophoresis performed by the Davis method (Davis, 1964) with some modification.

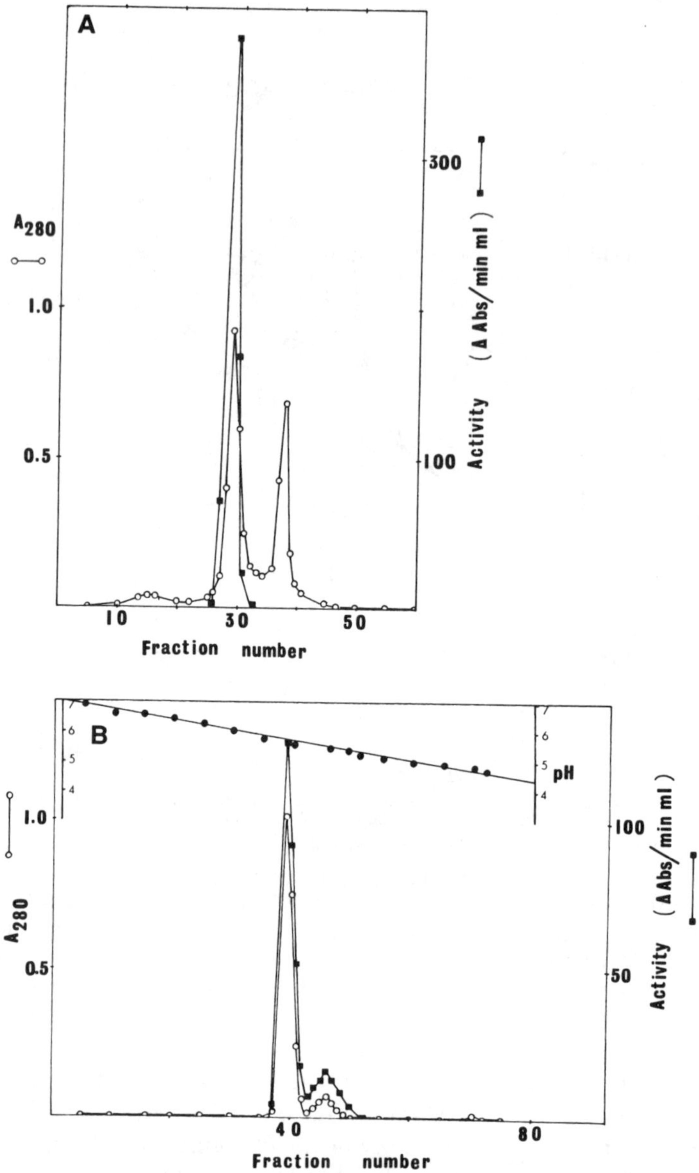

Fig. 5a. Purification of pea lipoxygenase by Sepharose 6B-100 gel filtration. Enzyme from the DEAE-cellulose column (Fig. 4(b), fractions 95-120) was precipitated by ammonium sulfate fractionation (50–90% saturation) and dissolved in 5 mM sodium phosphate buffer, pH 7·2, containing 30 mM NaCl. **Fig. 5b.** Purification of pea lipoxygenase by chromatofocusing chromatography. The enzyme fraction from the Sepharose 6B-100 column (fractions 26-31) was concentrated to 2 ml by ultrafiltration on a Millipore membrane (see Table 3). From Chen & Whitaker (1986) with permission.

Table 3. Purification of Lipoxygenase from Immature English Peas[a,b]

Treatment	*Units of activity 100 g^{-1}*	*Protein (absorbance units at 280 nm 100 g^{-1})*	*Specific activity (units of activity/ units of protein)*	*Purification[c] (-fold)*	*Recovery (%)*
Extract of acetone powder	31 100	6 370	4·88	1·00	100
25% $(NH_4)_2SO_4$ supnt	31 400	4 810	6·53	1·34	101
25–50% $(NH_4)_2SO_4$ ppt	26 600	1 680	15·8	3·24	85·5
Dialysis	23 200	766	30·3	6·21	74·6
DEAE-cellulose	8 910	41·4	215	44·1	28·6
Sepharose 6B-100	8 670	31·8	273	55·9	27·9
PBE[d]					
peak 1	2 510	2·59	969	199	8·07
peak 2	672	0·8	840	172	2·16

[a] From Chen & Whitaker (1986) with permission.
[b] 100 g of English peas.
[c] Based on specific activity of the original extract. The specific activity of each isoenzyme in the original extract is not known.
[d] Polybuffer exchange chromatography.

Duplicate gels for each sample were run; one was stained for protein with Coomassie Brilliant Blue R-250 and the other for activity by using the method of Guss *et al.* (1967) with modification. The results of the purification are shown in Table 3. The major lipoxygenase izoenzyme was purified 199-fold while a minor isozyme was purified 172-fold. The overall recovery of activity at the end was 10·2%.

MECHANISM OF ACTION OF LIPOXYGENASES

Relatively little is known about the mechanism of action of lipoxygenases other than soybean lipoxygenase-1. It is assumed, of course, that there is a common mechanism for all lipoxygenases in which iron is the cofactor, but this assumption is largely untested. A discussion of the lipoxygenase mechanism below requires attention to substrate specificity, nature of the products formed and kinetic aspects.

Substrate Specificity

The preferred substrates for soybean lipoxygenase-1 are linoleic and linolenic acids (Hamberg & Samuelsson, 1967; Holman *et al.*, 1969; Christopher *et al.*, 1972*b*). Required is a methylene-interrupted *cis,cis*-pentadiene moiety with the methylene group situated at the $\omega 8$ carbon and located between two double bonds (Fig. 6). Hamberg &

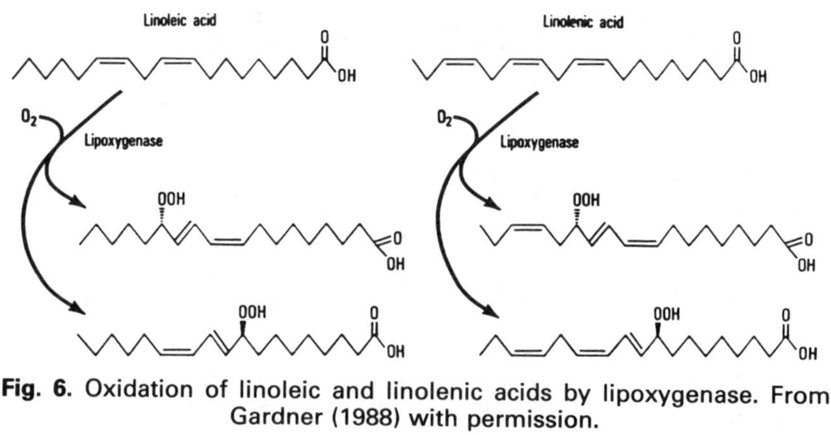

Fig. 6. Oxidation of linoleic and linolenic acids by lipoxygenase. From Gardner (1988) with permission.

Samuelsson (1967) and Holman *et al.* (1969) examined a series of positional isomers of octadecadienoic acids using soybean lipoxygenase-1. Although most of them had the required methylene-interrupted *cis,cis*-pentadiene moiety, with double bonds on either side of the methylene group, they were in fact poor substrates unless the methylene group was in the ω8 position. Therefore, there is direct evidence that soybean lipoxygenase-1 has specificity for the substrate directed toward the non-carboxyl terminal end (the methyl end) of the molecule.

Free fatty acid substrates are preferred by soybean lipoxygenase-1, which has a pH optimum of 9. Other isoenzymes of soybean, with pH optima between 6·5 and 7, and other higher plant lipoxygenases, also with pH optima between 6·5 and 7, have about equal activity on the free fatty acid, the methyl ester and the triglyceride (Christopher *et al.*, 1970, 1972*b*; Chen & Whitaker, 1986). This is illustrated in Table 4 for soybean lipoxygenase-1 and lipoxygenase isoenzyme-1 from English green peas (Chen & Whitaker, 1986). There is also a pH effect on the relative activities with free fatty acids, methyl esters and

Table 4. Substrate Specificity of Lipoxygenase Isoenzyme-1 from English Pea and Soybean Lipoxygenase[a]

Substrate[b]	*Detergent or solvent*	*Pea isoenzyme-1*		*Soybean lipoxygenase-1*	
		Activity (mM min⁻¹)	*%*[c]	*Activity (mM min⁻¹)*	*%*[c]
Linoleic acid	Tween-20	0·197	188	0·467	45·8
Linoleic acid	ethanol	0·105	100	1·02	100
Linoleic acid	ethanol + 0·05% BSA	0·048	46	0·277	27·2
Linoleyl alcohol	ethanol	0·073	70	0·0095	0·93
Linoleyl acetate	ethanol	0·030	29	0·0047	0·46
Methyl linoleate	Tween-20	0·054	51	0·0026	0·25
Methyl linoleate	ethanol	0·036	34	0·0066	0·65
Ethyl linoleate	ethanol	0·019	18	0·0036	0·35
Propyl linoleate	ethanol	0·023	22	0·0098	0·96
Trilinolein	ethanol	0·081	77	0·013	1·3

[a] From Chen & Whitaker (1986) with permission; 0·2 M phosphate buffer, pH 7·0, at 24°C.

[b] Substrate concentration was 0·2 mM in all cases.

[c] Relative to linoleic acid in ethanol as 100%.

triglycerides (Gardner, 1989). Chen (1983), Kühn *et al.* (1986) and Gardner (1988) suggested these differences may be a result of how the substrate is oriented into the active site of the enzyme (see 'Products formed').

Products Formed

Insertion of O_2 into the substrate is both chiral and positionally specific. As shown in Fig. 6 and Table 1, the hydroperoxide group can occur at either carbon-9 or carbon-13, depending on the lipoxygenase used. Also, in theory, both the R- and S-chiral orientation could occur. Hamberg & Samuelsson (1967) showed that removal of a hydrogen from the $\omega 8$ methylene of the substrate by soybean lipoxygenase-1 was the rate-limiting step and preceded the oxidation step. Hydrogen removal is chiral and occurs on the opposite side of the fatty acid chain from insertion of the O_2. Egmond *et al.* (1972) confirmed the chiral removal of hydrogen with both soybean lipoxygenase-1 and maize germ lipoxygenase. Soybean lipoxygenase-1, at pH 9, forms primarily 13S-LOOH, with a small amount of 9S-LOOH (Hamberg & Samuelsson, 1967). Chen & Whitaker (1986) reported 87·7% 13-LOOH and 12·3% 9-LOOH, respectively, for soybean lipoxygenase-1 at pH 9·0. Earlier reports indicated a ratio of 90:10 for 13-LOOH and 9-LOOH (Dolev *et al.*, 1967; Leu, 1974). There is also a small amount of 13R-LOOH and 9R-LOOH formed, presumably due to autoxidation products (Hamberg, 1971). Oxidation of linoleic acid by soybean lipoxygenase-1 at pH 6·6 gave an increased ratio of 9-LOOH to 13-LOOH and also increased the chiral preference for 9S-LOOH over 9R-LOOH (van Os *et al.*, 1979*a*). Soybean lipoxygenases-2 and -3, with pH optima near 7, give nearly equal amounts of the 13-LOOH and 9-LOOH isomers. Soybean lipoxygenase-2 appears to form 9R-LOOH, the only lipoxygenase so far shown to have this specificity (van Os *et al.*, 1979*b*).

Immature English green pea lipoxygenase isoenzyme-1 produced 50·3% 13-LOOH and 49·7% of 9-LOOH at pH 7·0 (Chen & Whitaker, 1986). These results agree with earlier reports that lipoxygenase isolated from dry English peas gave a 50:50 ratio of 13-LOOH and 9-LOOH (Hamberg, 1971; Roza & Francke, 1973). Oxidation of linoleic acid by maize germ lipoxygenase at pH 6·5–7·4 resulted in 82–89% 9S-LOOH (Gardner & Weisleder, 1970; Hamberg, 1971;

Gardner, 1975). Egmond *et al.* (1972), in a comparative study of soybean lipoxygenase-1 and maize germ lipoxygenase, found that soybean lipoxygenase-1 removed the pro-S hydrogen from carbon-11 methylene of linoleic acid and inserted O_2 on the opposite side of the fatty acid chain to give the 13S-LOOH. On the other hand, the maize germ lipoxygenase removed hydrogen from the pro-R position of the carbon-11 methylene group and inserted O_2 on the opposite side to give 9S-LOOH.

Chen (1983), Kühn *et al.* (1986) and Gardner (1988) suggested that the relative specificities of different lipoxygenases for free fatty acids, methyl esters and triglycerides, preference for forming 13-LOOH and/or 9-LOOH (or two isomers of arachidonic acid, for example), as well as the pH and temperature dependence of the specificity could be explained by how the substrate is oriented into the active site of the enzyme. Perhaps a head-to-tail orientation of substrates occurs at the active site of the enzyme, as shown in Fig. 7. Therefore, the hydrogen removals and oxygenations would be spatially identical. This postulation would explain the pH dependency of product formation, i.e., different affinities of the enzyme for the carboxylate versus the carboxylic acid form of the substrate. It could also explain why certain lipoxygenases (e.g. soybean lipoxygenase-1) recognize the ω end of the substrate and others (e.g. maize germ lipoxygenase) the carboxylic acid portion of the substrate and others (e.g. English pea lipoxygenase isoenzyme 1) appear to have no preference for the ω end.

As might be expected from the more complex physiological substrate, arachidonic acid (5,8,11,14-all-*cis*-eicosatetraenoic acid), the chiral and positional specificity of the animal lipoxygenases could be more complex (Kühn *et al.*, 1986) than with linoleic and linolenic acids. While the animal lipoxygenases do oxidize linoleic and linolenic acids, arachidonic acid is of more interest since it gives rise to the biologically important leukotrienes and other in-vivo lipoxygenase products. Twelve regular primary lipoxygenase products may be formed in principle from arachidonic acid, even though not all of them have been shown to occur (Schewe *et al.*, 1986).

Purified reticulocyte lipoxygenase, with arachidonic acid as substrate (Fig. 8), gives $15L_S$-hydroperoxy-5,8,11,13-*cis,cis,cis,trans*-eicosatetraenoic acid ($15L_S$-HPETE) and $12L_S$-hydroperoxy-5,8,10,14-*cis,cis,trans,cis*-eicosatetraenoic acid ($12L_S$-HPETE) in the ratio of about 15:1 (Byrant *et al.*, 1982). It converts linoleic acid to 13S-LOOH and 9S-LOOH in the ratio of about 10:1 (Kühn & Wiesner,

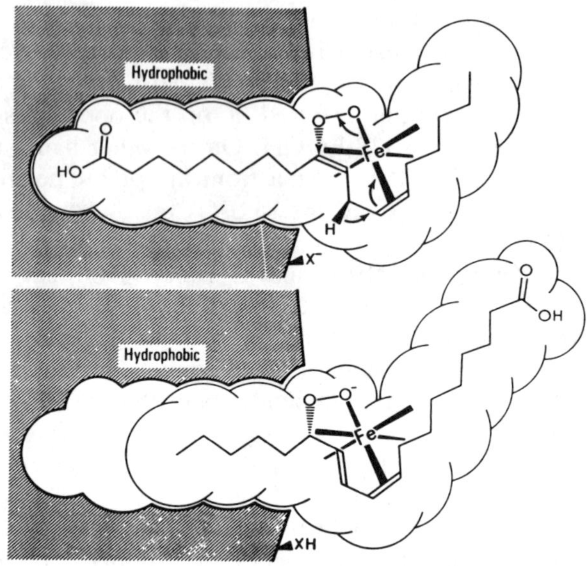

Fig. 7. Proposed oxidation mechanism of lipoxygenase, showing that oxidation at C-9 (top) and C-13 (bottom) as well as hydrogen removal from C-11 are spatially identical when the fatty acid chains are arranged head to tail at the active site. The model at the top shows initial attack of O_2 from behind the linoleic acid chain and electron transfer via the iron-active site; the model on the bottom shows the completed oxidation product, the hydroperoxydiene (anion). From Gardner (1988) with permission.

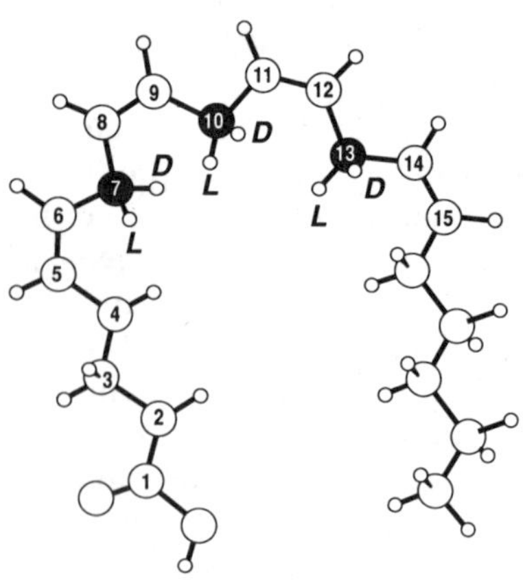

Fig. 8. Structure of arachidonic acid. The prochiral centers (double allylic carbon atoms) are hatched.

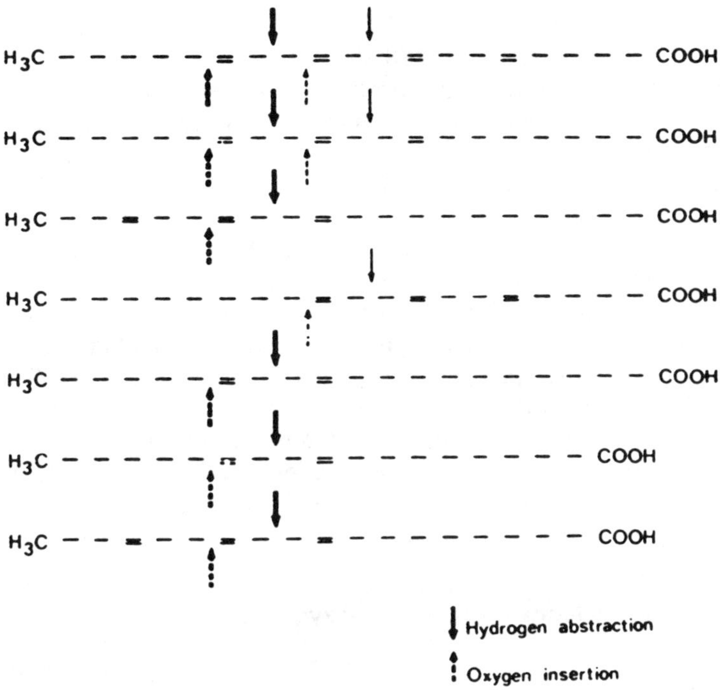

Fig. 9. Positional specificity of reticulocyte lipoxygenase. From Schewe *et al.* (1986) with permission.

unpublished results; see Schewe *et al.*, 1986). The positional specificity of reticulocyte lipoxygenase on several substrates is illustrated in Fig. 9. In all cases where the fatty acid serves as substrate, the major hydrogen extraction is at $\omega 8$ carbon and insertion of O_2 is at $\omega 6$ carbon. Removal of hydrogen and/or insertion of O_2 at other positions depends on the 'freshness' of the enzyme and the temperature of the reaction. When reticulocyte lipoxygenase was damaged by aerobic storage, 15-HPETE and 11-HPETE were formed from arachidonic acid in a ratio of nearly 1:1 (Schewe *et al.*, 1986; these authors suggest that the absence of formation of 11-HPETE may be a criterion for native preparations of reticulocyte lipoxygenase). When reaction of reticulocyte lipoxygenase with arachidonic acid was performed at 37°C, rather than 2°C, a variety of other products were formed, including the dihydroperoxy derivatives 5,15-dihydroperoxy-6,8,11,13-

eicosatetraenoic acid (5,15-DiHPETE) and 8,15-dihydroperoxy-5,9,11,13-eicosatetraenoic acid (8,15-DiHPETE) (Byrant *et al.*, 1985). These additional products are thought not to be formed by change in specificity of the enzyme, but rather by free radical intermediates that escape from the active site of the enzyme.

Partially purified leukocyte lipoxygenase produces $5L_S$-HPETE, $12L_S$-HPETE and $15L_S$-HPETE from arachidonic acid (Borgeat & Samuelsson, 1979; Yoshimoto *et al.*, 1982). It was suggested the three products most likely resulted from three different isoenzymes rather than representing a 'sloppy' enzyme. Blood platelet lipoxygenase produces $12L_S$-HPETE (Nugteren, 1975; Hamberg & Hamberg, 1980) while cyclooxygenase produces $11L_R$-HPETE in the first step (Gund & Shen, 1977; Kühn *et al.*, 1983).

Kühn *et al.* (1986) proposed a new nomenclature for the lipoxygenases whose stereo and chiral specificities are known (Table 5). A new nomenclature is required to include the animal lipoxygenases with preferred specificity for arachidonic acid. The lipoxygenases are

Table 5. Nomenclature of Various Lipoxygenases[a]

Source	*Nomenclature*	*Short form*
Soybean 1	$[C_{n-8}]$-$L_{pro\text{-}S}$:[+2]-L_S	15-LOX
Corn germ	$[C_{11}]$-$D_{pro\text{-}R}$:[−2]-D_S[b]	9-LOX[c]
Blood platelets	$[C_{n-11}]$-$L_{pro\text{-}S}$:[+2]-L_S	12-LOX
Reticulocytes	$[C_{n-8}]$-$L_{pro\text{-}S}$:[+2]-L_S[d]	15-LOX
	$[C_{n-11}]$-$L_{pro\text{-}S}$:[+2]-L_S[d]	12-LOX
Leukocytes	$[C_7]$-$D_{pro\text{-}S}$:[−2]-D_S[b,d]	5-LOX
	$[C_{n-11}]$-$L_{pro\text{-}S}$:[+2]-L_S[d]	12-LOX
	$[C_{n-8}]$-$L_{pro\text{-}S}$:[+2]-L_S[d]	15-LOX
Potato tuber	$[C_7]$-$D_{pro\text{-}S}$:[−2]-D_S[b,d]	5-LOX
Cyclooxygenase (first step)	$[C_{13}]$-$L_{pro\text{-}S}$:[−2]-L_R[b]	11-LOX
Leukotriene A_4 synthase (5,6-LTA_4)	$[C_{10}]$-$D_{pro\text{-}R}$:[−2][b] hydroperoxy fatty acid/epoxide synthase	

[a] From Kühn *et al.* (1986) with permission.
[b] For lipoxygenases with a [−] rearrangement of the fatty acid radical the positional specificity of the initial hydrogen removal has not been studied in detail. Therefore the position of the hydrogen removal with arachidonic acid is indicated.
[c] With linoleic acid as substrate.
[d] Conclusion from the stereochemistry of the dioxygenation product to the stereoselective hydrogen removal according to the antarafacial character of the lipoxygenase reaction.

named (chemically) by the following equation (eqn (1)):

$$[C_{n-x}(C_x)] - L(D)_{pro-S(R)} : [\pm 2(4)] - L(D)_{S(R)} \tag{1}$$

(polyenoic fatty acid: oxygen oxidoreductase)

To quote from Kühn *et al.* (1986):

The terms before the colon characterize the specificity of the initial hydrogen removal and those following the colon characterize the specificity of the dioxygen insertion.

1. Positional specificity of the initial hydrogen removal: $[C_{n-x}(C_x)]$; $n-x$ or x indicates the position of the carbon atom where the hydrogen removal takes place; (n is the total number of carbon atoms).

The designation$[C_{n-x}]$ or $[C_x]$ depends on which terminus serves as signal for the recognition of the site of the hydrogen removal. $[C_{n-x}]$ indicates that the place of hydrogen removal is determined by the distance of the prochiral center from the methyl terminus. In contrast, $[C_x]$ indicates that for these lipoxygenases the place of hydrogen removal is determined by the distance from the terminus possessing the highest oxidation number, for example, the carboxylic group.

2. Stereospecificity of hydrogen removal: $L(D)_{pro\text{-}S(R)}$; steric configuration of the hydrogen abstracted. The pro characterizes the carbon atom as a prochiral center.

3. Positional specificity of oxygen insertion: $[\pm 2(4)]$; rearrangement of the fatty acid radical formed during the hydrogen removal. The free radical electron is shifted by 2 or 4 carbon atoms in the direction of the carboxylic (−) or the methyl (+) end of the fatty acid. There oxygen insertion takes place.

4. Stereospecificity of oxygen insertion: $L(D)_{S(R)}$; steric configuration of the chiral center after dioxygen insertion.

As already discussed, the specification of the chiral and prochiral centers according to the Fischer convention [L/D] should be preferred. However, the specification according to the S/R system should be added as subscripts. In order to achieve uniformity according to this system it is proposed to use arachidonic acid, the biologically most important substrate, as model substrate.

Kinetics of Lipoxygenase-Catalyzed Reactions

Relatively little research has been done on the kinetics of lipoxygenase-catalyzed reactions, probably reflecting the complicated nature of the system. These complications include the low solubilities of both the fatty acid and O_2 substrates in aqueous systems, purity of the fatty acid, inhibition by the fatty acid substrate, aerobic and/or anaerobic reactions, effects of pH and temperature on positional and chiral specificities as discussed above, and inactivation of enzyme during reaction. Initial velocity determinations are essential.

Several methods can be used to measure lipoxygenase activity. Rate of product formation can be measured continuously in a recording spectrophotometer at 234 nm. The increase in absorbance is due to formation of conjugated double bonds in the fatty acid hydroperoxide. Activity can also be determined continuously with an O_2 electrode, measuring the rate of decrease in O_2 concentration. Other methods are less direct and/or not continuous. The rate of hydroperoxide formation can be measured by conversion of $Fe(CNS)_2$ to $Fe(CNS)_3$. However, aliquots must be removed from the reaction for color development. The colored $Fe(CNS)_3$ is unstable and absorbance must be read soon after mixing. One of the oldest methods is the coupled reaction involving added carotene (Balls *et al.*, 1943; Cosby & Sumner, 1945). The carotene color is bleached by the hydroperoxy radicals. However, there is not a linear or stoichiometric relationship between decrease in carotene absorbance and linoleate oxidation.

A kinetic lag phase is commonly observed on investigating the kinetics of the aerobic reaction. Haining & Axelrod (1958) found that high enzyme concentrations decreased the lag phase, as did a small amount of preformed hydroperoxides in the fatty acid substrate. Only the initial autoxidation products (LOOH) are effective in this regard. Some workers (Smith & Lands, 1972) postulated that the enzyme has a product-binding site and catalyzes its own inactivation during oxygenation of fatty acid substrates. Intermediate peroxy radicals are formed, which could be responsible for the enzyme inactivation, as is the case with polyphenol oxidase (Whitaker, 1972). Egmond *et al.* (1976) suggested that the initial lag period could be due to inhibition by the fatty acid substrate, presumably due to binding at the active site. This does not appear to be the case, since the lag period is inversely proportional to the amount of preformed LOOH in the fatty acid (Haining & Axelrod, 1958).

The insolubility of linoleic acid, other fatty acids or other fatty acid derivatives used as substrates is a problem, especially at pHs lower than 9. Tappel *et al.* (1952) reported that the kinetic behavior of soybean lipoxygenase-1 on methyl linoleate did not follow Michaelis-

Menten kinetics unless Tween-61 was added to the reaction system at pH 9·0. However, initial velocities at sodium linoleate concentrations below 3×10^{-4} M at pH 9·0 and 20°C followed Michaelis-Menten kinetics without Tween-61 and gave a K_m for linoleate of 2×10^{-5} M. An anomalous increase in reaction velocity was found in the linoleate concentration range of 3×10^{-4}–4×10^{-4} M. However, the Critical Micelle Concentration of linoleate is $1{\cdot}25 \times 10^{-5}$ M (Chen, 1983). Allen (1969) reported similar results as Tappel *et al.* (1952) for soybean lipoxygenase-1 at pH 9·0 and ionic strength of 0·2. K_m for linoleate was $2{\cdot}4 \times 10^{-5}$ M. It appears from the data of Allen (1969) and Tappel *et al.* (1952) that soybean lipoxygenase-1 can oxidize linoleate equally well whether the linoleate is in monomer or micellar form.

The K_m value of linoleic acid at high concentrations in the presence of detergent or solubilizer appears to be ~10× higher than K_m determined at low linoleic acid in the absence of detergent or solubilizer. How much is due to substrate inhibition at high concentration is not clear. Andrawis *et al.* (1982) found a K_m of $5{\cdot}5 \times 10^{-4}$ M for linoleic acid using Tween-20 and soybean lipoxygenase-2. Al-Obaidy & Siddiqi (1981) reported a K_m of $2{\cdot}8 \times 10^{-3}$ M for broad bean lipoxygenase in the presence of Tween-20. With partially purified soybean lipoxygenase, Koch *et al.* (1958) reported a K_m of $6{\cdot}5 \times 10^{-4}$ M for linoleic acid at pH 9·0 and 20°C with ethanol as solubilizer, and a K_m of $1{\cdot}44 \times 10^{-4}$ M for trilinolein using a mixture of acetone and ethanol (2:3) at pH 5·5 and 20°C.

Little work has been reported on the K_m for O_2. Tappel *et al.* (1952) found the apparent K_m for O_2 was 3×10^{-5} M and $2{\cdot}9 \times 10^{-4}$ M at $3{\cdot}6 \times 10^{-4}$ M and $7{\cdot}2 \times 10^{-3}$ M linoleic acid, respectively, using soybean lipoxygenase-1. On the other hand, Finazzi-Agro *et al.* (1973) reported the same K_m for O_2 at different concentrations of linoleic acid.

Two detailed reports of the effect of linoleic acid and O_2 on the initial velocity of the soybean lipoxygenase-1 catalyzed reactions are available. Egmond *et al.* (1976) investigated the steady-state kinetics at pH 9 over the range of $5{\cdot}4 \times 10^{-6}$–140×10^{-6} M linoleic acid and 5×10^{-6} M O_2, with no detergents present. The K_m was 2×10^{-5} M for linoleic acid and $3{\cdot}5 \times 10^{-6}$ M for O_2. Chen & Whitaker (1986) investigated the steady-state kinetics at pH 7·0 for soybean lipoxygenase-1 over the range of $0{\cdot}44 \times 10^{-3}$–$2{\cdot}5 \times 10^{-3}$ M linoleic acid and $0{\cdot}66 \times 10^{-4}$–$2{\cdot}64 \times 10^{-4}$ M O_2 in the presence of Tween-20. K_m for linoleic acid was $1{\cdot}63 \times 10^{-3}$ M and for O_2 K_m was $1{\cdot}8 \times 10^{-4}$ M. The data of Chen & Whitaker (1986) indicate that soybean lipoxygenase-1

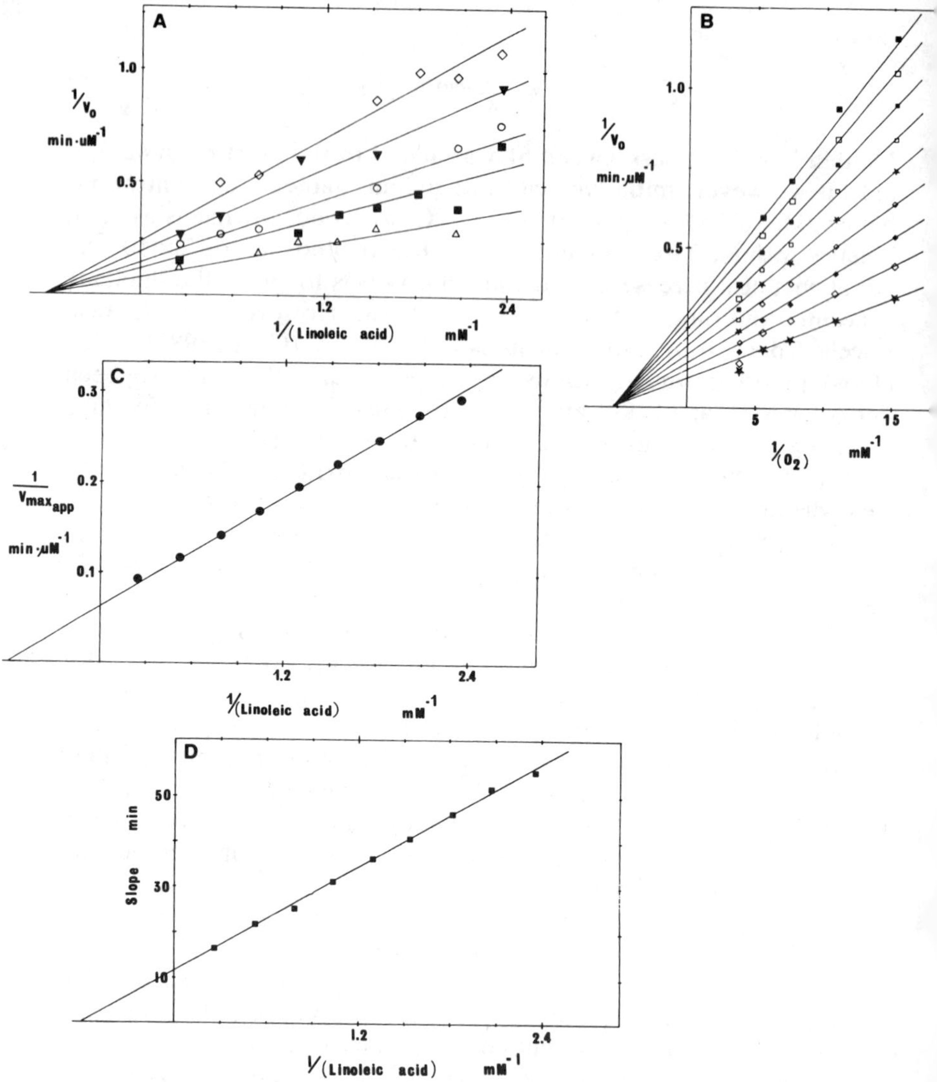

Fig. 10a. Effect of concentration of linoleic acid and oxygen on initial velocities with soybean lipoxygenase-1. The O_2 concentration in each reaction was determined with an O_2 electrode: ◇, 0·066 mM O_2; ▼, 0·0925 mM O_2; ○, 0·132 mM O_2; ■, 0·185 mM O_2; △, 0·264 mM O_2. **Fig. 10b.** Effect of concentrations of linoleic acid and oxygen on initial velocities with soybean lipoxygenase-1 catalyzed reaction. Data were replotted from Fig. 10a. The concentrations of linoleic acid and symbols are ■, 0·43 mM; □, 0·48 mM; ▪, 0·55 mM; ▫, 0·65 mM; ★, 0·77 mM; ○, 0·96 mM; ●, 1·29 mM; △, 1·93 mM; ▼, 2·42 mM linoleic acid. **Fig. 10c.** Replot of the *y* intercepts of the data of Fig. 10b versus 1/[linoleic acid]. **Fig. 10d.** Replot of the slopes of the data of Fig. 10b versus 1/[linoleic acid]. From Chen & Whitaker (1986) with permission.

follows a Bi Bi Ordered Sequential Mechanism, not a Ping-Pong Mechanism (Fig. 10). Evidence for Ordered versus Random Mechanism is that the hydrogen must be removed from the $\omega 8$ carbon prior to oxygenation (Hamberg & Samuelsson, 1967). The same kinetic data were obtained for immature English green pea lipoxygenase isoenzyme-1. K_m for linoleic acid was $2{\cdot}67 \times 10^{-3}$ M and K_m for O_2 was $3{\cdot}6 \times 10^{-4}$ M. Again, the data support a Bi Bi Ordered Sequential Mechanism for the pea lipoxygenase (Fig. 11).

PATHWAY OF LIPID PEROXIDATION CATALYZED BY LIPOXYGENASES

The generally accepted pathway of lipid oxidation by lipoxygenase is that proposed by de Groot *et al.* (1975) (Fig. 12). The complete reaction is composed of four parts: (a) the activation of the enzyme; (b) the aerobic pathway; (c) the anaerobic pathway; and (d) the non-enzymic pathway. It is suggested that the iron prosthetic group of the native ('resting') enzyme is in the inactive $E{-}Fe^{II}{-}O_2$ form. Activation of the enzyme requires a small amount of the hydroperoxy fatty acid (shown as 13-L-ROOH) to oxidize the iron to $E{-}Fe^{III}$. $E{-}Fe^{III}$ stereospecifically abstracts a hydrogen from the lipid to form $E{-}Fe^{II} \cdots R^{\cdot}$ which, in the aerobic pathway, adds O_2 to give $E{-}Fe^{II} \cdots ROO^{\cdot}$ and then rearranges to $E{-}Fe^{III} \cdots ROO^{-}$. The product $E{-}Fe^{III} \cdots ROO^{-}$ is protonated to $E{-}Fe^{III} \cdots ROOH$, ROOH dissociates (13-L-ROOH in Fig. 11) and the $E{-}Fe^{III}$ is ready to oxidize another fatty acid molecule. Gardner (1988) has modified the proposed pathway (Fig. 12) to include the chemical structures (Fig. 13). Under anaerobic conditions, the $E{-}Fe^{II} \cdots R^{\cdot}$ in a lipohydroperoxidase-type reaction, catalyzes a number of oxidation, rearrangement and polymerization steps (Fig. 13). The free radical $ROO^{\cdot}$ may dissociate from the active site of the enzyme and undergo non-enzymic reaction with the polyunsaturated fatty acid to give racemic (S/R) mixtures of products. This non-enzymic reaction is most likely responsible for the observation of small concentrations of 9R-LOOH in most lipoxygenase-catalyzed reactions (above).

PHYSIOLOGICAL ROLES OF LIPOXYGENASES

Several physiological roles for lipoxygenase in higher plants have been proposed. Lieberman & Mapson (1964) proposed that lipoxygenase

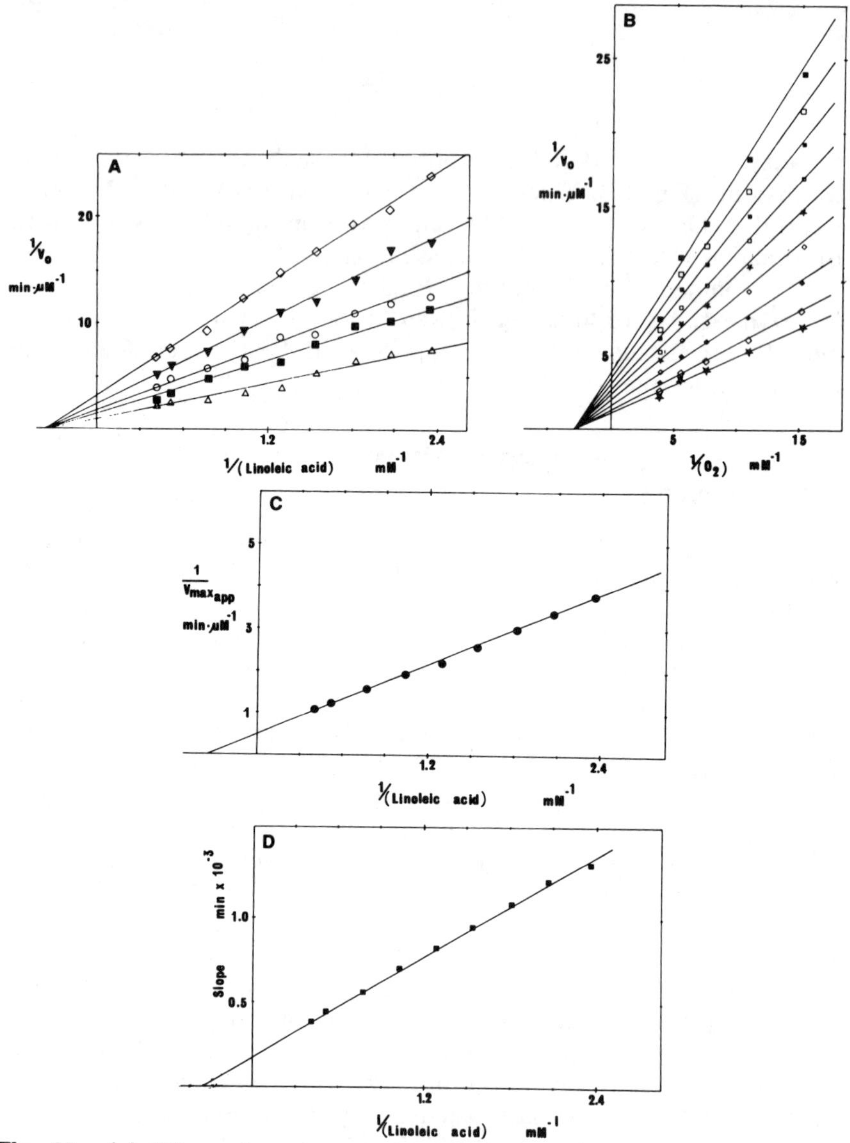

Fig. 11a. (a) Effect of concentration of linoleic acid and or oxygen on initial velocities with isolated pea lipoxygenase isoenzyme-1. The O_2 concentration in each reaction was determined with an O_2 electrode. The O_2 concentrations were the same as in Fig. 10. **Fig. 11b.** Effect of concentration of linoleic acid and oxygen on initial velocities with isolated pea lipoxygenase isoenzyme-1. Data were replotted from Fig. 11a. The concentrations of linoleic acid and symbols were as in Fig. 10. **Fig. 11c.** Replot of the *y* intercepts of the data of Fig. 11b versus 1/[linoleic acid]. **Fig. 11d.** Replot of the slopes of the data of Fig. 11b versus 1/[linoleic acid]. From Chen & Whitaker (1986) with permission.

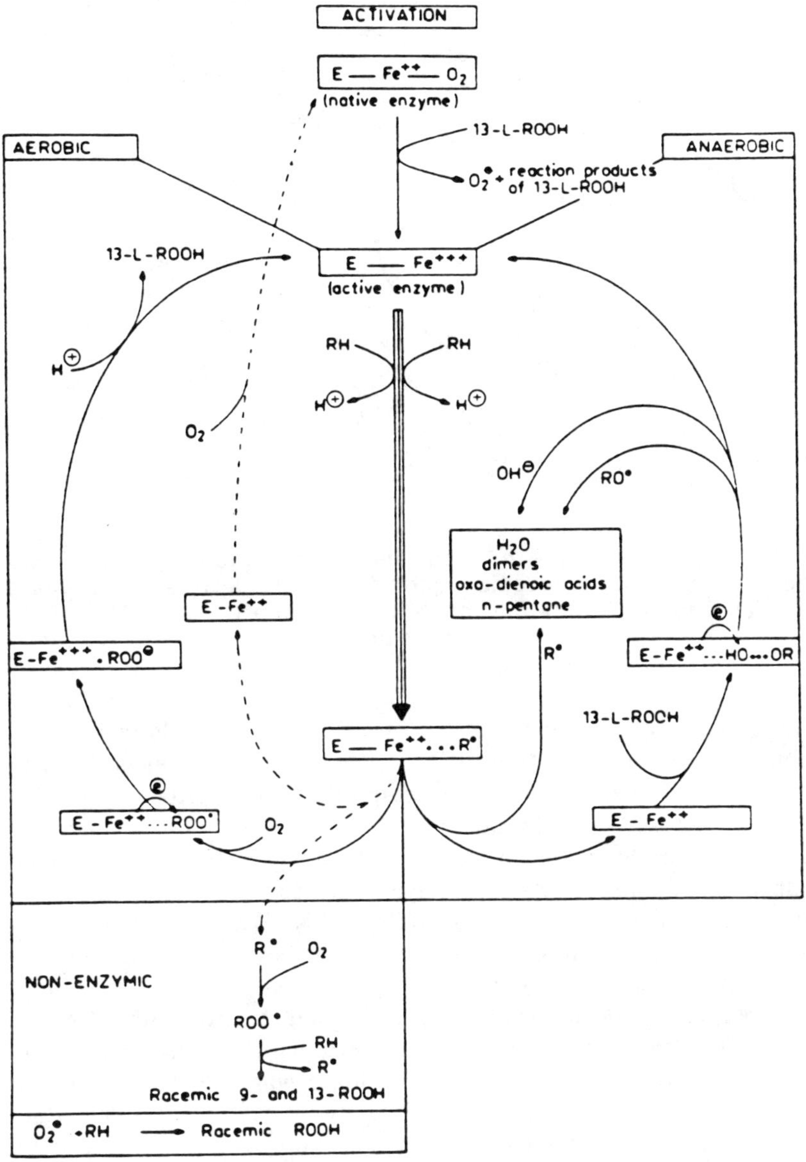

Fig. 12. Proposed mechanism of lipoxygenase-1 action. From de Groot *et al.* (1975) with permission.

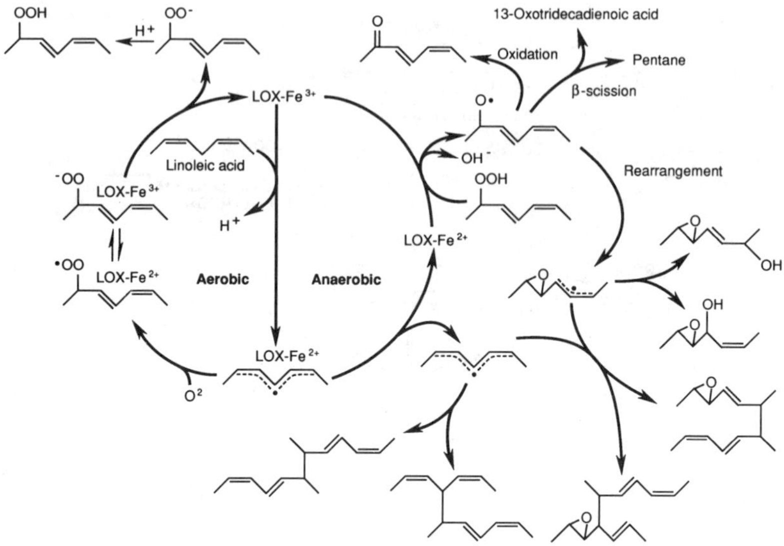

Fig. 13. Proposed scheme for lipoxygenase oxidation of linoleic acid under aerobic and anaerobic conditions. Fatty acid structures are abbreviated to show only C-8 through C-14. From Gardner (1988) with permission.

plays a key role in production of ethylene from linoleic acid. Wooltorten *et al.* (1965) found that an increase in lipoxygenase activity paralleled the rise in both the respiration and ethylene in the climacteric phase of apple ripening. Galliard *et al.* (1968) found an absolute requirement for linolenic acid and ascorbate for ethylene production. We now know that ethylene arises primarily from methionine (Yang, 1985) and not from fatty acids. Wang & Yang (1987) have reported a physiological role of lipoxygenase in ethylene production from 1-aminocyclopropane-1-carboxylic acid. Firn & Friend (1972) reported that violaxanthin, in a coupled reaction with lipoxygenase and linoleic acid, formed xanthoxin, a plant growth inhibitor. A role for lipoxygenase as a terminal oxidase was proposed by Craig (1936). However, Fritz *et al.* (1958) showed that plant tissue only incorporated small amounts of $^{18}O_2$. The lipoxygenase level can be substantially reduced in soybeans without physiological effects and some higher plants have very little lipoxygenase activity (Axelrod, 1974). A further point to bear in mind is that the highest lipoxygenase levels are found in legume and cereal seeds, not the actively metabolizing green leaves.

Some physiological roles for the animal lipoxygenases are well known. Some lipoxygenases are involved in the biosynthesis of bioactive oxyeicosanoids, such as the leukotrienes. Lipoxygenase appears to play a key role in the maturation of reticulocytes (Schewe *et al.*, 1986). A possible regulatory role in the activation and aggregation of platelets has been proposed by some workers (Aharony *et al.*, 1984) but is not accepted by others. The product 12-HETE may play a role in the modulation of the interactions with other cells (Schewe *et al.*, 1986). Arachidonate 15-lipoxygenase in leukocytes may be important in biosynthesis of lipoxins (Schewe *et al.*, 1986).

IMPORTANCE OF LIPOXYGENASES IN FOOD SCIENCE AND TECHNOLOGY

Higher plant lipoxygenases are important in food stability and quality (Eriksson, 1987). While most of the reactions are deleterious, some are beneficial. One of the first uses of lipoxygenase in food processing was the bleaching of carotenoids in wheat dough. This is generally done by addition of a small quantity of soybean flour, high in lipoxygenase activity (Haas & Bohn, 1934). Faubion & Hoseney (1981) recently reviewed the use of lipoxygenase in wheat flour. Because of their neutral pH optima and high peroxidizing activity, lipoxygenases-2 and −3 are more important than lipoxygenase-1 in this role. Endogeneous lipoxygenase is deleterious in pasta manufacture because the yellow color of carotenoid is a quality indicator. Prior heat treatment of the flour before adding water and linoleic acid (Matsuo *et al.*, 1970), adding ascorbic acid (Walsh *et al.*, 1970) or kneading in a vacuum (Burov *et al.*, 1974) prevented carotenoid loss.

Lipoxygenase, in the form of soybean flour, is also added to dough to improve the rheological properties, which results in larger loaf volumes and improved texture (Sullivan *et al.*, 1936). Lipoxygenase also increases the ability of the dough to withstand overmixing. There are several hypotheses to explain this role. The most accepted hypothesis is that oxidation of the thiol groups of gluten by the lipid hydroperoxide formed is responsible for the improvement. This hypothesis is re-enforced in that lipoxygenase can replace the need for use of the oxidizing reagents, bromates and iodates (Koch, 1956; Smith *et al.*, 1957; Tsen & Hlynka, 1962, 1963). Some workers also believe lipoxygenase assists in the release of bound lipid. Lipoxygenase also appears to have a marked effect in retarding bread staling (Frazier, 1979). On the negative side, lipoxygenase added to bread dough does cause some loss of polyunsaturated fatty acids from the steryl esters, triglycerides, diglycerides, galactolipids and phosphatidylcholine (Morrison & Panpaprai, 1975). All classes of lipids were peroxidized in soybean flour-supplemented doughs (Daniels *et al.*, 1970) and there was a decrease in the linoleic acid content of dough triglycerides (Daniels *et al.*, 1968) indicating that lipoxygenases-2 and −3 probably are responsible.

Lipoxygenase is probably responsible for the degradation of chlorophyll in underblanched green beans, green peas and other leguminous foods (Lee & Wagenknecht, 1958; Williams *et al.*, 1986). Whitaker and co-workers have shown that the loss of green color of blanched, frozen English green peas correlates with the amount of lipoxygenase activity left (Williams *et al.*, 1986).

Lipoxygenase has been shown to be responsible for off-odor formation in soybeans (Chen, A. O., pers. comm. 1989) and in English green peas, green beans and corn (Williams *et al.*, 1986; Velasco *et al.*, 1989). These latter results were achieved by adding lipoxygenase purified from these plants back to a blanched puree of the vegetable in the same amount in which it originally occurred. Only addition of lipoxygenase caused the same changes associated with unblanched, pureed vegetables. Peroxidase, catalase or lipase added back did not cause off-odor formation. Oxidized soy phosphatidylcholine is thought to be responsible for bitterness in soy products (Sessa *et al.*, 1976) and in oat and wheat flours (Biermann *et al.*, 1980).

Lipoxygenases are responsible for some of the desirable flavors and aromas of some vegetables and fruits. In tomatoes, the formation of 2-hexenal from polyunsaturated fatty acids is important (Kazeniac & Hall, 1970). In fruits, the initial products of lipid oxidation by lipoxygenase are converted by other enzymes to alcohols and acids, which then form esters (Croteau, 1978).

REFERENCES

Aharony, D., Smith, J. B. & Silver, M. J. (1984). Platelet arachidonate lipoxygenase. In *The Leukotrienes,* ed. L. W. Cahkwin & D. M. Bailey. Academic Press, New York, pp. 103–23.

Allen, J. C. (1969). Octadeca-9,12-dienyl sulfate ion; a new, water soluble substrate for lipoxygenase. *J. Chem. Soc. D,* **16,** 906–7.

Al-Obaidy, H. M. & Siddiqi, A. M. (1981). Properties of broad bean lipoxygenase. *J. Food Sci.,* **46,** 622–5, 629.

Andrawis, A., Pinsky, A. & Grossman, S. (1982). Isolation of soybean lipoxygenase-1 by affinity chromatography. *Phytochemistry,* **21,** 1523–5.

Andre, E. & Hou, K.-W. (1932*a*). The presence of a lipoid oxidase in soy bean, *Glycine soja,* Lieb. *Compt. Rend. Acad. Sci., Paris,* **194,** 645–7.

Andre, E. & Hou, K.-W. (1932*b*). The lipoid oxidases of *Glycine soja* [Sieb] and *Phaseolus vulgaris* L. *Compt. Rend. Acad. Sci., Paris,* **195,** 172–4.

Antis, P. J. P. & Friend, J. (1974). Lipoxygenase in dark-grown *Pisum sativum. Phytochemistry,* **13,** 567–73.

Arens, D., Seilmeier, W., Weber, F., Kloos, G. & Grosch, W. (1973). Purification and properties of a carotene co-oxidizing lipoxygenase from peas. *Biochim. Biophys. Acta,* **327,** 295–305.

Averman, L. Y., Popov, M. P., Dubtsov, G. G. & Samsonov, M. M. (1971). Lipoxygenase activity of wheat grain and flour. *Prikl. Biokhim. Mikrobiol.,* **7,** 678–81.

Axelrod, B. (1974). Lipoxygenases. *Adv. Chem. Ser.,* **136,** 324–48.

Balls, A. K., Axelrod, B. & Kies, M. W. (1943). Soybean lipoxidase. *J. Biol. Chem.,* **149,** 491–504.

Biermann, V. U., Wittman, A. & Grosch, W. (1980). Occurrence of bitter hydroxy fatty acids in oats and wheat. *Fette Seifen Anstrichm.,* **82,** 236–40.

Bohn, R. M. & Haas, L. W. (1928). Cited in *Chemistry and Methods of Enzymes,* 3rd edn, ed. J. B. Sumner & G. F. Somers. Academic Press, New York, p. 311.

Borgeat, P. & Samuelsson, B. (1979). Arachidonic acid metabolism in polymorphonuclear leukocytes: Effects of ionophore A23187. *Proc. Natl. Acad. Sci. U.S.A.,* **76,** 2148–52.

Borgeat, P., Hamberg, M. & Samuelsson, B. (1976). Transformation of arachidonic acid and homo-γ-linolenic acid by rabbit polymorphonuclear leukocytes. *J. Biol. Chem.,* **251,** 7816–20.

Boyd, D. H. J. & Adams, G. A. (1955). Assay method for lipoxidase in animal tissue. *Can. J. Biochem. Physiol.,* **33,** 191–8.

Burov, L. A., Medvedev, G. M. & Ilias, A. (1974). Lipoxygenase, carotenoids, and the color of macaroni products. *Khlebopek. Konditer. Promst.,* **1974,** 25–6.

Byrant, R. W., Bailey, J. M., Schewe, T. & Rapoport, S. M. (1982). Positional specificity of a reticulocyte lipoxygenase. *J. Biol. Chem.,* **257,** 6050–55.

Byrant, R. W., Schewe, T., Bailey, J. M. & Rapoport, S. M. (1985). Leukotriene formation by a purified reticulocyte lipoxygenase enzyme. Conversion of arachidonic acid and 15-hydroperoxyeicosatetraenoic acid to 14,15-leukotriene A_4. *J. Biol. Chem.,* **260,** 3548–55.

Chan, H. W.-S. (1973). Soya-bean lipoxygenase: An iron-containing dioxygenase. *Biochim. Biophys. Acta,* **327,** 32–5.

Chan, H. W.-S. (1987). The mechanism of oxidation. In *Autoxidation of*

Unsaturated Lipids, ed. H. W.-S. Chan. Academic Press, New York and London, pp. 1–16.

Chan, J. T. & Black, H. S. (1976). Distribution of cholesterol-5α,6α-epoxide formation and its metabolism in mouse skin. *J. Invest. Dermat.*, **66,** 112–16.

Chen, A. S.-M. (1983). Purification and characterization of pea lipoxygenase; relationship to quality deterioration in frozen peas. PhD dissertation, University of California, Davis.

Chen, A. O. & Whitaker, J. R. (1986). Purification and characterization of a lipoxygenase from immature English peas. *J. Agric. Food Chem.*, **34,** 203–11.

Chisom, G. W. (1985). Soy lipoxygenase. In *Flavor Chemistry of Fats and Oils*, ed. D. B. Min & T. H. Smouse. American Oil Chemists' Society, pp. 175–87.

Christopher, J., Pistorius, E. & Axelrod, B. (1970). Isolation of an isozyme of soybean lipoxygenase. *Biochim. Biophys. Acta,* **198,** 12–19.

Christopher, J. P., Pistorius, E. K. & Axelrod, B. (1972*a*). Isolation of a third isozyme of soybean lipoxygenase. *Biochim. Biophys. Acta,* **284,** 54–62.

Christopher, J. P., Pistorius, E. K., Regnier, F. E. & Axelrod, B. (1972*b*). Factors influencing the positional specificity of soybean lipoxygenase. *Biochim. Biophys. Acta,* **289,** 82–7.

Colas, A. & Chargelegue, A. (1974). Lipasic and lipoxygenasic activity of wheats and products of milling. *Ann. Technol. Agric.*, **23,** 323–34.

Coleman, M. H. (1978). A model system for the formation of N-nitrosopyrrolidine in grilled or fried bacon. *J. Food Technol.*, **13,** 55–69.

Cosby, E. L. & Sumner, J. B. (1945). Lipoxidase. *Arch. Biochem.*, **8,** 259–64.

Craig, F. N. (1936). A fat oxidation system in *Lupinus albus. J. Biol. Chem.*, **114,** 727–46.

Croteau, R. (1978). Biosynthesis of flavor components: Volatile carbonyl compounds and monoterpenoids. In *Postharvest Biology and Biotechnology*, ed. H. O. Hultin & M. Milner. Food and Nutrition Press, Westport, CT, pp. 400–432.

Daniels, N. W. R., Richmond, J. W., Russell Eggitt, P. W. & Coppock, J. B. M. (1968). Studies on the lipids of flour. IV. Factors affecting lipid binding in breadmaking. *J. Sci. Food Agric.*, **20,** 129–36.

Daniels, N. W. R., Wood, P. S., Russell Eggitt, P. W. & Coppock, J. B. M. (1970). Lipids of flour. *V*. Effect of air on lipid binding. *J. Sci. Food Agric.*, **21,** 377–84.

Davis, B. J. (1964). Disc electrophoresis. II. Method and application to human serum proteins. *Ann. N. Y. Acad. Sci.*, **121,** 404–27.

Dixon, R. A. F., Jones, R. E., Diehl, R. E., Bennett, C. D., Kargman, S. & Rouzer, C. A. (1988). Cloning of the cDNA for human 5-lipoxygenase. *Proc. Natl. Acad. Sci. USA,* **85,** 416–20.

Dolev, A., Rohwedder, W. K. & Dutton, H. J. (1967). Mechanism of lipoxidase reaction. I. Specificity of hydroperoxidation of linoleic acid. *Lipids,* **2,** 28–32.

Douillard, R. & Bergeron, E. (1979). Foliar lipoxygenase activity: Chloroplast localization and age variations in young plants. *Physiol. Veg.*, **17,** 457–76.

Douillard, R. & Bergeron, E. (1981). Lipoxygenase activities of young wheat leaves. *Physiol. Plant.*, **51,** 335–8.

Egmond, M. R., Vliegenthart, J. F. G. & Boldingh, J. (1972). Stereospecificity of the hydrogen abstraction at carbon atom n-8 in the oxygenation of linoleic acid by lipoxygenases from corn germs and soya beans. *Biochem. Biophys. Res. Commun.*, **48,** 1055–60.

Egmond, M. R., Brunori, M. & Fasella, P. M. (1976). The steady-state kinetics of the oxygenation of linoleic acid catalysed by soybean lipoxygenase. *Eur. J. Biochem.*, **61,** 93–100.

Erikkson, C. E. (1979). Review of the biosynthesis of volatiles in fruits and vegetables since 1975. In *Progress in Flavour Research,* ed. D. G. Land & H. E. Nursten. Applied Science Publishers, London, pp. 159–74.

Eriksson, C. E. (1987). Oxidation of lipids in food systems. In *Autoxidation of Unsaturated Lipids,* ed. H. W.-S. Chan. Academic Press, New York and London, pp. 207–31.

Eriksson, C. E. & Svensson, S. G. (1970). Lipoxygenase from peas, purification and properties of the enzyme. *Biochim. Biophys. Acta,* **198,** 449–59.

Faubion, J. M. & Hoseney, R. C. (1981). Lipoxygenase: Its biochemistry and role in breadmaking. *Cereal Chem.*, **58,** 175–80.

Finazzi-Agro, A., Avigliano, L., Veldink, G. A., Vliegenthart, J. F. G. & Boldingh, J. (1973). The influence of oxygen on the fluorescence of lipoxygenase. *Biochim. Biophys. Acta,* **326,** 462–70.

Firn, R. D. & Friend, J. (1972). Enzymatic production of the plant growth inhibitor, xanthoxin. *Planta,* **103,** 263–6.

Fleming, J., Thiele, B. J., Chester, J., O'Prey, J., Janetzki, S., Aitken, A., Anton, I. A., Rapoport, S. M. & Harrison, P. R. (1989). The complete sequence of the rabbit erythroid cell-specific 15-lipoxygenase mRNA: Comparison of the predicted amino acid sequence of the erythrocyte lipoxygenase with other lipoxygenases. *Gene,* **79,** 181–8.

Frazier, P. J. (1979). Lipoxygenase action and lipid binding in breadmaking. *Bakers Dig.*, **53**(6), 8–10, 12–13, 16, 18, 20, 29.

Fritz, G. & Beevers, H. (1955). Lipoxygenase and the oxygen absorption of homogenates from corn seedlings. *Arch. Biochem. Biophys.*, **55,** 436–46.

Fritz, G. J., Miller, W. G., Burris, R. H. & Anderson, L. (1958). Direct incorporation of molecular oxygen into organic material by respiring corn seedlings. *Plant Physiol.*, **33,** 159–61.

Führling, D. (1975). Lipoxygenase and linoleic acid hydroperoxidase isomerase in barley (*Hordeum vulgare*). PhD thesis, University of Berlin.

Fukuba, H. (1952). The optimum pH of lipoxidase. *J. Agr. Chem. Soc. Japan,* **26,** 170–72.

Galliard, T., Hulme, A. C., Rhodes, M. J. C. & Wooltorton, L. S. C. (1968). Enzymic conversion of linolenic acid to ethylene by extracts of apple fruits. *FEBS Letters,* **1,** 283–6.

Gardner, H. W. (1975). Isolation of a pure isomer of linoleic acid hydroperoxide. *Lipids,* **10,** 248–52.

Gardner, H. W. (1983). Effect of lipid hydroperoxides on food components. In *Xenobiotics in Foods and Feeds,* ed. J. W. Finley & D. E. Schwass.

ACS Symp. Ser. 234, American Chemical Society, Washington, DC, pp. 63–84.
Gardner, H. W. (1988). Lipoxygenase pathways in cereals. In *Advances in Cereal Science and Technology,* ed. Y. Pomeranz. Volume IX, American Association of Cereal Chemists, St Paul, MN, pp. 161–215.
Gardner, H. W. (1989). Soybean lipoxygenase-1 enzymatically forms both (9S)- and (13S)-hydroperoxides from linoleic acid by a pH-dependent mechanism. *Biochim. Biophys. Acta,* **1001,** 274–81.
Gardner, H. W. & Weisleder, D. (1970). Lipoxygenase from *Zea mays*: 9-D-hydroperoxy-*trans*-10,*cis*-12-octadecadienoic acid from linoleic acid. *Lipids,* **5,** 678–83.
Goetze, A. M., Fayer, L., Bouska, J., Bornemeier, D. & Carter, G. W. (1985). Purification of a mammalian 5-lipoxygenase from rat basophilic leukemia cells. *Prostaglandins,* **29,** 689–701.
Goldstein, A. H., Anderson, J. O. & McDaniel, R. G. (1980). Cyanide-insensitive and cyanide-sensitive O_2 uptake in wheat. I. Gradient-purified mitochondria. *Plant Physiol.,* **66,** 488–93.
Groot, J. J. M. C. de, Veldink, G. A., Vliegenthart, J. F. G., Boldingh, J., Wever, R. & van Gelder, B. F. (1975). Demonstration by EPR spectroscopy of the functional role of iron in soybean lipoxygenase-1. *Biochim. Biophys. Acta,* **337,** 71–9.
Grosch, W. (1987). Reactions of hydroperoxides. Products of low molecular weight. In *Autoxidation of Unsaturated Lipids,* ed. H. W.-S. Chan. Academic Press, New York and London, pp. 95–139.
Gund, P. & Shen, T. Y. (1977). A model for the prostaglandin synthetase cyclooxygenation site and its inhibition by antiinflammatory aryl acetic acids. *J. Med. Chem.,* **20,** 1146–52.
Guss, P. L., Richardson, T. & Stahmann, M. A. (1967). The oxidation-reduction enzymes of wheat. III. Isoenzymes of lipoxidase in wheat fractions and soybeans. *Cereal Chem.,* **44,** 607–10.
Guss, P. L., Macko, V., Richardson, T. & Stahmann, M. A. (1968). Lipoxygenase in early growth of wheat. *Plant Cell Physiol.,* **9,** 415–22.
Haas, L. W. & Bohn, R. M. (1934). Bleaching bread dough. US Patents 1,957,33–37.
Haining, J. L. & Axelrod, B. (1958). Induction period in the lipoxidase-catalyzed oxidation of linoleic acid and its abolition by substrate peroxide. *J. Biol. Chem.,* **232,** 193–202.
Hale, S. A., Richardson, T., Von Elbe, J. H. & Hagedorn, D. J. (1969). Isoenzymes of lipoxidase. *Lipids,* **4,** 209–15.
Hamberg, M. (1971). Steric analysis of hydroperoxides formed by lipoxygenase oxygenation of linoleic acid. *Anal. Biochem.,* **43,** 515–26.
Hamberg, M. & Hamberg, G. (1980). On the mechanism of the oxygenation of arachidonic acid by human platelet lipoxygenase. *Biochem. Biophys. Res. Commun.,* **95,** 1090–97.
Hamberg, M. & Samuelsson, B. (1967). On the specificity of the oxygenation of unsaturated fatty acids catalyzed by soybean lipoxidase. *J. Biol. Chem.,* **242,** 5329–35.
Hamberg, M. & Samuelsson, B. (1974). Prostaglandin endoperoxides. III.

Novel transformations of arachidonic acid in human platelets. *Proc. Natl. Acad. Sci. U.S.A.*, **71**, 3400–404.

Hansson, G., Malmsten, C. & Raadmark, O. (1983). The leukotrienes and other lipoxygenase products. In *Prostaglandins and Related Substances*, ed. C. R. Pace-Asciak & E. Granström. Elsevier, Amsterdam, pp. 127–69.

Hauge, S. M. (1935). Evidence of enzymatic destruction of the vitamin A value of alfalfa during the curing process. *J. Biol. Chem.*, **108**, 331–6.

Hauge, S. M. & Aitkenhead, W. (1931). The effect of artificial drying upon the vitamin A content of alfalfa. *J. Biol. Chem.*, **93**, 657–65.

Haydar, M. & Hadziyev, J. (1973). A study of lipoxygenase in pea seeds and seedlings. *J. Sci. Food Agric.*, **24**, 1039–53.

Heimann, W. & Klaiber, V. (1977). On lipoxygenase and enzymes which decompose linoleic acid hydroperoxides in rye. *Z. Lebensm. Unters. Forsch.*, **165**, 131–6.

Heimann, W., Dresen, P. & Schreier, P. (1973*a*). The lipoxygenase–'lipoperoxidase'-system in cereals. Isolation of two protein complexes with lipoxygenase and linoleic acid hydroperoxide activity from oats and soybeans. *Z. Lebensm. Unters. Forsch.*, **152**, 147–51.

Heimann, W., Dresen, P. & Klaiber, V. (1973*b*). Formation and decomposition of linoleic acid hydroperoxides in cereals. Quantitative determination of the reaction products. *Z. Lebensm. Unters. Forsch.*, **153**, 1–5.

Ho, P. P. K., Walters, C. P. & Sullivan, H. R. (1977). A particulate arachidonate lipoxygenase in human blood platelets. *Biochem. Biophys. Res. Commun.*, **76**, 398–405.

Holman, R. T. (1947). Crystalline lipoxidase. II. Lipoxidase activity. *Arch. Biochem. Biophys.*, **15**, 403–13.

Holman, R. T. (1949). Spectrophotometric studies of the oxidation of fats. VIII. Coupled oxidation of carotene. *Arch. Biochem. Biophys.*, **21**, 51–7.

Holman, R. T., Egwim, P. O. & Christie, W. W. (1969). Substrate specificity of soybean lipoxygenase. *J. Biol. Chem.*, **244**, 1149–51.

Ida, S., Masaki, Y. & Morita, Y. (1983). The isolation of multiple forms and product specificity of rice lipoxygenase. *Agric. Biol. Chem.*, **47**, 637–41.

Kazeniac, S. J. & Hall, R. M. (1970). Flavor chemistry of tomato volatiles. *J. Food Sci.*, **35**, 519–30.

Kies, M. W. (1947). Complex nature of soybean lipoxidase. *Fed. Proc.*, **6**, 267.

Kikugawa, K., Tsukuda, K. & Kurechi, T. (1980). Studies on peroxidized lipids. I. Interaction of malondialdehyde with secondary amines and its relevance to nitrosamine formation. *Chem. Pharm. Bull.*, **28**, 3323–31.

Klein, B. P. (1976). Isolation of lipoxygenase from split pea seeds, snap beans, and peas. *J. Agric. Food Chem.*, **24**, 938–42.

Koch, R. B. (1956). Mechanism of fat oxidation. *Bakers Dig.*, **30**, 48.

Koch, R. B., Stern, B. & Ferrari, C. G. (1958). Linoleic acid and trilinolein as substrates for soybean lipoxygenase(s). *Arch. Biochem. Biophys.*, **78**, 165–79.

Kühn, H., Wiesner, R., Schewe, T. & Rapoport, S. M. (1983). Reticulocyte lipoxygenase exhibits both n-6 and n-9 activities. *FEBS Letters*, **153**, 353–6.

Kühn, H., Heydeck, D., Wiesner, R. & Schewe, T. (1985). The positional specificity of wheat lipoxygenase; The carboxylic group as signal for the recognition of the site of the hydrogen removal. *Biochim. Biophys. Acta*, **830**, 25–9.

Kühn, H., Schewe, T. & Rapoport, S. M. (1986). The stereochemistry of the reactions of lipoxygenases and their metabolites. Proposed nomenclature of lipoxygenases and related enzymes. *Adv. Enzymol.*, **58**, 273–311.

Kurechi, T., Kikugawa, K. & Ozawa, M. (1980). Effect of malondialdehyde on nitrosamine formation. *Food Cosmet. Toxicol.*, **18**, 119–22.

Lee, F. A. & Wagenknecht, A. C. (1958). Enzyme action and off-flavor in frozen peas. II. The use of enzymes prepared from garden peas. *Food Research*, **23**, 584–90.

Leu, K. (1974). Formation of isomeric hydroperoxides from linoleic acid by lipoxygenase. *Lebensm. Wiss. Technol.* **7**, 82–85; *Chem. Abstr.*, **81**, 87115k (1974).

Lieberman, M. & Mapson, L. W. (1964). Genesis and biogenesis of ethylene. *Nature*, **204**, 343–5.

Lillard, D. A. (1978). Chemical changes involved in the oxidation of lipids in foods. *ACS Symp. Ser.*, **75**, 68–80.

Lim, M. H., Velasco, P. J., Pangborn, R. M. & Whitaker, J. R. (1989). Enzymes involved in off-aroma formation in broccoli. *ACS Symp. Ser.*, **405**, 72–83.

Matsumoto, T., Funk, C. D., Radmark, O., Hoog, J.-O., Jornvall, H. & Samuelsson, B. (1988). Molecular cloning and amino acid sequence of human 5-lipoxygenase. *Proc. Natl. Acad. Sci. USA*, **85**, 26–30.

Matsuo, R. R., Bradley, J. W. & Irvine, G. N. (1970). Studies on pigment destruction during spaghetti processing. *Cereal Chem.*, **47**, 1–5.

Morrison, W. R. & Panpaprai, R. (1975). Oxidation of free and esterified linoleic and linolenic acids in bread doughs by wheat and soya lipoxygenases. *J. Sci. Food Agric.*, **26**, 1225–36.

Navez, A. E. (1929). Cited in Craig, J. B. (1936). A fat oxidation system in *Lupinus albus*. *J. Biol. Chem.*, **114**, 727–46.

Nicolas, J., Autran, M. & Drapron, R. (1982). Purification and some properties of wheat germ lipoxygenase. *J. Sci. Food Agric.*, **33**, 365–72.

Nugteren, D. H. (1975). Arachidonate lipoxygenase in blood platelets. *Biochim. Biophys. Acta*, **380**, 299–307.

Os, C. P. A. van, Vente, M. & Vliegenthart, J. F. G. (1979*a*). A NMR shift method for determination of the enantiomeric composition of hydroperoxides formed by lipoxygenase. *Biochim. Biophys. Acta*, **574**, 103–11.

Os, C. P. A. van, Rijke-Schilder, G. P. M. & Vliegenthart, J. F. G. (1979*b*). 9-LR-Linoleyl hydroperoxide, a novel product from the oxygenation of linoleic acid by type-2 lipoxygenases from soybeans and peas. *Biochim. Biophys. Acta*, **575**, 479–84.

Peng, S. K., Taylor, C. B., Tham, P. & Mikkelson, (1978). Effect of auto-oxidation products from cholesterol on aortic smooth muscle cells. *Arch. Pathol. Lab. Med.*, **102**, 57–61.

Pistorius, E. K. & Axelrod, B. (1973). Ferric iron as a cofactor of lipoxygenase. *Fed. Proc.*, **32**, 544.

Pistorius, E. K. & Axelrod, B. (1974). Iron, an essential component of lipoxygenase. *J. Biol. Chem.*, **249,** 3183–6.

Rapoport, S. M., Schewe, T., Wiesner, R., Halangk, W., Ludwig, P., Janicke-Höhne, M., Tannert, C., Hiebsch, C. & Klatt, D. (1979). The lipoxygenase of reticulocytes. Purification, characterization and biological dynamics of the lipoxygenase. Its identity with the respiratory inhibitors of the reticulocytes. *Eur. J. Biochem.*, **96,** 545–61.

Reynolds, P. A. (1982). Characterization of pea seed lipoxygenase enzymes. *Diss. Abst. Int. B.*, **42,** 3623.

Roza, M. & Francke, A. (1973). Soybean lipoxygenase: An iron-containing enzyme. *Biochim. Biophys. Acta*, **327,** 24–31.

Schewe, T., Halangk, W., Hiebsch, C. & Rapoport, S. M. (1975). Lipoxygenase in rabbit reticulocytes which attacks phospholipids and intact mitochondria. *FEBS Letters*, **50,** 149–52.

Schewe, T., Wiesner, R. & Rapoport, S. M. (1981). Lipoxygenase from rabbit reticulocytes. *Methods Enzymol.*, **71,** 430–41.

Schewe, T., Rapoport, S. M. & Kühn, H. (1986). Enzymology and physiology of reticulocyte lipoxygenase: Comparison with other lipoxygenases. *Adv. Enzymol.*, **58,** 191–217.

Schormüller, J., Weber, J., Höxer, B. & Grosch, W. (1969). Guajacollinolsäurehydroperoxid- oxydoreduktase in sojabohnen. Versuche zur abtrennung des enzyms von der H_2O_2-peroxydase und der lipoxygenase. *Z. Lebensm. Unters. Forsch.*, **139,** 357–64.

Schroeder, D. H. (1968). Some properties of soybean lipoxygenase, PhD thesis, Purdue University.

Sessa, D. J., Warner, K. & Rackis, J. J. (1976). Oxidized phosphatidylcholines from defatted soybean flakes taste bitter. *J. Agric. Food Chem.*, **24,** 16–21.

Shamberger, R. J. (1977). Antioxidants and cancer. VII. Presence of malonaldehyde in beef and other meats and its epidemiological significance. *Trace Subst. Environ. Health*, **11,** 36–43.

Shamberger, R. J. (1980). Is peroxidation important in the cancer process? In *Autoxidation in Food and Biological Systems*, ed. M. G. Simic & M. Karel. Plenum Press, New York, pp. 639–49.

Shastry, B. S. & Raghavendra Rao, M. R. (1975). Studies on the lipoxygenase from rice bran. *Cereal Chem.*, **52,** 597–603.

Shibata, D., Steczko, J., Dixon, J. E., Hermodson, M., Yazdanparast, R. & Axelrod, B. (1987). Primary structure of soybean lipoxygenase-1. *J. Biol. Chem.*, **262,** 10080–85.

Shibata, D., Steczko, J., Dixon, J. E., Andrews, P. C., Hermodson, M. & Axelrod, B. (1988). Primary structure of soybean lipoxygenase L-2. *J. Biol. Chem.*, **263,** 6816–21.

Siddiqi, A. M. & Tappel, A. L. (1956). Catalysis of linoleate oxidation by pea lipoxidase. *Arch. Biochem. Biophys.*, **60,** 91–9.

Smith, D. E., Buren, J. P. van & Andrews, J. S. (1957). Some effects of oxygen and fat upon the physical and chemical properties of flour doughs. *Cereal Chem.*, **34,** 337–49.

Smith, L. L. & Kulig, M. J. (1975). Sterol metabolism. XXXIV. On the derivation of carcinogenic sterols from cholesterol. *Cancer Biochem. Biophys.*, **1,** 79–84.
Smith, W. L. & Lands, W. E. M. (1972). Oxygenation of unsaturated fatty acids by soybean lipoxygenase. *J. Biol. Chem.*, **247,** 1038–47.
Sullivan, B., Near, C. & Foley, G. H. (1936). The role of the lipids in relation to flour quality. *Cereal Chem.*, **13,** 318–31.
Sumner, R. J. (1943). Lipoid oxidase studies. Method for the determination of lipoxidase activity. *Ind. Eng. Chem., Anal. Ed.*, **15,** 14–15.
Sumner, J. B. & Sumner, R. J. (1940). The coupled oxidation of carotene and fat by carotene oxidase. *J. Biol. Chem.*, **134,** 531–3.
Tappel, A. L. (1953). Linoleate oxidation catalysts occurring in animal tissues. *Food Research,* **18,** 104–8.
Tappel, A. L., Boyer, P. D. & Lundberg, W. O. (1952). The reaction mechanism of soy bean lipoxidase. *J. Biol. Chem.*, **199,** 267–81.
Tauber, H. (1940). Unsaturated fat oxidase. *J. Amer. Chem. Soc.*, **62**, 2251.
Taylor, C. B., Peng, S.-K., Werthessen, N. T., Tham, P. & Lee, K. T. (1979). Spontaneously occurring angiotoxic derivatives of cholesterol. *Amer. J. Clin. Nutr.*, **32,** 40–57.
Theorell, H., Holman, R. T. & Åkeson, A. (1947). A note on the preparation of crystalline soybean lipoxidase. *Arch. Biochem.*, **14,** 250–52.
Tsen, C. C. & Hlynka, I. (1962). The role of lipids in oxidation of doughs. *Cereal Chem.*, **39,** 209–19.
Tsen, C. C. & Hlynka, I. (1963). Flour lipids and oxidation of sulfhydryl groups in dough. *Cereal Chem.*, **40,** 145–53.
Velasco, P. J., Lim, M. H., Pangborn, R. M. & Whitaker, J. R. (1989). Enzymes responsible for off-flavor and off-aroma in blanched and frozen-stored vegetables. *Biotech. Appl. Biochem.*, **11,** 118–27.
Vernooy-Gerritsen, M., Bos, A. L. M., Veldink, G. A. & Vliegenthart, J. F. G. (1983). Localization of lipoxygenases-1 and -2 in germinating soybean seeds by an indirect immunofluorescence technique. *Plant Physiol.*, **73,** 262–7.
Vernooy-Gerritsen, M., Leunissen, J. L. M., Veldink, G. A. & Vliegenthart, J. F. G. (1984). Intracellular localization of lipoxygenases-1 and -2 in germinating soybean seeds by indirect labeling with protein A-colloidal gold complexes. *Plant Physiol.*, **76,** 1070–79.
Vick, B. A. & Zimmerman, D. C. (1982). Levels of oxygenated fatty acids in young corn and sunflower plants. *Plant Physiol.*, **69,** 1103–8.
Walker, G. C. (1963). The formation of free radicals during the reaction of soybean lipoxidase. *Biochem. Biophys. Res. Commun.*, **13,** 431–4.
Walsh, D. E., Youngs, V. L. & Gilles, K. A. (1970). Inhibition of durum wheat lipoxidase with L-ascorbic acid. *Cereal Chem.*, **47,** 119–25.
Wang, T.-T. & Yang, S. F. (1987). The physiological role of lipoxygenase in ethylene formation from 1-aminocyclopropane-1-carboxylic acid in oat leaves. *Plant Physiol.*, **84,** 277–81.
Wardale, D. A. & Galliard, T. (1975). Subcellular localization of lipoxygenase and lipolytic acyl hydrolase enzymes in plants. *Phytochemistry,* **14,** 2323–9.

Wardale, D. A. & Galliard, T. (1977). Further studies on the subcellular localization of lipid-degrading enzymes. *Phytochemistry,* **16,** 333–8.

Whitaker, J. R. (1972). *Principles of Enzymology for the Food Sciences.* Marcel Dekker, New York.

Williams, D. C., Lim, M. H., Chen, A. O., Pangborn, R. M. & Whitaker, J. R. (1986). Blanching of vegetables for freezing—which indicator enzyme to choose. *Food Technol.,* **40,** 130–40.

Wooltorton, L. S. C., Jones, J. D. & Hulme, A. C. (1965). Genesis of ethylene in apples. *Nature,* **207,** 999–1000.

Yabuuchi, S. (1976). Occurrence of a new lipoxygenase isoenzyme in germinating barley embryos. *Agric. Biol. Chem.,* **40,** 1987–92.

Yamamoto, A. & Fujii, Y. (1980). Partial purification and study of some properties of rice germ lipoxygenase. *Agric. Biol. Chem.,* **44,** 443–5.

Yang, S. F. (1985). Biosynthesis of ethylene. In *Current Topics in Plant Biochemistry and Physiology,* Vol. 4, ed. D. P. Randall. Interdisciplinary Plant Biochemistry and Physiology Program, University of Missouri–Columbia, Missouri, pp. 126–38.

Yenofsky, R. L., Fine, M. & Liu, C. (1988). Isolation and characterization of a soybean (*Glycine max*) lipoxygenase-3 gene. *Mol. Gen. Genet.,* **211,** 215–22.

Yoshimoto, T., Miyamoto, Y., Ochi, K. & Yamamoto, S. (1982). Arachidonate 12-lipoxygenase of porcine leukocyte with activity for 5-hydroxyeicosatetraenoic acid. *Biochim. Biophys. Acta,* **713,** 638–46.

6

Polyphenol Oxidase

Jerzy Zawistowski,[a,c] Costas G. Biliaderis,[a] & N. A. Michael Eskin[b]

[a]Department of Food Science, [b]Department of Foods & Nutrition, [c]Agri-Food Biotechnology Laboratory, Manitoba Research Council, University of Manitoba, Manitoba, Canada

INTRODUCTION

Polyphenol oxidase (PPO) has received continuous attention from food chemists and processors since its discovery over 90 years ago. This is because it is involved in the enzymatic browning (Eskin, 1990) of many edible plant products, especially fruits and vegetables, mushrooms, as well as crustaceans such as shrimps, lobsters and crabs.

Different names have been associated with this enzyme including tyrosinase, cresolase, catecholase, diphenolase and phenolase. Of these, it is PPO that generally reflects the ability of this enzyme to utilise many different phenolic compounds as substrates. The Commission on Enzymes of the International Union of Biochemistry (IUB) revised the nomenclature and placed PPO into two key categories as EC 1.14.18.1-monophenol monooxygenase, also known as tyrosinase (this category was previously reserved for EC 1.10.3.1 *o*-diphenol oxidase, and EC 1.10.3.2 *p*-diphenol oxidase or laccase), and EC 1.10.3.2 *o*-diphenol:O_2 oxidoreductase (Mayer, 1987). This classification differentiates only between two reactions, cresolase and catecholase, of the same enzyme (Mayer, 1987). In this chapter we will be using PPO to reflect both IUB entries. We have excluded other enzymes such as laccase (EC 1.10.3.1) and peroxidase (EC 1.11.1.7), which may be considered PPOs as they may convert phenols to *o*-diphenols and *o*-diphenols to *o*-quinones.

HISTORICAL

Early research on enzymatic browning identified two types of activity referred to as tyrosinase and laccase. Laccase differed from tyrosinase or PPO by its ability to oxidise *p*-diphenols and is not inhibited by carbon monoxide (Mayer & Harel, 1979). Laccase has been identified in some higher plants, and a number of fungi. The first report of the formation of a blue pigment by an extract of mushroom, *Boletus luciferus,* was by Schoenbein in 1856. Some 30 years later, Yoshida (1886), observed the presence of laccase in the latex of the lac tree. This enzyme was also identified in several mushroom varieties by Bertrand (1896). Subsequent work by Bourquelot & Bertrand (1895, 1896) identified several mushroom varieties, *Russula fortens* and *R. nigricans,* which turned red and finally dark brown or black during the course of oxidation. A crystalline substance separated out of the juice of *R. nigricans* which formed a melanin-like product in the presence of the oxidase system from the juice. This substance was identified by Bertrand (1896) as the amino acid tyrosine. Since tyrosine was inactive towards laccase, Bertrand (1896) was convinced he had identified a new oxidase system, later named tyrosinase, capable of catalysing the aerobic oxidation of a variety of monohydroxy phenolic compounds including *p*-hydroxyphenyl amine, *p*-hydroxyphenyl methyl amine, *p*-cresol and phenol (Bertrand, 1908).

The first serious study of this enzyme followed the large-scale production of potato tyrosinase by Kubowitz (1938). It was the first enzyme purified by affinity chromatography (Lerman, 1953). Tyrosinase was also the first enzyme recognized as being able to incorporate molecular oxygen into an organic (phenolic) molecule (Mason *et al.*, 1955), and one of the first examples of a suicide enzyme inactivation (Abeles & Maycock, 1976). Considerable research has been published on PPOs since then, as discussed in several recent reviews (Vámos-Vigyázó, 1981; Robb, 1984; Mayer, 1987; Eskin, 1990).

OCCURRENCE AND LOCALISATION

PPO is widely distributed in the plant and animal kingdom. This enzyme is not restricted to any particular organ in plants but in higher animals it is confined to skin, hairs, feathers and eyes (Robb, 1984). Thus foods that may contain PPO are derived mainly from plants, mushrooms and crustaceans.

The distribution of PPO in edible plant products has been reviewed by several authors (Mayer & Harel, 1979; Vámos-Vigyázó, 1981; Mayer, 1987). Furthermore, this enzyme has been recently isolated and studied in a number of food commodities not previously reported including wheat flour (Marsh & Galliard, 1986) and pasta (Temesvari, 1986), Jerusalem artichoke (Zawistowski *et al.*, 1986), chilli (Bajaj & Kaur, 1986), sapota (Prabha & Patwardhan, 1986), carambola or starfruit (Tengku *et al.*, 1986), Japanese persimmon (Taira *et al.*, 1987), legumes (Udayasekhara Rao & Deosthale, 1987), cocoa beans (Kuznetsova *et al.*, 1987; Kim, 1988), hassaku fruit (Fujita & Tono, 1988), sago palm (Okamoto *et al.*, 1988), cherimoya (Sauchez de Medina *et al.*, 1987; Martinez-Cayuela *et al.*, 1988*a*, 1989), soybean (Toiguchi *et al.*, 1989), cucumber (Miller *et al.*, 1990), wild rice (Owusu-Ansah, 1989), strawberry (Wesche-Ebeling & Montgomery, 1990), Shiitake mushrooms (Yamaguchi *et al.*, 1988) as well as pink and white shrimps (Simpson *et al.*, 1987, 1988; Al-Jassir, 1989; Chu & Shaw, 1989; Shaw *et al.*, 1989), spiny lobster (Ferrer, 1988; Ferrer *et al.*, 1989*a,b*) and golden crab (Marshall *et al.*, 1984).

Of PPOs isolated from a great number of fruits and vegetables, grapes (Ruiz-Hernandez, 1983; Sapis *et al.*, 1984; Cheynier *et al.*, 1988; Chin & Chen, 1988; Lamikanra, 1988; Sánchez-Ferrer, 1988, 1989; Valero *et al.*, 1988), apples (Prabha & Patwardhan, 1986;

Coseteng & Lee, 1987; Czapski *et al.*, 1988; Janovitz-Klapp *et al.*, 1989; Sapers *et al.*, 1989*b*; Zemel *et al.*, 1990), and potato enzymes (Batistuti & Lourenco 1985; Randhawa *et al.*, 1986; Matheis, 1987*a,b*; Ha & Lee, 1988; Sapers *et al.*, 1989*a*) continue to receive the highest attention because of their contribution to the quality of wine, juice or processed potato tubers, respectively.

The level of PPO in plants is dependent on the species, cultivar, maturity and age (Vámos-Vigyázó, 1981). Differences in the PPO content have been found between cultivars of apples (Walker, 1962; Janovitz-Klapp *et al.*, 1989), grapes (Sapis *et al.*, 1983), avocado (Sharon-Raber & Kahn, 1983), eggplants (Flick *et al.*, 1977), yam tubers (Anosike & Ayaebene, 1981), sweet potato (Walter & Purcell, 1980) and potato (Sapers *et al.*, 1989*a*). The level of this enzyme in some potato cultivars, such as Atlantic, has been reported to be almost four times lower than for Russet Burbank (Sapers *et al.*, 1989*a*). Food processors, recently deprived of the application of sulphites to inhibit enzymatic browning, are now interested in varieties with lower PPO content, which are less susceptible to this type of discoloration.

The PPO activity is very low in young plants, often undetectable. This is because PPO exists as a particulate enzyme tightly bound to a membrane. In green sugar beet leaves, PPO is present in the chloroplast and linked at various points to some membrane structure through peptide bonds of non-aromatic amino acids (Mayer, 1966). It has been shown that breakage of this structure is necessary to release PPO (Mayer & Friend, 1960). In freshly harvested apples, PPO exists almost exclusively as two particulate fractions in the chloroplasts and mitochondria (Harel *et al.*, 1964). The maximum activity of the particulate forms of apple PPO occurs prior to the fruit reaching full maturity and decreases in ripe apples with a concomitant increase in the soluble form of PPO (Harel *et al.*, 1966). A considerable increase in soluble PPO form activity has also been reported during the ripening of olives (Ben-Shalom *et al.*, 1977).

The level of soluble forms of PPO increases with the onset of the senescence of plants (Meyer & Biehl, 1980; Trémolières & Bieth, 1984), and aging of mushrooms (Goodenough, 1978) as well as following wounding (Bastin, 1968). The biochemical mechanisms associated with this phenomenon remain to be elucidated, although several workers suggested that changes in total PPO activity during aging or after injury may be correlated with de-novo protein synthesis. Goodenough (1978) showed that increased PPO activity occuring during post-harvest storage of mushrooms can be prevented by inhibitors of protein synthesis. In addition, the increase in PPO level after cutting Jerusalem artichoke tubers could be prevented by inhibition of RNA synthesis required for the production of this enzyme (Bastin & Ünlüer, 1972). Contradictory evidence provided by Meyer & Biehl (1980) suggests that the increase of PPO activity in spinach leaves is not due to de-novo synthesis but to the release of the active enzyme from a thylakoid-bound latent form caused by the aging-related loss of structural integrity of the membrane.

Although PPO is not restricted to any particular part of the plant, analysis of Jerusalem artichoke tubers showed the outer peels contained the majority of PPO activity, while the middle core exhibited only 10% of the total activity (Zawistowski *et al.*, 1986). The highest enzymatic activity has been reported in the peels of grapes (Ivanov, 1966), some apple cultivars (Voigt & Noske, 1966), cucumber (Miller *et al.*, 1990) and other fruits (Vámos-Vigyázó, 1981). Furthermore, the highest activity of mushroom PPO has been shown to be localised in the stalk, gills and cup epidermis (Moore *et al.*, 1988). Epidermis or cuticle was also found to be the primary location for PPO in lobsters (Ferrer *et al.*, 1989*b*). It is also important to note that the distribution of PPO in plants, mushrooms and crustaceans varies with the species, age and maturity (Walker, 1975; Vámos-Vigyázó, 1981; Moore & Flurkey, 1989; Ferrer *et al.*, 1989*b*). Prabha & Patwardhan (1986) reported that during fruit ripening the activity of PPO generally increased in the peel tissues of apple, avocado and mango, whereas it decreased in the pulp of apples, increased in the pulp of avocado and was absent in mango pulp.

PPO is an intracellular enzyme reported to be located in chloroplasts, mitochondria, microsomes, peroxisomes, and cellular plasma (Kato *et al.*, 1976; Mayer & Harel, 1979; Henry *et al.*, 1982). Recent cytochemical studies, however, provide evidence that PPO is solely a plastid enzyme, localised in a diverse series of plastids: root plastids, potato tuber amyloplasts, epidermal plastids, carrot tissue culture plastids, etioplasts, chromoplasts as well as chloroplasts of many different species (Vaughn & Duke, 1984*a*). Moreover, studies on PPO with tentoxin- (a tetrapeptide that specifically affects plastid development and induces severe chlorosis) treated plants revealed a lack of PPO activity by means of cytochemical, electrophoretic and spectrophotometric procedures (Vaughn & Duke, 1981*a*, 1984*b*; Duke & Vaughn, 1982).

Biological Role for PPO

The role of PPO in plants is ambiguous, because in healthy green tissues this enzyme exists in the latent form, which hampers the study *in vivo*. Several hypotheses, however, have been put forward regarding the functions of PPO. Early studies by James (1953) suggested PPO was a terminal oxidase in respiration. This hypothesis was supported by findings that latent PPO is bound to the thylakoid membrane where photosynthesis takes place. It was later postulated that PPO has a role in pseudocyclic phosphorylation or the 'Mehler Reaction', a non-cyclic electron transport system involving oxygen instead of $NADP^+$ as the terminal acceptor (Tolbert, 1973). An increase in the photosynthetic evolution of oxygen occurred when PPO was inhibited by KCN (Siegenthaler & Vaucher-Bonjour, 1971). Meyer & Biehl (1980) also noted an increase in oxygen evolution when latent PPO was activated during aging of isolated spinach chloroplasts. These studies correlated the loss of Mehler reaction with activation of PPO (Vaughn & Duke, 1984*a*). Activation of PPO occurs during infection or mechanical injury resulting in quinone formation and consequently insoluble polymers. The plant tissues 'impregnated' with these polymers provide a barrier for preventing the spread of infection in the plant (Rubin & Artskhoskaya, 1960). In this capacity the enzyme affords resistance to plant infections by viruses, bacteria, fungi as well as from mechanical damage. Some of the intermediates formed during polymerisation of phenolics could also inactivate or bind plant viruses and limit their potency in their infestation of plants (Pierpoint, 1966). Melanins and quinones were later found by Pierpoint *et al.* (1977) to be bacteriostatic and able to partially inactivate a potato virus. Production of quinones has been claimed to be essential for resistance of apple to infection by *Venturia inequalis* (Overeem, 1976) and helps to prevent spread of infection by *Botrytis cinerea* on broad bean (Deverall, 1961).

Another possible role for PPO is in root formation and development (Al-Barazi & Schwabe, 1984). These researchers located considerable PPO activity in both the secondary xylem and pith, adjacent to the sites of root initiation in *Pistacia vera* cuttings.

The role of PPO in the biosynthesis of plant phenolics, particularly phenylpropanoids and indirect involvement in lignin formation, has also been reported (Robb, 1984). However, inconclusive results were obtained to fully support this theory (Vaughn & Duke, 1984*a*).

Extraction

Extraction of PPOs from foods of plant origin is associated with many problems, making isolation extremely difficult (Mathew & Parpia, 1971). The major drawback to PPO isolation is the occurrence of both enzyme and substrate side by side in intact cells. PPO is located exclusively in the plastids of healthy plant tissues and is apparently not even activated until it crosses the plastid envelope. In contrast, the majority of phenolic compounds in higher plants are located in the vacuole distinct from the plastids (Czaninski & Catesson, 1972; Vaughn & Duke, 1981*b*; Vaughn & Duke, 1984*a*). Once the plastid and vacuole contents mix, following cellular disruption, the internal organization is damaged. As a consequence, endogenous phenolics are enzymatically oxidised to quinones which polymerize to polyphenols, responsible for enzymatic browning or tanning (Balasingam & Ferdinand, 1970). Polyphenols are multidentate ligands and cross-link protein molecules including PPO via hydrophobic and hydrogen-bonding interactions to yield aggregates and/or precipitates (Loomis, 1974; McManus *et al.*, 1985). Moreover, phenols readily transformed to quinones, may react with nucleophilic groups (e.g. SH and NH_2) on the proteins to form covalent products (Matheis & Whitaker, 1984; McManus *et al.*, 1985). It has also been reported that phenolics may inactivate PPO (Sanderson, 1965; Loomis, 1974), or in contrast, react with proteins and produce additional multiple forms which are not present in the intact tissue (Smith & Montgomery, 1985). Thus it is important to carefully select methods of extraction and isolation of PPO which minimise or even eliminate any changes in enzyme structure and activity during its purification.

Many attempts have been made to minimise phenolic–protein interactions including comminution and homogenisation in liquid nitrogen or in nitrogen atmosphere (Balasingham & Ferdinand, 1970; Benjamin & Montgomery, 1973; Thomas & Janave, 1973; Coombs *et al.*, 1974). Extraction steps have also been carried out at −20 to −30°C (Vámos-Vigyázó, 1981). Early researchers used lead acetate as a precipitant for polyphenols (Nelson & Dawson, 1944), but its use is limited as it does not precipitate all plant phenols (Loomis & Battaille, 1966) and resulted in considerable loss of enzyme itself (Mathew & Parpia, 1971). More recently, insoluble polyvinyl pyrrolidone (PVP) has been used for the removal of phenols during separation of subcellular fractions and in the initial purification steps of the enzyme from plant tissues (Chan & Yang, 1971; Montgomery & Sgarbieri, 1975; Cash *et al.*, 1976; Galeazzi *et al.*, 1981; Roudsari *et al.*, 1981). This technique, however, has its disadvantages as it requires a large amount of PVP, an acidic pH during the extraction (maximum affinity for plant phenols occurs at pH 3·5), and a time-consuming pre-treatment procedure prior to extraction (Andersen & Sowers, 1968; Smith & Montgomery, 1985). In addition, PVP was also found to inhibit enzyme PPO activity to some extent (Harel *et al.*, 1964; Anderson, 1968). Other polymers have also been successfully used for the removal of phenols, e.g. ion exchange resins (Cornwell & Wrolstad, 1981; Batistuti & Lourençe, 1985; Smith & Montgomery, 1985), polycaprolactam (Sanderson, 1964*a*), and polyethylene glycol (Badran & Jones, 1965; Benjamin & Montgomery, 1973; Rivas & Whitaker, 1973; Park & Luh, 1985). Polyethylene glycol (PEG), because of its solubility in acetone, is often used in combination with the latter for the preparation of acetone powder (Loomis, 1968; Flurkey & Jen, 1978; Wissemann & Lee, 1980; Owusu-Ansah, 1989). Acetone precipitation followed by buffer extraction is one of the methods most often used as the initial step in isolation of PPO (Clayton, 1959; Palmer, 1963; Walker, 1964; Dizik & Knapp, 1970; Wong *et al.*, 1971; Padron *et al.*, 1975; Cash *et al.*, 1976; Gross & Coombs, 1976; Satjawatcharaphong *et al.*, 1983; Oda *et al.*, 1989). This method offers certain advantages in obtaining a crude enzyme preparation with a high yield of activity (Flurkey & Jen, 1978) and free of artifacts (Kahn, 1975). In order to prevent inactivation during buffer extraction, reducing agents or quinone couplers such as ascorbic acid (Abukharma & Woolhouse, 1966; Rivas & Whitaker, 1973; Zenin & Park, 1978; Sharma & Ali, 1980; Janovitz-Klapp *et al.*, 1989), cysteine (Stelzig *et al.*, 1972; Batistuti & Lourenco, 1985), sodium metabisulfite (Stokes *et al.*, 1968), sodium diethyldithiocarbamate (Pierpoint, 1966), sodium tetrathiolate (Park & Luh, 1985) or thiourea (Van Driessche *et al.*, 1984) have been included.

In addition to the difficulties discussed, other problems must be overcome during the extraction of PPO. The soluble enzyme is present in relatively small amounts so that purification factors in excess of 100 are frequently reported indicating less than 1% of the total protein (Robb, 1984). Kertesz & Zito (1962) estimated that even in mushroom samples, generally regarded as the best source of PPO, the concentration of this enzyme is of the order of 0·004% (40 mg/kg).

Although PPO occurs in its active form in most foods (Vámos-

Vigyázó, 1981), the latent forms of this enzyme exist in broad bean, *Vicia faba* L. (Kenten, 1957; Robb *et al.*, 1964; Hutcheson *et al.*, 1980; Moore & Flurkey, 1990; Sánchez-Ferrer *et al.*, 1990), bananas (Palmer, 1963), spinach (Sato & Hasegawa, 1976), avocado (Kahn, 1977), mushrooms (Yamaguchi *et al.*, 1970), grapes (Sánchez-Ferrer *et al.*, 1989), and crustaceans (Savagaon & Sreenivasan, 1978; Marshall *et al.*, 1984; Ferrer *et al.*, 1989*a,b*). PPO can be released from its latency by different treatments including acid and base shock (Kenten, 1957, 1958), anionic detergents (Swain *et al.*, 1966; Flurkey 1986; Moore & Flurkey, 1990), proteolytic enzymes (Savagaon & Screenivasan, 1978; Marshall *et al.*, 1984; King & Flurkey, 1987), urea (Swain *et al.*, 1966; Lerner *et al.*, 1972), fatty acids (Siegenthaler & Vaucher-Bonjour, 1971; Hutcheson *et al.*, 1980; Golbeck & Cammarata, 1981), fungal glycoproteins, bacterial lipopolysaccharides and β-1,3-glucans (Söderhäll & Unestam, 1979).

Activation of latent PPO was first achieved by brief exposure of broad bean extracts to acidic (pH 3·0–3·5) or alkaline (pH 11·0–11·5) conditions, by incubating in the presence of ammonium sulphate at pH 5·0 (Kenten, 1957), or by treatment with anionic wetting agents at pH 6·0 (Kenten, 1958). Although the nature of latent PPO has not yet been elucidated, it has been suggested that activation may result from removal of an enzyme inhibitor from a PPO–inhibitor complex. The latency of spinach chloroplast PPO was attributed, in part, to the formation of an inactive complex between the active protein (PPO) and a low molecular weight polypeptide. Dissociation of this inactive complex by various detergents such as sodium deoxycholate, sodium dodecylsulphate (SDS), and digitonin, resulted in activation of the enzyme. Other researchers (Robb *et al.*, 1964; Swain *et al.*, 1966) studying latency of broad bean PPO, reported that mild denaturing treatments such as used by Kenten (1957, 1958) or exposure to dilute detergent solutions (Manoxal OT) brought about activation of the enzyme with concomitant changes of its physico-chemical behaviour (e.g. increased electrophoretic mobility, higher affinity for copper-chelating agents). These studies showed that the activation of latent PPO involved rearrangement of the protein tertiary structure, rendering the active site more accessible to substrates, rather than dissociation of an enzyme–inhibitor complex. Recent studies by Moore & Flurkey (1990) also argued against the Kenten theory and indicated only limited conformational changes of protein during activation of latent broad bean PPO with SDS. A similar mechanism was proposed earlier by Golbeck & Cammarata (1981) for activation of spinach PPO by linolenic acid.

Of several proteolytic enzymes, trypsin was identified as a major activator of latent plant and crustacean PPO. The precise mechanism, however, is not fully understood and proposed theories are controversial. For shrimp and lobster PPO, the mechanism of activation involved limited proteolysis of the enzyme to form active isoenzymes (Savagaon & Sreenivasan, 1978; Ferrer *et al.*, 1989*a*). Limited proteolysis has also been shown to activate spinach (Golbeck & Cammarata, 1981), broad bean (King & Flurkey, 1987), and carrot (Söderhäll *et al.*, 1985) PPOs. These results are in contrast to the work of Tolbert (1973), who observed that spinach PPO can be activated even with denatured trypsin preparations.

Isolation of PPO may be hampered by the presence of endogenous proteolytic enzymes which could generate artifactual multiple and interconvertible forms of PPO. This was evident when protease inhibitors (phenylmethylsulfonyl fluoride and Trasylol) were shown to decrease the number of isoenzyme forms obtained upon purification of peach (Flurkey & Jen, 1980) and date PPO (Sachde *et al.*, 1989*a*).

Several plant PPOs have been found to be bound tightly to membranes (Mayer & Harel, 1979; Vaughn & Duke, 1984*a*). Solubilisation involves the release from or breakage of a membrane structure following treatment with detergents (Harel *et al.*, 1965; Stelzig *et al.*, 1972), proteolytic enzymes (Mayer, 1966), and lipases (which hydrolyse membrane lipids and disrupt membrane integrity) (Benson, 1964), or the use of hypotonic extractors (Meyer & Biehl, 1980) and sonification (Hutcheson *et al.*, 1980; Golbeck & Cammarata, 1981). Several problems, however, may be encountered including modification of the enzyme during solubilisation. For example, an almost complete loss of hydroxylase activity has been reported by several researchers following solubilisation (Harel *et al.*, 1965; Harel & Mayer, 1971). A second problem is due to interference by chlorophyll liberated when detergents were used to extract the enzyme. This could be overcome by removing chlorophyll by chromatographic methods or using more drastic treatments such as acetone powder. Solubilisation of PPO without liberation of chlorophyll was achieved using cationic detergents such as cetyl trimethylammonium bromide (Vaughan *et al.*, 1975) and more recently by temperature-induced phase separation with a non-ionic detergent, Triton X-114 (Sánchez-Ferrer *et al.*, 1989). The latter method has the additional advantage of removing phenolic compounds along with chlorophyll.

Purification

Numerous publications have dealt with the purification of PPO; the *Food Science and Technology Abstracts* (1970–1990) listed 71 reports on the purification of this enzyme. Nevertheless, for reasons discussed in the 'Extraction' section, only a few PPOs have been purified to apparent homogeneity. These include PPOs from wheat (Interesse *et al.*, 1981, 1983), grapes (Kidron *et al.*, 1977), potatoes (Balsingam & Ferdinand, 1970), tea leaves (Gregory & Bendall, 1966), spinach (Vaughan *et al.*, 1975; Oda *et al.*, 1989), pears (Rivas & Whitaker, 1973), peach (Wong *et al.*, 1971; Flurkey & Jen, 1980) and artichoke (Zawistowski *et al.*, 1988*a,b*).

Published procedures on the purification of PPO have varied according to the enzyme source and degree of purity attained (Vámos-Vigyázó, 1981). The most common first step of purification is the removal of inactive protein by precipitation with protamine sulphate (Balasingam & Ferdinand, 1970), calcium sulphate (Patil & Zucker, 1965), or fractionation with ammonium sulphate (being the most widely used) followed by dialysis or chromatography on Sephadex G-25 (Vámos-Vigyázó, 1981). Further purification frequently included adsorption, ion exchange and gel filtration chromatography, or a combination of some of these methods; hydrophobic chromatography (Mayer, 1987) and preparative isoelectric focussing have also been used (Interesse *et al.*, 1980).

Adsorption chromatography with celite was found to be selective for PPO, as this adsorbent is relatively specific for copper-containing proteins (Mathew & Parpia, 1971). Adsorption on calcium phosphate has been applied to apple (Stelzig *et al.*, 1972), and wheat PPO (Interesse *et al.*, 1983), while hydroxyapatite chromatography by itself (Kidron *et al.*, 1977), or following elution from carboxymethyl (CM) cellulose (Vaughan *et al.*, 1975), or diethylaminoethyl (DEAE) cellulose columns were employed for purification of spinach (Oda *et al.*, 1989), pear (Flurkey & Jen, 1980) and peach (Rivas & Whitaker, 1973) PPO, respectively.

DEAE-cellulose proved to be a popular ion exchanger for the purification of a variety of PPOs, although other gel matrices in combination with DEAE cellulose, Sephadex (Wong *et al.*, 1971), Sepharose (Wichers *et al.*, 1984; Zawistowski *et al.*, 1988*a*), and Toyopearl (polyvinyl beads) (Fujita & Tono, 1988) have also been employed. Ion exchange chromatography has become the favored method, especially in combination with gel filtration chromatography on Sephadex (Mayer & Harel, 1979), or more recently on Sepharose (Batistuti & Lourenco, 1985; Jayaraman *et al.*, 1987; Fujita & Tono, 1988; Okamoto *et al.*, 1988), Sephacryl S-200 (Hsu *et al.*, 1984), and Ultrogel AcA 34 (Flurkey & Jen, 1980) columns.

Several attempts were made to purify different PPOs by affinity chromatography including those of mushroom (Gutteridge & Robb, 1973; O'Neill *et al.*, 1973; Menon & Haberman, 1975), housefly larvae (Shimoda *et al.*, 1975), and mammalian tissue (Jimbow *et al.*, 1975, 1981). Affinity chromatography was first introduced by Lerman (1953) to purify mushroom PPO, although it is difficult to conclude whether true affinity chromatography was involved. The low capacity of design absorbents and the oxidation of ligands (phenolic substrates) during operation of the column precludes its widespread use (Robb, 1984). For example, attempts to purify Jerusalem artichoke PPO by affinity chromatography using L-DOPA Sepharose 4B-CL column were unsuccessful due to the rapid darkening of the gel caused by autoxidation of DOPA (Zawistowski, 1987).

Purification of PPO by hydrophobic chromatography on Phenyl Sepharose CL-4B was recently reported (Mayer, 1987). This method is particularly attractive as the active site of PPO has a hydrophobic character which binds well to hydrophobic media such as phenyl-Sepharose, at high ionic strength (Robb, 1984). It is difficult to evaluate the effectiveness of this method as only a few applications on PPO systems have been published (Jen & Flurkey, 1979; Wissemann & Lee, 1980; Janovitz-Klapp *et al.*, 1989).

A relatively new method, referred to as 'immobilised metal affinity chromatography', which uses copper conjugated to iminodiacetic acid Sepharose 6B, was used to purify Jerusalem artichoke PPO (Zawistowski *et al.*, 1987*a*). This type of chromatography is based on the ability of proteins to form coordination bonds with a metal due to the presence of surface chelating amino acids–metal combining sites (Porath & Olin, 1983). Amino acid residues possessing an ionisable proton can display ligand properties. Nitrogen donors include the imidazole group of histidine, the ε-amino group of lysine and the guanidine group of the arginine. The carboxyl groups of aspartate and glutamate and the hydroxyl group of tyrosine can function as oxygen donors, while the thiol group of cysteine acts as a sulphur donor for co-ordination with metal ions. The nitrogen atom of the peptide linkage and the *N*-terminal amino group can also co-ordinate metal ions (Conlon & Murphy, 1976; Hemdan & Porath, 1985). Coordination binding between proteins and the immobilised metal is pH-dependent, and proteins are usually desorbed simply by reducing the pH, or by decreasing the salt concentration of the eluent (Lonnerdal & Keen, 1982). More selective desorption was achieved by including competing displacers (ligands) in the eluent buffer, such as glycine, histamine, histidine and imidazole (Kikuchi & Watanabe, 1981; Andersson, 1984). This method is simple and permits purification of PPO with high yield within a relatively short time period. Although this type of chromatography has been used to purify other plant enzymes (Redinbaugh & Campbell, 1985; Zawistowska *et al.*, 1988), its use in purifying PPO is so far limited to artichoke enzyme (Zawistowski *et al.*, 1987*a*).

METHODS OF PPO DETERMINATION

Accurate determination of PPO activity continues to be a challenge in spite of the numerous methods described in the literature (Mayer & Harel, 1979; Vámos-Vigyázó, 1981; Robb, 1984). This is because quinones formed during the course of the enzymatic reaction undergo secondary reactions with themselves, unreacted substrates, oxygen, other constituents and the enzyme. These reactions lead to the formation of complex polymeric molecules interfering with the reliability of the assay.

The existence of other enzymes with similar molecular properties, such as laccase and peroxidase in plant tissues, may be mistaken for PPO activity during its determination. Enzymes show overlapping substrate specificities by being able to convert phenols to quinones (Harel & Mayer, 1970; Robb, 1984). It is, therefore, very important to differentiate between PPO and the other enzymes to avoid confusion. A number of methods to distinguish between these enzymes have been proposed (Mayer *et al.*, 1964; Lerner *et al.*, 1971; van Loon, 1971; Kahn, 1985*b*) and reviewed by Mayer & Harel (1979) and Mayer (1987).

Monophenolase Activity

In contrast to mammalian tyrosinase, which is relatively specific for tyrosine, plant monophenolases are characterised by a broad substrate specificity. Consequently *p*-cresol, tyrosine, tyramine and *p*-coumaric acid are the most frequently used monohydroxyphenol substrates. As mentioned earlier, the hydroxylation reaction which is frequently associated with a very long induction period, can be reduced or entirely eliminated by the addition of small amounts of *o*-dihydroxyphenols which act as the electron donors (Vámos-Vigyázó, 1981). Furthermore, reducing agents such as ascorbate, $NADH^+$, cysteine, and dimethyltetrahydropteridine can all be used to overcome this lag phase (Vaughan & Butt, 1970; Golan-Goldhirsh & Whitaker, 1984).

The simplest and least accurate methods of assaying monophenolase activity of PPO involve measuring the increase in absorbance caused by the appearance of quinones (Sánchez-Ferrer *et al.*, 1988), or by determination of oxygen uptake (Dawson & Magee, 1955; Kertesz & Zito, 1962). As time progresses, however, the oxidase activity can make an increasing contribution to the overall activity (Robb, 1984). A sensitive and more accurate method for assaying mammalian tyrosinase was developed based on determination of tritiated water liberated from [3,5^{3}H] tyrosine in the course of hydroxylation (Pomerantz, 1966; Pomerantz & Li, 1970). Although this radioassay has been modified for use in the determination of monophenolase activity of food PPO (Kahn & Pomerantz, 1980) it has not been widely adopted in spite of its sensitivity (Mayer & Harel, 1979). An alternative method with comparable sensitivity to the radioassay procedure was introduced by Goodenough *et al.* (1983). This method was used to monitor the hydroxylation of *p*-coumaric acid and phloridizin (flavonoid) by apple PPO using HPLC. The reactions were conducted in the presence of either $NADH^+$ or ascorbate to stop further oxidation of the hydroxylation products. The quantitation of peak size from HPLC was used to determine an apparent K_m value for both compounds.

Diphenolase Activity

The types of assays for measuring diphenolase activity have been described in the literature. The chronometric method measures the rate of ascorbate disappearance in a PPO/phenolic substrate/ascorbate coupled system (Dawson & Magee, 1955). Although easy to perform, this method is unsuitable for kinetic studies where it is more convenient to determine the rate of loss of reducing agents (ascorbate,

265 nm = 15, 300 M^{-1} cm^{-1}; NADH, 340 nm = 6, 220 M^{-1} cm^{-1}) spectrophotometrically (Robb, 1984). The manometric method described by Gregg & Nelson (1940) and Nelson & Mason (1970) is no longer performed because of the time, technical difficulties, and the large sample size required. The spectrophotometric methods, by far, have been the most widely used for measuring diphenolase activity of PPO. The simplest assay records the final colour yield when the enzyme is incubated with a suitable substrate such as catechol or methylcatechol. Although this procedure has proved adequate for comparative studies of the browning levels of different fruits and vegetables, it is fraught with errors (Walker, 1975). The most accurate spectrophotometric assay is to monitor PPO activity using coupled reactions with the formed quinones. These quinones, in aqueous solution, which may condense with themselves or with other molecules including enzymatic proteins, are usually trapped by the coupling agents such as *o*-phenylenediamine (Racusen, 1970), Besthorn's hydrazone (Pifferi & Baldassari, 1973; Mazzocco & Pifferi, 1976) or reduced back to phenolic compounds by addition of ascorbate (El-Bayoumi & Frieden, 1957). In some cases the initial rate of formation of quinone is determined and in others the disappearance of the coupling reagents is followed (Mayer & Harel, 1979). Polarographic measurement of oxygen uptake, as described by Harel *et al.* (1964) and Yamaguchi *et al.* (1969), however, is the method of choice for its convenience, simplicity and accuracy (Mayer *et al.*, 1966). The rate of oxygen utilisation by the reaction catalysed by PPO is monitored in a reaction vessel equipped with an oxygen electrode, usually a Clark type (Harel *et al.*, 1964; Estabrook, 1967; Yamaguchi *et al.*, 1969). The amount of oxygen consumed reflects the amount of phenolic substrate being oxidised and is proportional to the enzyme concentration. The ratio of oxygen consumed/substrate oxidised changes with the time of the reaction and depends on the substrate structure and concentration, pH, temperature and the buffer used (Mayer & Harel, 1979). Furthermore, the reaction velocity is dependent on oxygen concentration (Yamaguchi *et al.*, 1969). Thus it is important to measure only the initial rate of oxygen uptake using strictly controlled conditions.

Polyclonal antibodies to tyrosinase have recently been used to develop a competitive enzyme-linked immunosorbent assay (ELISA) for the quantitative analysis of the murine enzyme (Ishiguro *et al.*, 1986). Although the ELISA test is a specific, sensitive and simple technique, due to the heterogeneity and lack of standards for PPO, quantitation of food PPOs by this method is difficult. Nevertheless, Flurkey and his associates (Flurkey, 1986; Lanker *et al.*, 1988) attempted to use polyclonal as well as monoclonal antibodies to PPO in conjunction with immunoblotting or with the Ouchterlony test to detect and study the cross-reactivity of PPO in a wide range of higher plants.

Reactions Catalysed

PPO is a copper-containing enzyme capable of catalysing two different types of reactions (Fig. 1); the hydroxylation of monophenols to *o*-diphenols and the dehydrogenation of *o*-diphenols to *o*-quinones (Whitaker, 1972; Matheis & Whitaker, 1984). The latter products subsequently undergo a series of non-enzymatic reactions to yield brown-black melanin pigments (Fig. 2). These two enzymatic activities are referred to as monophenolase (hydroxylase or cresolase) and diphenolase (catecholase or oxidase), respectively (Nelson & Dawson, 1944; Robb, 1984). The ratio of diphenolase : monophenolase activities

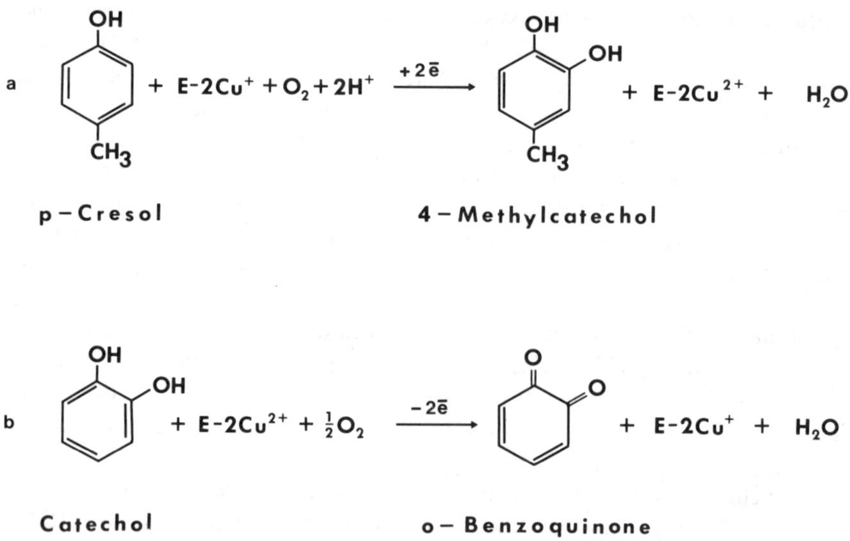

Fig. 1. Reactions catalysed by PPO. (a) Hydroxylation of monophenol to *o*-diphenol. (b) Dehydrogenation of *o*-diphenol to *o*-quinone.

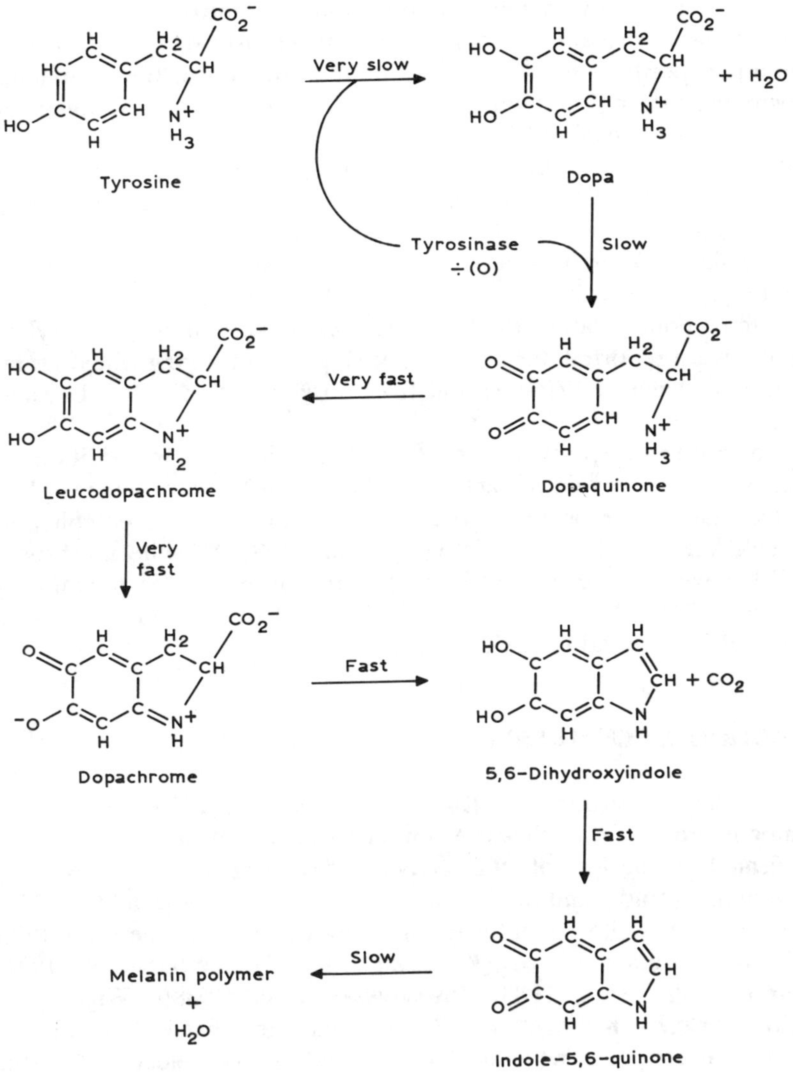

Fig. 2. Mechanism of melanin formation from tyrosine (Mason, 1948). Although deviations from this pathway are known to occur in melanin formation, this metabolic path adequately represents the majority of product formation.

varies with the plant source, although diphenolase was generally found to be 5–10 times higher than monophenolase (Vámos-Vigyázó, 1981). Studies on Jerusalem artichoke PPO by Zawistowski *et al.* (1986) found the specific activity to be 10–50 times greater for the *o*-diphenol substrates, catechol and chlorogenic acid, compared to monophenols, tyrosine and *p*-cresol.

A number of fruits such as mango (Joshi & Shiralkar, 1977), cherry (Lanzarini *et al.*, 1972), banana (Palmer, 1963), and pear (Rivas & Whitaker, 1973) are devoid of monophenolase activity. PPO preparations which lack or have attenuated hydroxylase ability are common but preparations exhibiting only monophenolase activity have been reported. Food-related PPO isoenzymes having only hydroxylase activity were reported during the later stages of maturing wheat grains (Taneja & Sachar, 1974*a,b*) and from sorghum (Stafford & Dresler, 1972).

The substrate specificity of PPO is ambiguous. In addition to hydroxylase and oxidative activities, PPO activity from several plant sources has been reported to oxidise tri-, tetra and other polyphenols (Vámos-Vigyázó, 1981; Zawistowski *et al.*, 1988*a*), while mushroom PPO behaves as a peroxidase in the absence of oxygen (Strothkamp & Mason, 1974). Moreover Boyer *et al.* (1986) discovered that mushroom PPO exhibited ferroxidase activity.

REACTION MECHANISM

The classical experiment by Kubowitz (1938) established copper as a component of PPO which was essential for its overall activity. This was confirmed by the loss of PPO activity when copper was removed by potassium cyanide and was restored when copper was added. This experiment was subsequently reported by other researchers for PPO isolated from many sources (Kertesz & Zito, 1965; Wong *et al.*, 1971; Beltramini & Lerch, 1983; Zawistowski *et al.*, 1988*b*; Wigfield & Goltz, 1990*a,b*). Kubowitz (1938) also concluded that PPO contained a binuclear copper at its active site involved in a two-electron transfer in which the valence state of copper is changed (Fig. 1). Nevertheless, Kobowitz's theory was challenged by a number of researchers. For example, PPO isolated from *Neurospora crassa* was reported to contain only one copper atom per molecule and thus unable to participate in two-electron transfers (Fling *et al.*, 1963; Gutteridge &

Robb, 1975). A copper content of one atom per functional unit was also found in PPO from broad bean (Robb *et al.*, 1965), mushroom (Kertesz & Zito, 1965), spinach beet (Vaughan *et al.*, 1975) and grapes (Kidron *et al.*, 1977). Other studies, however, suggested that the original theory proposed by Kubowitz (1983) was correct. Jolley *et al.* (1974) showed a stoichiometric relationship existing between one molecule of hydrogen peroxide and two copper atoms in the reaction of mushroom PPO with hydrogen peroxide indicating that in mushroom the enzyme contained a copper pair *per* active site. These findings were supported by EPR (Schoot Uiterkamp & Mason, 1973) and magnetic susceptibility studies (Makino *et al.*, 1974) of nitric oxide derivatives of mushroom PPO. Reinvestigation of the molecular structure of *N. crassa* PPO by Lerch (1976) contradicted earlier work and found the enzyme to indeed contain two copper atoms per functional unit of mol. wt 46 000.

The state of copper in PPO has also been the subject of controversy. Kertesz & Zito (1962) reported that all copper in a fresh preparation of mushroom enzyme occurs in the cuprous state and only this form is catalytically active. This was supported by the fact that the enzymatic copper could be recovered in the form of cuprous ions (Brooks & Dawson, 1966; Kertesz, 1966). PPO was inhibited by carbon monoxide (Kertesz, 1966; Vanneste & Zuberbühler, 1974), and by the absence of a cupric EPR spectrum of the active enzyme (Kertesz *et al.*, 1972). Conflicting evidence, however, was provided by Dressler & Dawson (1960) who showed that mushroom PPO exchanged enzymatic copper against $^{64}Cu^{2+}$ during oxidation of substrates. This pointed to a divalent copper associated with the enzyme. Further studies on the oxygenation of mushroom PPO revealed that reaction with hydrogen peroxide involved different states of the enzyme, resting and oxygenated forms associated with cupric copper as well as cuprous deoxygenated form.

Spectroscopic and chemical studies of the different states of mushroom PPO have all indicated that the enzyme is structurally similar to hemocyanin; both bind carbon monoxide and also reversibly with oxygen to produce oxygenated forms (Vanneste & Mason, 1966). The absorption spectra of the oxygenated forms exhibited similar maxima in the visible region at 345 and 600 nm with extinction coefficients of approximately 9000 and 600 $\text{M}^{-1}\,\text{cm}^{-1}$, respectively (Jolley *et al.*, 1972). Moreover, the EPR spectra of both enzymatic deoxyforms after derivatisation with nitric oxide were strikingly similar (Schoot Uiterkamp & Mason, 1973). An absorption spectrum with a maximum at 340 nm was also observed by Vaughan & McIntyre (1975) for an oxygenated spinach beet PPO form. More recent studies by Himmelwright *et al.* (1980) and Lerch's group (Lerch, 1981; Pfiffner & Lerch, 1981; Huber & Lerch, 1986; Lerch *et al.*, 1986) reveal great similarities between PPO of *N. crassa* and hemocyanin with respect to a binuclear copper at the active site. The proposed active site had two antiferromagnetically spin-coupled copper atoms (commonly referred to as 'type 3' copper) in close proximity (around 3·5 Å) responsible for the lack of an EPR signal. One of the copper atoms appeared to be bound to three histidine residues (His-188, His-193 and His-289), as measured by dye-sensitised photo-oxidation, while the other copper was bound to a single histidine residue (His-306) based on reaction inactivation experiments [Fig. 3(a)]. All four histidine residues are indicated in Fig. 3(b) showing the amino acid sequence of the *Neurospora* enzyme. In addition, an intramolecular thioether bond linking cysteine (Cys-94) and histidine (His-96) residues, but not binding copper atoms, may be involved in regulation of enzymatic activity (Lerch, 1981).

The structure of the active site depends on the oxidation state of copper. In the resting state of PPO (more recently referred to as the met form), two tetragonal Cu(II) ions are bridged by an exogenous ligand (L) and an endogenous protein ligand (R). The latter ligand (R) provides an effective pathway for superexchange between the two Cu(II)s which in turn accounts for the absence of an EPR signal. Treatment with nitric oxide ruptures the endogenous ligand (R) increasing the distance between the copper atoms to 6 Å and thereby results in EPR-detectable derivatives. In the oxy form, the exogenous ligand (L) is replaced by a peroxide linkage. Based on the available data on the nature of the copper site and protein structure, the overall reaction mechanism for *Neurospora* PPO for both monophenolase and diphenolase activity is schematically shown in Fig. 4.

The first step in the cycle was chosen somewhat arbitrarily since a dispute remains regarding which substrate, *o*-diphenol or oxygen, binds PPO first (Mayer & Harel, 1979; Janovitz-Klapp *et al.*, 1990). Deoxyenzyme (1) was found to function as a pseudoperoxidase (Strothkamp & Mason, 1974), and binds reversibly with oxygen. Two-electron transfer from cuprous ions to oxygen is associated with this step and as a result an oxyform (2) is produced. The geometrical configuration of the binuclear copper site of the oxyenzyme, with a

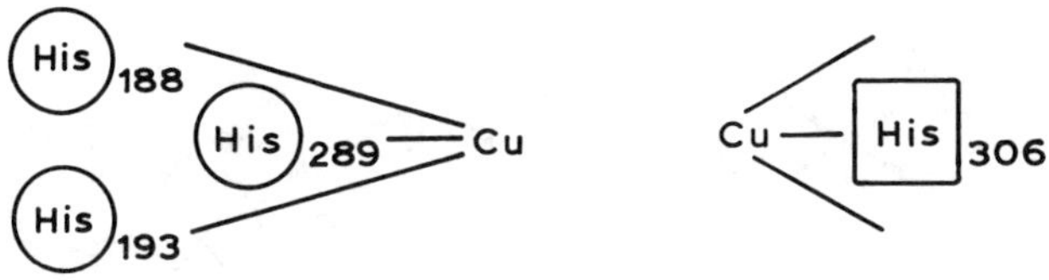

(a)

1
AC-S T D I K F A I T G V P T T P S S N G A V P L R R E L R D L

31
Q Q N Y P E Q F N L Y L L G L R D F Q G L D E A K L D S Y Y

61
Q V A G I H G M P F K P W A G V P S D T D W S Q P G S S G F

91
G G Y C T H S S I L F I T W H R P Y L A L Y E Q A L Y A S V

121
Q A V A Q K F P V E G G L R A K Y V A A A K D F R A P Y F D

151
W A S Q P P K G T L A F P E S L S S R T I Q V V D V D G K T

181
K S I N N P L H R F T F H P V N P S P G N F S A A W S R Y P

211
S T V R Y P N R L P G A S R D E R I A P I L A D E L A S L R

241
N N V S L L L L S Y K D F D A F S Y N R W D P N T N P G D F

271
G S L E A V H N E I H D R T G G N G H M S S L E V S A F D P

301
L F W L H H V N V D R L W S I W Q D L N P N S F M T P R P A

331
P Y S T F V A Q E G E S Q S K S T P L E P F W D K S A A N F

361
W T S E Q V K D S I T F G Y A Y P E T Q K W K Y S S V K E Y

391
Q A A I R K S V T A L Y G S N V F-OH

(b)

Fig. 3. Amino acid sequence of *Neurospora* tyrosinase (Lerch, 1978). One letter amino acid abbreviations [see IUB-IUPAC (1968)]. Cysteinyl residue 94 is linked to histidyl residue 96 via a thioether bridge. Asterisks indicate the histidyl residues lost after dye-sensitised photooxidation of the apoenzyme.

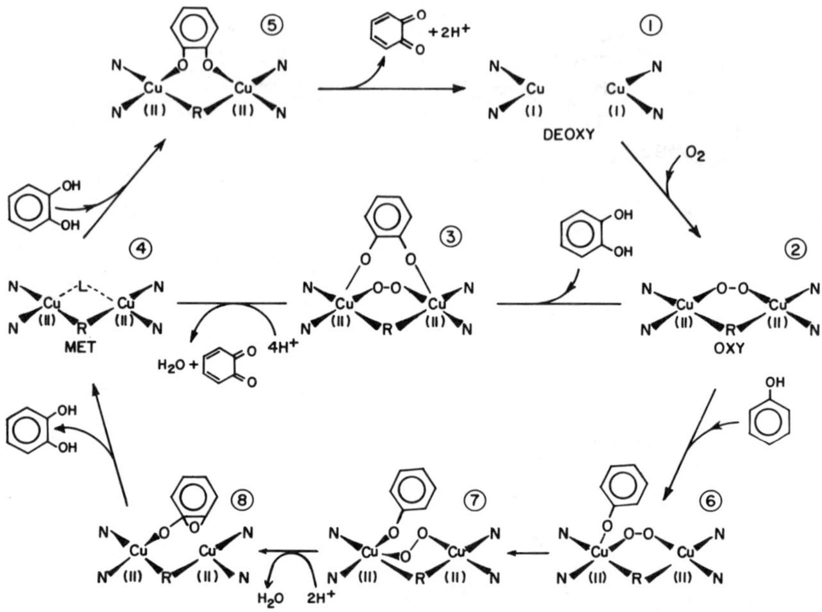

Fig. 4. Reaction mechanism of monophenolase and diphenolase activities of PPO. N = protein ligand. R = endogenous protein bridge. L = exogenous ligand.

metal–metal distance of about 3·6 Å, allows for co-ordination of the *o*-diphenolic oxygens or monophenolic oxygen, accounting for diphenolase and monophenolase activities, respectively. In the case of diphenolase activity, the oxyenzyme reacts with *o*-diphenol to form an intermediate (3) which breaks down to a more stable metenzyme (4) with the release of an *o*-quinone molecule. This step is associated with another two-electron transfer from *o*-diphenol to peroxide oxygens. The next step involves the reduction of bicupric metenzyme with a second molecule of *o*-diphenol to form bicuprous deoxyenzyme (1) via an intermediate (5). Thus, during one cycle of diphenolase turnover (Fig. 1) two molecules of *o*-diphenols are oxidized to two molecules of *o*-quinones with concomitant four-electron reduction of oxygen yielding two molecules of water (Lerch, 1981).

In the case of monophenolase activity, the oxyform (2) binds a monophenol to one of its copper atoms (6) and undergoes rearrangement to a trigonal–bipyramidal intermediate (7). This triggers disruption of the peroxide linkage and hydroxylation of a monophenol, possibly via an epoxide intermediate (8), as suggested by Robb (1984). The *o*-diphenol is then released from the enzyme with the production of the binuclear cupric metform (4). Whether the *o*-diphenol is released in the free form or undergoes an immediate oxidation to *o*-quinone is still uncertain, although McIntyre & Vaughan (1975) demonstrated that the conversion of coumaric acid to caffeic acid as a final product of spinach PPO, is possible.

The next sequence of reactions, outlined above, couples both monophenolase and diphenolase activities where the metenzyme is reduced to binuclear cuprous deoxyenzyme (1) by *o*-diphenol. Monophenolase activity is characterised by a lag phase which can be abolished by the addition of *o*-diphenol or other reductants acting as the electron donors. Robb (1984) suggested an activation of hydroxylation occurs during step (4)–(5). The mechanism outlined above is generally accepted to explain the catalytic action of *Neurospora* PPO (Lerch, 1981; Robb, 1984; Whitaker, 1985). Whether it is applicable to other PPOs is uncertain as their molecular structures are still unresolved (Mayer, 1987).

HETEROGENEITY AND MOLECULAR STRUCTURE

Despite the inherent difficulties associated with the isolation and purification of PPO due to the presence of endogenous phenolics and proteases, attempts have been made to characterise the enzyme from various sources. The comprehensive reviews of Mayer & Harel (1979), Vámos-Vigyázó (1981), Robb (1984) and Mayer (1987) summarised earlier work on molecular properties, substrate specificities and inhibitor action of fungal (including mushroom), mammalian and many plant enzymes. As stated earlier, only in a few studies has PPO been purified to apparent homogeneity. Any reports on enzyme heterogeneity, amino acid composition and copper content must be treated with caution since they reflect different conditions of extraction and varying degrees of purification.

Gel electrophoresis, isoelectric focussing, immunological analysis and gel filtration of many PPO preparations indicate that the enzyme exists in multiple forms. Enzyme multiplicity has been attributed to association–dissociation phenomena of similar and/or different sub-units, in-vivo post-translational modification including limited proteolysis, polymerisation with phenolics, glycosylation, and proteolytic action during isolation/purification of the enzyme, as well as interaction between enzyme and non-enzymatic proteins (Jolley *et al.*, 1969*a*; Mayer & Harel, 1979; Flurkey & Jen, 1980; Vámos-Vigyázó, 1981; Robb, 1984; Smith & Montgomery, 1985; Flurkey, 1989; Kupper *et al.*, 1989). These forms may also be interconvertible (depending on ionic strength, pH, enzyme concentration and presence of dissociating agents, etc.) and differ in their substrate specificity, pH optima, thermal stability and response to inhibitors or chemical reagents.

In the case of mushroom (*Agaricus bispora*) tyrosinase, the quaternary structure consists of two types of polypeptide chains, a heavy (H, 43 K) and a light (L, 13·4 K), which differ in their amino acid composition (Jolley *et al.*, 1969*b*; Strothkamp *et al.*, 1976). In aqueous solutions, the most predominant form (120 K) has the structure L_2H_2, while the active form of the enzyme (exhibiting both cresolase and catecholase activity) has the subunit composition L_2H. Electrophoresis data also suggest that the high molecular weight forms of the enzyme are aggregates in which interpolypeptide disulphide linkages are not involved. The monomeric *N. crassa* enzyme (46 K) has been shown to originate from a much larger precursor polypeptide sequence (75 K); the mature enzyme differs from the proenzyme by a long carboxyl terminal extension (213 amino acids) that is present only in the proenzyme. By analogy with other structurally related proteins, Kupper *et al.* (1989) suggested that this extension, cleaved by a protease with chymotrypsin-like activity, is intimately involved in the control of the enzymic activity of the *N. crassa* tyrosinase. Using an in-vitro translation system and poly(A) + *m*RNA isolated from *Vicia faba* leaves, Flurkey (1985) reported that the synthesised translation product for broad bean PPO was a 45 K polypeptide.

In a subsequent study, Flurkey (1989) found four polypeptides (43, 45, 60 and 61·5 K) in a purified PPO preparation from broad bean leaves. Because all four proteins had a similar amino-terminal sequence, post-translational modification of the original polypeptide was suggested to account for their presence; this modification may involve glycosylation (Robb *et al.*, 1965) and covalent attachment of phenolic groups (Bendall & Gregory, 1963).

Reported molecular weights of multiple forms of PPO, as determined by SDS electrophoresis or gel filtration, extend over a wide range, 12 000–400 000 (Mayer & Harel, 1979; Robb, 1984; Vámos-Vigyázó, 1981); apparent enzyme multiplicity often reflects association–dissociation phenomena, proteolysis as well as technique and solvent conditions employed for determination of molecular weights. Often the pattern of multiple forms was found to vary within the same species or even cultivars depending on the time of harvest, storage conditions, etc. Thus, while Harel & Mayer (1968) observed three forms of catechol oxidase in apple fruit (30–40, 60–70 and 120–130 K), Janovitz-Klapp *et al.* (1989) found a single PPO component (mol. wt 46 000) for another apple cultivar. Studies on PPO from Jerusalem artichoke tubers, using SDS-PAGE, revealed four distinct forms (32, 40, 86 and 120 K), while gel filtration indicated only two eluting species (34 and 42 K) (Zawistowski *et al.*, 1988*a*). Moreover, the purified enzyme from spinach leaves showed a mol. wt of 40 K by SDS-PAGE and 150 K (corresponding to a tetramer of the basic subunit) by gel filtration (Oda *et al.*, 1989). For this same enzyme, Lieberei *et al.* (1981) previously reported a mol. wt of 240 K by Sepharose 2B chromatography and 43 K by electrophoresis. Multiplicity of molecular weights has also been observed for PPO from avocado (14, 28, 56, 112 and 400 K) (Dizik & Knapp, 1970), potato tubers (aggregates of 2, 4, 8 and 16 of a 36-K subunit) (Balasingam & Ferdinand, 1970) and sugar-cane (32 and 130 K) (Coombs *et al.*, 1974), as assessed by gel filtration chromatography; wild rice (35, 42, 48 and 116 K) (Owusu-Ansah, 1989); kiwifruit (15, 20, 25 and 45 K) (Park & Luh, 1985), and wheat (23·5 and 30 K) (Interesse *et al.*, 1983) as revealed by SDS-PAGE. Other reported molecular weights for PPOs include grapes, 80 K by sedimentation (Kidron *et al.*, 1977); using gel filtration, yam tubers, 115 K (Anosike & Ayaebene, 1981), egg plant, 79 K (Roudsari *et al.*, 1981), sago palm, 40 K (Okamoto *et al.*, 1988); using SDS-PAGE, spinach leaves, 43 K (Lieberei *et al.*, 1981), white (*Penaeus setiferus*) and pink (*P. duorarum*) shrimp, 30 and 40 K, respectively (Simpson *et al.*, 1988). These results seem to indicate that the molecular weight of the basic subunit of PPO from plants and other sources ranges between 30 and 45 K.

Proteolytic degradation of the enzyme leads to smaller molecular weight species that retain activity (Flurkey & Jen, 1980). An active PPO from cotton leaves with a very low molecular weight (5·3 K), presumably due to proteolysis was reported by Mayer & Harel (1979). Moreover, some high molecular weight forms have been shown to convert into smaller molecular weight species when treated with urea (Vámos-Vigyázó, 1981). Ferrer *et al.* (1989*a*) recently proposed that activation of PPO from lobster cuticle involves limited proteolysis (by

Table 1. Amino Acid Composition (% Mole) of Phenoloxidases from Various Sources

Amino acid	*Neurospora*[a]	*Jerusalem artichoke*[b]	*Wheat*[c]	*Grape*[c]	*Spinach beet*[c]	*Potato*[c]	*Mushroom*[a]	
							H	*L*
Alanine	8·1	11·0	8·5	8·2	7·1	7·8	6·6	4·3
Arginine	4·9	2·2	3·9	3·7	3·0	3·9	4·3	4·3
Aspartic acid	10·8	7·1	10·3	12·8	12·6	10·7	12·4	13·7
Half-cystine	0·2	1·7	1·5	0·1–0·4	2·2	1·4	0·5	0·7
Glutamic acid	8·3	10·2	11·4	9·2	5·5	10·0	11·3	7·9
Glycine	6·2	12·6	9·1	7·2	7·9	8·1	5·5	11·5
Histidine	2·5	0·5	1·8	2·0	2·4	2·0	2·8	1·4
Isoleucine	3·2	4·3	3·9	5·6	4·3	5·5	5·1	5·7
Leucine	8·1	8·4	7·3	6·9	9·2	9·1	7·6	6·5
Lysine	4·2	4·8	5·7	5·9	5·5	6·5	4·0	4·3
Methionine	0·7	1·2	1·5	—	2·2	2·0	2·5	1·4
Phenylalanine	5·9	4·4	3·9	3·9	5·2	4·6	5·8	5·0
Proline	7·6	6·0	6·9	8·5	8·6	5·5	8·3	6·5
Serine	10·3	8·4	8·5	7·9	7·3	7·1	5·3	7·9
Threonine	5·4	7·5	5·7	5·9	5·9	5·5	6·1	7·2
Tyrosine	4·9	2·5	3·8	5·9	2·5–2·8	3·4–3·6	4·8	3·6
Valine	5·6	7·2	6·1	6·9	7·9	6·8	5·5	6·5
Tryptophan	2·9	N.D	1·0	N.D	0·6	N.D	1·5	1·4

[a] Robb (1984).
[b] Zawistowski *et al.* (1988*b*).
[c] Interesse *et al.* (1983).

endogenous protease or by trypsin) of a high molecular weight inert pro-enzyme; the released active forms had molecular weights of 64 (tryptin-activated) and 62·5 K (endogenous protease-activated).

Lieberei *et al.* (1981) using immunological methods concluded that several electrophoretically distinct PPO species were multiple forms rather than true isoenzymes. All PPO forms seemed to contain at least one common epitope recognized by antibodies against a single electrophoretic form. In addition, polyclonal and monoclonal antibodies against broad bean PPO reacted with both active and inactive forms of PPO from other plant sources as assessed by immunoblotting (Lanker *et al.*, 1988). Thus, considerable structural similarities exist among plant PPOs with respect to subunit composition. Isoelectric focussing also reveals the heterogeneous nature of the enzyme with isoelectric points varying within the range 3·3–9·6 (Vámos-Vigyázó, 1981; Robb, 1984); microheterogeneity is even exhibited by individual forms of PPO isolated by ion exchange chromatography (Zawistowski *et al.*, 1988*b*).

Published amino acid analysis data of phenolases from several plant sources as well as from *N. crassa* and the L and H subunits of mushroom PPO are summarised in Table 1. There are similarities between the enzymes as shown by their content of hydrophobic and sulphur-containing amino acids. Moreover, all plant PPOs except artichoke exhibit similarities in their content of basic amino acids (Arg, His, Leu). With respect to the prosthetic group, the current consensus is that the functional unit of the *Neurospora* and mushroom enzymes contains a pair of copper ions per single polypeptide chain. The reported copper contents of a variety of plant and fungal diphenolases range between 0·2 and 0·32% (Vámos-Vigyázó, 1981; Interesse *et al.*, 1983). These values were often found to correspond to stoichiometric amounts of less than one atom of copper per subunit. However, as pointed out by Mayer & Harel (1979), literature data on the copper content must be viewed with caution, since partial removal of copper from the enzyme can occur during purification and/or catalytic action.

SUBSTRATE SPECIFICITY

Many reports on PPOs were concerned with the specificity of the enzyme towards phenolic compounds. Although the subject has been reviewed by Mayer & Harel (1979) and Vámos-Vigyázó (1981) a few general comments as well as some recent developments in this area will be discussed.

PPOs from higher plants and fungi can act on a wide range of phenolic compounds (both mono- and *o*-diphenols). For the same substrates, large differences in kinetic parameters (K_m and V_{max}) are found depending on the source (genus, cultivar and even tissue the enzyme was extracted from) and purity of the enzyme (Vámos-Vigyázó, 1981). Several enzymes of plant origin are lacking cresolase activity (Mayer & Harel, 1979; Whitaker, 1985; Oda *et al.*, 1989; Owusu-Ansah, 1989). It is not certain, however, whether such preparations truly represent the 'native' form of the enzyme or loss of cresolase activity is attributed to changes in protein structure during isolation (Sharma & Ali, 1980; Walter & Purcell, 1980; Batistuti & Lourenco, 1985). The affinity of PPO towards a given substrate may even vary among the multiple forms of a particular preparation. Matheis & Belitz (1977) observed that only high molecular weight forms of potato tuber PPO are capable of oxidising monophenols. Similarly, Zawistowski *et al.* (1988*a*) found that while the high molecular weight aggregated forms of Jerusalem artichoke PPO exhibited activity towards monohydroxy-, *o*-dihydroxy- and polyhydroxy-phenols, the low molecular subunits were specific for *o*-dihydroxyphenols.

The most important natural phenolic substrates of PPO found in fruits and vegetables are catechins, cinnamic acid esters (e.g. chlorogenic acid), 3,4-dihydroxy phenylalanine (DOPA) and tyrosine. The type and position of substituents present on mono- and *o*-dihydroxy phenols are also important determinants of substrate reactivity. Monophenols are hydroxylated only if they have a $-CH_3$ group in the para position. Moreover, the rate of oxidation of *o*-diphenols increases with increasing power of an electron donating group and decreases as the electron withdrawing ability of the substituent in para position increases. Hence, various attempts have been made to relate kinetic parameters (K_m, K_{cat}, K_{cat}/K_m) of oxidation of several 4-substituted catechol derivatives with the 'substituent constant' (σ) values according to the Hammett equation. For purified mushroom tyrosinase (Duckworth & Coleman, 1970) reasonably linear relationships were shown between σ and log K_m or log K_{cat} (K_m and K_{cat} decreased in the order of $H > SCN > COCH_3 > CHO > CN > NO_2$), while for spinach leaf PPO (Oda *et al.*, 1989), K_{cat} and K_{cat}/K_m obeyed the Hammett relationship. Similar observations were made for lettuce, apple and cherry PPOs (Mayer & Harel, 1979). For substrates containing —COOH groups, Oda *et al.* (1989) suggested that there might be non-productive binding of the substrate to the active site (through the OH moiety of the —COOH group) which can alter the reaction rate. For such substrates K_{cat}/K_m is a better descriptor of enzyme-substrate affinity (Vámos-Vigyázó, 1981). The affinity of PPOs from various plant sources (Mayer & Harel, 1979; Hasegawa & Maier, 1980; Vámos-Vigyázó, 1981; Interesse *et al.*, 1983; Fujita & Tono, 1988; Zawistowski *et al.*, 1988*a*; Janovitz-Klapp *et al.*, 1989; Oda *et al.*, 1989; Sachde *et al.*, 1989*b*), mushroom (Duckworth & Coleman, 1970), lobsters (Ferrer *et al.*, 1989*a*) and shrimp (Simpson *et al.*, 1988) for various phenolic substrates is relatively low as evidenced by the K_m values which are generally within 0·5–10·0 mM. Much higher affinities (K_m ranging from 0·01 to 0·02 mM), however, were reported for PPO from potato tubers (Batistuti & Lourenco, 1985) and bananas (Mayer & Harel, 1979). The affinity of PPO to oxygen is also relatively low with K_m values ranging from 0·1 to 0·5 mM depending on the phenolic substrate being oxidised (Mayer & Harel, 1979).

Substrate inhibition at high concentrations has been reported for PPO from several plant sources. At substrate concentrations greater than 10 mM, kinetic plots of chlorogenic acid oxidation by PPO from potato (Abukharma & Woolhouse, 1966) and Jerusalem artichoke tubers (Zawistowski *et al.*, 1988*b*) showed a strong inhibitory effect. Similar responses were observed for wheat PPO with 4-methylcatechol (Interesse *et al.*, 1983). This phenomenon is often rationalised by assuming that at high substrate concentrations, both active [ES] and inactive [ES_2] enzyme–substrate complexes are formed (Koller & Neukom, 1969). In the case of the latter it has been suggested that a second substrate molecule bound ineffectively near the active center of the enzyme retards its catalytic action (Trémolières & Bieth, 1984).

The kinetic studies of Janovitz-Klapp *et al.* (1990) on purified apple PPO, where 4-methylcatechol and benzoic acid were used as substrate and competitive inhibitor, respectively, suggested that apple PPO combines with oxygen first prior to binding the phenolic substrate. In this regard, it is worth noting that kinetic data pertaining to the mechanism of PPO action are scarce and rather controversial. Earlier studies on concentration effects of oxygen and phenolics on reaction kinetics lead to several mechanisms. Gregory & Bendall (1966) considered a ping-pong mechanism, whereas Ingraham (1957) and

Matheis & Belitz (1977) suggested an ordered mechanism where oxygen binds first. In contrast, other studies (Duckworth & Coleman, 1970; Gutteridge & Robb, 1975; Lerner & Mayer, 1976) pointed to a random mechanism or an ordered one where oxygen does not bind first. The reason for such discrepancies might be due to a number of factors affecting the accuracy of obtaining 'true' kinetic data. These are inactivation of PPO during its catalysis (Golan-Goldhirsh & Whitaker, 1984); oxygen consumption by non-enzymatic processes involving secondary products of the reaction; the presence of both mono- and diphenolase activities (i.e. *o*-diphenol has to be considered as a substrate and a product), and the complications associated with spectrophotometric measurements of PPO activity (Janovitz-Klapp *et al.*, 1990). Janovitz-Klapp *et al.* (1990) also conducted kinetic studies with systems containing two phenolic compounds and derived an equation to predict oxygen consumption from the relative concentrations and kinetic parameters of each individual phenolic substrate present. This approach could be interesting in studying enzymatic browning of phenolics in mixtures as they are actually found in fruits and vegetables.

Inhibitors of PPO

Inhibitors of PPO and thereby the control of enzymatic browning has been the subject of continuing interest and a challenge to the food processor. Enzymatic browning is detrimental to appearance, organoleptic properties, nutritional quality and occasionally safety of the product (Friedman & Molnar-Perl, 1990). A wide range of compounds are known to inhibit PPO, although their efficiency in controlling browning reactions is dependent on the nature and concentration of the inhibitor, source of PPO, substrate availability (O_2, phenolics), pH and temperature (Mayer & Harel, 1979; Vámos-Vigyázó, 1981). The mode of action also varies and in many cases it is difficult to assign a single specific mechanism of action for an inhibitor. For example, some inhibitors act simultaneously on the enzyme, substrate and some products. Moreover, aside from any economic feasibility, the use of effective browning inhibitors in food processing is restricted due to toxicity problems, effects on taste, flavor, texture and wholesomeness of the food product.

With respect to inhibitors acting directly on the enzyme, several modes of action have been recognized. Since PPO is a metalloprotein it can be inhibited by chelating agents such as cyanide, carbon monoxide, sodium diethyl-dithiocarbamate (DIECA), tropolone, 2-mercaptobenzothiazole, azide, potassium methyl xanthate and EDTA (Mayer & Harel, 1979; Vámos-Vigyázó, 1981; Kahn & Andrawis, 1985). Inorganic ions, including halides, are also known to inhibit PPOs from many sources (Lerner, 1953; Peñafiel *et al.*, 1984; Martinez *et al.*, 1986). Investigations on frog epidermis tyrosinase showed that inhibition by halides, which increases with a decrease in pH, is due to their interaction with the copper at the active center of the enzyme (Peñafiel *et al.*, 1984). The pH dependence has been explained by considering that halides compete with a protonable ligand (LH^+, a histidine residue) for the metal ion. In a comparative study of tyrosinases from different sources by Martinez *et al.* (1986), they suggested that the degree of tyrosinase inhibition by halide anions is correlated not only with the strength of Cu- - -X^- interaction but also the accessibility of the exogenous halide ligands to the enzyme active site. Inhibition of PPO by these agents is of limited commercial value since most of these compounds are toxic with the exception of NaCl. However, sodium chloride, useful in preventing darkening of freshly peeled fruits and vegetables (Ponting, 1960), is generally a weak inhibitor at low to moderate concentrations.

Benzoic acid and some substituted cinnamic acids were mostly found as competitive inhibitors of PPOs from cherries, apples, pears, apricots and potatoes, as reviewed by Vámos-Vigyázó (1981). The inhibitor constant (K_i) generally increases in the order of cinnamic acid < *p*-coumaric < ferulic < *m*-coumaric < *o*-coumaric < benzoic. In addition, compounds which do not resemble the substrate structure but are slowly oxidised, such as 4-nitrocatechol, *p*-nitrophenol and 4-chlorocatechol, have been identified as competitive inhibitors of PPO (Mayer & Harel, 1979).

Compounds showing competitive inhibition with oxygen are non-competitive towards the phenolic substrate; similarly, competitive inhibitors with respect to the phenolics are non-competitive or uncompetitive towards oxygen. Hence, inhibition of mushroom tyrosinase by benzoic acid and cyanide ions shows that the former is competitive with catechol and non-competitive with oxygen, while the latter is competitive with oxygen and non-competitive with catechol (Duckworth & Coleman, 1970). In a kinetic investigation of apple

PPO, benzoic acid was found to be a competitive inhibitor towards 4-methylcatechol and uncompetitive with respect to oxygen (Janovitz-Klapp *et al.*, 1990). Competitive inhibition towards both substrates (oxygen, phenol) has been reported for diacetyl and 2,3-naphthalenediol (Mayer & Harel, 1979). The latter compound, being a highly specific inhibitor for PPO does not exert any action on laccase and therefore is used to distinguish between the two enzymes (Mayer *et al.*, 1964).

In the presence of chlorogenic acid as the substrate, the action of cinnamic, *p*-coumaric and ferulic acids was non-competitive on a particulate form of PPO from apples and competitive on its solubilised counterpart (Walker & Wilson, 1975). Conformational changes brought about upon solubilisation of the enzyme have been attributed for the different inhibitory action on the two enzyme forms. The properties of an inhibitor (type of inhibition and K_i) also depend on whether an *o*-diphenol or monophenol is used as the substrate. The different responses of the two activities have been interpreted to originate from the existence of two distinct active sites in PPO for mono- and diphenols (Macrae & Duggleby, 1968). In addition, the K_i values for an inhibitor may vary with respect to isoenzymes of the same PPO system. Hence, for coumaric acid (non-competitive towards chlorogenic acid), Rivas & Whitaker (1973) reported K_i values of 0·38 and 0·50 mM for the A and B PPO isoenzymes from pears, respectively. The corresponding K_i values for benzoic acid (competitive inhibitor) were 0·04 and 0·11 mM for the A and B PPO isoenzymes. Overall, as pointed out by several authors (Mayer & Harel, 1979; Janovitz-Klapp *et al.*, 1990), kinetic data of inhibitor studies on PPO and interpretations derived therefrom must be viewed carefully and critically due to the multi-substrate nature of catalysis as well as other factors (discussed in the section on 'Substrate Specificity'), which may further complicate the observed reaction kinetics.

Inhibition of PPO is also caused by reducing agents among which sulphites, SO_2, ascorbic acid (AA), erythorbic acid and thiol-containing compounds are the most frequently studied. Some of these inhibitors do not simply act by reducing *o*-quinones back to *o*-diphenols but also react with quinones to yield colorless compounds or exert a direct inhibitory effect on the enzyme (Walker, 1964; Embs & Markarkis, 1965; Pierpoint, 1966; Golan-Goldhirsh & Whitaker, 1984; Kahn & Andrawis, 1986; Dudley & Hotchkiss, 1989; Ferrer *et al.*, 1989*a*; Friedman & Molnar-Perl, 1990*a,b*). The effect of reducing agents could be temporary, because they themselves are irreversibly oxidised during the process; e.g. AA is readily oxidised by endogenous enzymes or by iron- or copper-catalysed oxidation. Moreover the use of certain reducing agents can lead to oxidation products with off-flavors and thus limit their application to foods.

Sulphites (sodium sulphite, sodium bisulphite, sodium and potassium metabisulphite), although the most effective inhibitors of PPO, are currently under re-examination because of public health concerns; some individuals, particularly asthmatics, show strong allergic responses towards these agents (Brown, 1985; Fan & Book, 1987). While earlier reports on inactivation of potato PPO by sulphites showed dependence on pH and time of exposure (Amla & Francis, 1961; Muneta & Wang, 1977), recent studies by Molnar-Perl & Friedman (1990*b*) indicated that inhibition of apple PPO by these agents was independent of pH. In addition to reaction with intermediate quinones forming sulphoquinones, sulphites have been reported to irreversibly inhibit PPO from mushroom (Embs & Markakis, 1965; Golan-Goldhirsh & Whitaker, 1984) and lobster (Ferrer *et al.*, 1989*a*).

A dual mode of action as a quinone reductant and direct inhibition of the enzyme has been assigned to AA (Golan-Goldhirsh & Whitaker, 1984). Electron spin resonance studies by Hsu *et al* (1988) demonstrated that Cu^{2+} of mushroom PPO is reduced to Cu^{+} by AA. Furthermore, the possibility that AA, as well as other reductants, inactivates PPO via formation of free radicals close to the active site, which could then oxidise histidine, was recently suggested by Andrawis & Kahn (1990). These authors previously proposed that inactivation of mushroom PPO by H_2O_2 also proceeds by production of $OH^{\cdot}$ bound to the enzyme, with the latter attacking an adjacent histidine and oxidising it (Andrawis & Kahn, 1985). Although AA, as a natural inhibitor, is the first choice for replacing sulphites, its instability to oxidation makes it less effective than other inhibitors at equivalent concentrations (Hsu *et al.*, 1988; Sapers *et al.*, 1989*b*; Molnar-Perl & Friedman, 1990*a*). Ascorbic acid derivatives, having greater stability against O_2, have recently been examined as potential inhibitors. Studies by Sapers *et al.* showed that AA-2-PO_4 and AA-2-$(PO_4)_3$ were effective in preventing browning at cut surfaces of apples but relatively ineffective in apple juice. Cinnamate and benzoate, used in combination with AA and its derivatives, enhanced their anti-browning effect in juice but induced browning on apple slices. Combinations of AA with Sporix, an acidic polyphosphate which acts as a chelator, proved highly effective in preventing browning in both apple juice and on cut surfaces (Sapers *et al.*, 1989*b*). AA-derivatives, AA-2-PO_4 and AA-2-SO_4, were both examined in a model system containing mushroom PPO and found to be ineffective inhibitors unless phosphatase or sulphatase enzymes were included to cleave the esters.

Thiol compounds, including L-cysteine, glutathione, dithiothreitol, mercaptoethanol and thiourea, are effective inhibitors of many PPO systems including apple (Walker, 1964; Molnar-Perl & Friedman, 1990b), pear (Montgomery, 1983), Jerusalem artichoke tubers (Zawistowski *et al.*, 1987*b*), cherimoye (Martinez-Cayuela *et al.*, 1988*b*), banana, avocado and mushroom (Kahn, 1985*a*). *N*-Acetyl-L-cysteine and reduced glutathione were recently shown to be as effective as sulphites in preventing browning of both apples and potatoes as well as protein-containing foods (Molnar-Perl & Friedman, 1990*a,b*). Because of their lower flavor threshold compared to cysteine (Molnar-Perl & Friedman, 1990*b*), more applied research is needed to determine whether these compounds offer a practical alternative to sulphites. Thiol compounds, being strong nucleophiles, suppress enzymic browning mainly via formation of colorless addition products with *o*-quinones, i.e. the quinone intermediate is preferentially trapped as a sulphur adduct instead of polymerizing to melanin (Roberts, 1959; Pierpoint, 1966). The structure of some compounds has been recently elucidated by NMR Spectroscopy. For example, the adduct of cysteine with catechol appears to be S-(2,3-dihydroxyphenol)-cysteine (Dudley & Hotchkiss, 1989) while glutathione and caffeoyl-L-tartaric acid produce 2-*S*-glutathionylcaffeoyltartaric acid (Cheynier *et al.*, 1986).

The inhibitory effect of honey on PPO and on the browning of apple slices, grape juice and in model systems was attributed to a small peptide with a molecular weight of 600 (Oszmianski & Lee, 1990). While several inhibitors in the form of small peptides have been reported in the literature (Harel *et al.*, 1967; Madhosingh, 1975; Mayer & Harel, 1979) the use of such 'natural inhibitors' has not been explored. Treatment with a dilute solution of a sulphydryl plant protease, ficin, was found to be effective in retarding shrimp melanosis (Taoukis *et al.*, 1990). In a previous study, cysteine was found to be ineffective in preventing oxidation of DOPA by shrimp enzymes (Simpson *et al.*, 1988).

As an alternative to sulphites, Langdon (1987) recommended dipping fruits and vegetables in a mixed-acid solution (citric and ascorbic acids plus potassium sorbate to inhibit yeast and mold growth) to control enzymatic browning. Citric acid exerts a double inhibitory effect by lowering the pH and by chelating copper; as such it also stabilises AA against oxidation. This treatment could be even more effective when combined with modified atmosphere packaging where most O_2 is removed. In a recent report by Molnar-Perl & Friedman (1990*b*), however, mixed acid solutions appeared of limited value in retarding browning of potato and apple slices. A dye-sensitised photo-oxidation was investigated as a non-thermal means to inactivate mushroom tyrosinase in a model system (Lowum & Parkin, 1989). The rate of inactivation was dependent on pH, O_2, temperature, protein concentration, ionic strength and the nature and concentration of the dye. Photo-oxidation is believed to proceed via a free-radical mechanism involving active oxygen species which oxidise key amino acids of the enzyme. Lerch (1981) first suggested that the 'suicide-type' inactivation of tyrosinase is caused by $OH^{\cdot}$ bound to the enzyme as Cu^{2+}—$OH^{\cdot}$ which could oxidise His-306 at the active site and lead to loss of activity.

pH Optimum and Thermal Stability

The pH optimum of PPO from many plant sources, including apple, pear, peach, grape, banana, avocado, potato and mushroom, is generally between 5·0 and 7·0, with the exception of the sweet cherry PPO (pH optimum 4·0–4·5), as reviewed by Vámos-Vigyázó (1981). Within this pH range also lie the pH optima of several PPOs characterised in the last decade from kiwifruit, pH 6·8–7·3 (Park & Luh, 1985); dates, pH 6·0–7·0 (Sachde *et al.*, 1989*b*); eggplant, pH 4·0 (Fujita & Tono, 1988); spinach leaves, pH 6·5 (Oda *et al.*, 1989); wild rice, pH 7·8 (Owusu-Ansah, 1989); Jerusalem artichoke tubers, pH 6·0 (Zawistowski *et al.*, 1988*a*); shrimp, pH 6·5–8·0 (Simpson *et al.*, 1987, 1988); lobster, pH 6·5–8·0 (Ferrer *et al.*, 1989*a*). It is also noteworthy that two pH optima have been found for some PPO preparations, with the activity being greater at the higher pH, e.g. grapes, 5·0 and 7·3 (Harel & Mayer, 1971); wheat, 5·3 and 5·9 (Interesse *et al.*, 1980); apples, 5·2 and 7·3 (Shannon & Pratt, 1967). PPO is often reported to be inactive at pHs below 4·0, thus providing a method for controlling browning reactions by lowering the pH (McCord & Kilara, 1983; Zemel *et al.*, 1990). Other features which recur from various reports in the literature are that the pH–activity profiles could vary with the cultivar and stage of maturity, enzyme purity, nature of the phenolic substrate and isoenzyme form.

Although dependent on the source, PPO is generally considered as an enzyme of low thermostability. Heat stability was reported to differ among cultivars and multiple forms of PPO from the same source as well as between fruit tissue homogenates and their respective juices (Vámos-Vigyázó, 1981). Thermal treatments of relatively short duration at 70–90°C are sufficient to substantially reduce or completely eliminate the PPO activity in products of plant origin. In contrast, the enzyme is relatively stable under storage at subzero temperatures (Vámos-Vigyázó, 1981).

APPLICATIONS OF PPO

The major concern in the food industry is to prevent the development of enzymatic browning prior to the processing of fruits and vegetables (Eskin, 1990). This is accomplished by inhibiting PPO activity using some of the inhibitors discussed in an earlier section. PPO, however, plays a critical role in the fermentation of black tea, and in recent years tyrosinase has been recognized for its potential medical applications.

Tea Manufacture

Tea, one of the world's popular beverages, is produced from the tender, rapidly growing shoots of the tea plant (tea flush) or tender young leaves (Wickremasinghe, 1978; Yaminishi, 1978). It is classified according to the process of manufacture as fermented tea (black tea) and non-fermented tea (green tea). PPO plays a key role during fermentation of black tea by oxidising polyphenols (Sanderson, 1964*a,b*). The major group of polyphenols in tea are the catechins which include, (+)-catechin, (−)-epicatechin, (+)-gallocatechin, (−)-epigallocatechin, (−)-epicatechin gallate and (−)-epigallocatechin gallate. During fermentation the catechins are oxidised to the corresponding *o*-quinones which then undergo secondary oxidation to form theaflavin (TF) and theaflavin gallates (TFG). The latter compounds are both yellow pigments which undergo oxidation to thearubigins (TR), dark brown pigments responsible for the color of

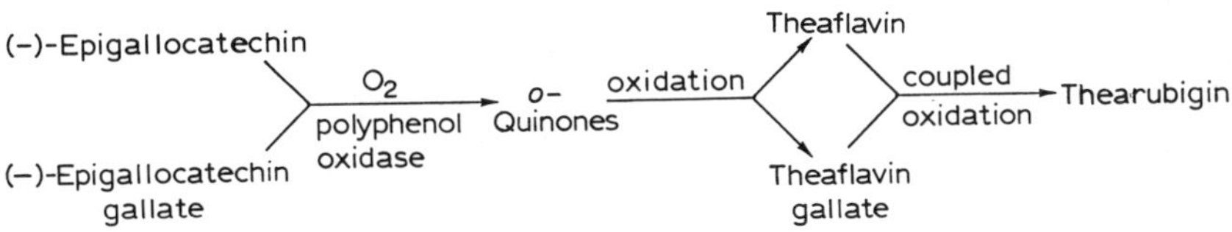

Fig. 5. Oxidative transformations of (−)-epigallocatechin and its gallate during tea fermentation (Eskin, 1990).

black tea (Sanderson, 1972; Cloughley, 1980; Graham, 1983). These oxidative reactions are summarised in Fig. 5.

The quality of tea, based on sensory evaluation of colour and bitterness, has been correlated with total theaflavin content (Hilton & Ellis, 1972). Theaflavins, thearubigins and caffeine are all essential ingredients in high-quality teas. Early work by Roberts (1952) and Sanderson (1964*b*) positively correlated PPO activity with the quality of tea. Owuor & Othieno (1987) showed that the recommended plucking standard of two leaves and a bud resulted in the highest level of caffeine, crude fiber and theaflavins. The high content of theaflavins was attributed to PPO which increases in activity from the bud to the mature leaf stages (Basu & Dev Choudhury, 1984). Oxidative degradation of the phloroglucinol rings of theaflavins by peroxidase resulted in a loss of tea quality. Van Lelyveld & de Rooster (1986) reported much higher levels of PPO in high-quality hybrid tea clones compared to a poor-quality seedling tea. Their results, in Table 2, also showed higher levels of peroxidase in the high-quality hybrid tea clones, the significance of which remains unclear.

Using model systems, Bajaj *et al.* (1987) showed PPO oxidised theaflavin monogallates much faster than theaflavin digallates. Nevertheless, Thanaraj & Seshadri (1990) found that theaflavin

Table 2. Specific Activities of MT12 and Seedling Tea Leaves[a,b]

Clone	*Peroxidase*[c]	*Polyphenol oxidase*[d]
M12	1·821	0·055
Seedling	0·735	0·020

[a] Van Lelyveld & de Rooster (1986). Reprinted with permission from J. Hortic. Sci., **61**, 545.
[b] Specific activity in Δ OD/min per mg protein.
[c] Significant at $P < 0{\cdot}05$.
[d] Significant at $P < 0{\cdot}01$.

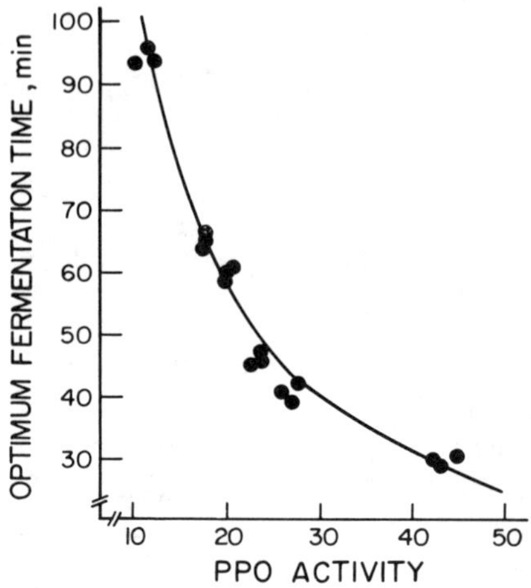

Fig. 6. Relationship between PPO activity of fresh shoots and optimum fermentation time of six clones (Thanaraj & Seshadri, 1990).

digallates were a better indicator of tea quality than total theaflavins. These researchers reported variations in PPO activity, total phenolic compounds and catechins in different clones and shoots in relation to black tea quality (*Camellia sinensis* (L) O Kuntze). The fast-fermenting clone had three to five times more PPO activity than the slow-fermenting clone. The optimum fermentation time was inversely related to PPO activity of six clones as shown in Fig. 6. A marked decrease in fermentation time accompanied the initial increase in PPO activity, although beyond 25 units only minor changes occurred. An increase in theaflavin content parallelled the rise in PPO activity reaching a maximum at 25 units of activity.

Medicinal Applications

Tyrosinase has received considerable attention because of its ability to produce black melanin or eumelanin. This insoluble heteropolymer, composed of 5,6-dihydroxyindole and 5,6-dihydroxy-indole-2-carboxylic acid units, finds uses as UV absorber, cation exchanger, in amorphous semiconductors and novel biopolymers capable of binding drugs (Bell & Wheeler, 1986; Ito, 1986). The therapeutic use of melanocytotoxic substances against cancer has also been reported. In mammals the active form of tyrosinase is located in a specialised cytoplasmic granule, the melanosome, present in melanocytes. The mechanism involves oxidation of melanocytes (Riley, 1970; Riley *et al.*, 1975). Dewey *et al.* (1977) found that 4-hydroxyanisole, a substrate for tyrosinase, caused regression of Harding-Passey melanomas when transplanted in mice, while Morgan *et al.* (1981) and Morgan (1984) showed it caused a regression of localised malignant melanoma. Studies by Doherty *et al* (1985) showed mushroom tyrosinase metabolised 1-naphthol to 1,2-naphthoquinone and covalently bound species. The mechanism proposed was *o*-hydroxylation of 1-naphthol to 1,2-dihydroxynaphthalene followed by oxidation to the corresponding quinone. The covalently bound species may alter the kinetics of this enzyme, which suggested that 1-naphthol or structurally related derivatives may be of therapeutic value for the treatment of melanomas high in tyrosinase activity.

Biotechnology and Melanin Formation

Della-Cioppa *et al.* (1990) successfully cloned the gene for tyrosinase together with a putative protein from *Streptococcus antiobioticus* into *Eschericha coli.* The production of melanin by the recombinant *E. coli* cells in agar plates and liquid culture required the presence of copper, tyrosine and the protein. Clones with high tyrosinase levels produced little melanin in the absence of the protein. The protein acted as a *trans* activator or as a metallothionein-like protein delivering copper to the apotyrosinase (Bernan *et al.*, 1985; Lee *et al.*, 1988). This development provides an alternative method for obtaining large quantities of melanin, compared to limited amounts of melanin available from natural sources (squid sacs or bovine retinas) or from the synthetic oxidation of DOPA.

REFERENCES

Abeles, R. H. & Maycock, A. L. (1976). Suicide enzyme inactivators. *Acc. Chem. Res.*, **9**, 313–18.

Abukharma, D. A. & Woolhouse, H. W. (1966). The preparation and properties of *o*-diphenol:oxygen oxidoreductase from potato tubers. *New Phytol.*, **65**, 477–8.

Al-Barazi, Z. & Schwabe, W. W. (1984). The possible involvement of polyphenol-oxidase and the auxin-oxidase system in root formation and development in cuttings of *Pistacia vera*. *J. Hortic. Sci.*, **59**, 453–61.

Al-Jassir, M. S. H. (1989). Polyphenol oxidase inhibition by glucose oxidase in pink shrimp (*Pandalus borealis*). *Dissert. Abst. Int., B,* **49,** 2433–4.

Amla, B. L. & Francis, F. J. (1961). Effects of pH of dipping solutions on the quality of prepeeled potatoes. *Am. Potato J.,* **38,** 121–30.

Andersen, R. A. & Sowers, J. A. (1968). Optimum conditions for bonding of plant phenols to insoluble polyvinylpyrrolidone. *Phytochemistry,* **7,** 293–301.

Anderson, J. W. (1968). Extraction of enzyme and subcellular organelles from plant tissues. *Phytochemistry,* **7,** 1973–88.

Andersson, L. (1984). Fractionation of human serum proteins by immobilized metal affinity chromatography. *J. Chromatogr.,* **315,** 167–74.

Andrawis, A. & Kahn, V. (1985). Inactivation of mushroom tyrosinase by hydrogen peroxide. *Phytochemistry,* **24,** 397–405.

Andrawis, A. & Kahn, V. (1990). Ability of various chemicals to reduce copper and to inactivate mushroom tyrosinase. *J. Food Biochem.,* **14,** 103–15.

Anosike, E. O. & Ayaebene, A. O. (1981). Purification and some properties of polyphenol oxidase from the yam tubers, *Discorea bulbifera. Phytochemistry,* **20,** 2625–8.

Badran, A. M. & Jones, D. E. (1965). Polyethylene glycols—tannins interaction in extracting enzymes, *Nature,* **206,** 622–4.

Bajaj, K. L. & Kaur, G. (1986). Polyphenol oxidase activity and chemical composition of some important Indian varieties of chillies (*Capsicum annuum* L.). *Tropical Sci.,* **26,** 121–8.

Bajaj, K. L., Anan, T., Tsushida, T. & Ikegaya, K. (1987). Effect of (−)-epicatechin on oxidation of theaflavins by polyphenol oxidase from tea leaves. *Agric. Biol. Chem.,* **51,** 1767–72.

Balasingam, K. & Ferdinand, W. (1970). The purification and properties of a ribonucleoenzyme, *o*-diphenol oxidase, from potatoes. *Biochem. J.,* **118,** 15–23.

Basu, R. P. & Dev Choudhury, M. N. (1984). How plucking and pruning affect quality of plains teas. *Two and a Bud,* **31,** 19–21.

Bastin, M. (1968). Effect of wounding on the synthesis of phenols, phenol oxidase, and peroxidase in the tuber tissue of Jerusalem artichoke. *Can. J. Biochem.,* **46,** 1339–43.

Bastin, M. & Ünlüer, O. (1972). Effect of actinomycin D on the formation of enzymes in Jerusalem artichoke tuber slices. *Planta,* **102,** 357–61.

Batistuti, J. P. & Lourenço, E. J. (1985). Isolation and purification of polyphenol oxidase from a new variety of potato. *Food Chem.,* **18,** 251–63.

Bell, A. A. & Wheeler, M. H. (1986). Biosynthesis and functions of fungal melanins. *Ann. Rev. Phytopathol.,* **24,** 411–51.

Beltramini, M. & Lerch, K. (1983). The reconstitution reaction of *Neurospora* apotyrosinase. *Biochem. Biophys. Res. Com.,* **110,** 313–19.

Bendall, D. S. & Gregory, R. P. F. (1963). Purification of phenol oxidases. In *Enzyme Chemistry of Phenolic Compounds,* ed. J. B. Pridham. Pergamon Press, Oxford, pp. 7–24.

Benjamin, N. O. & Montgomery, M. W. (1973). Polyphenol oxidase of *Royal Ann* cherries: purification and characterization. *J. Food Sci.,* **38,** 799–806.

Ben-Shalom, N., Kahn, V., Harel, E. & Mayer, A. M. (1977). Catechol oxidase from green olives: properties and partial purification. *Phytochemistry*, **16**, 1153–8.

Benson, A. A. (1964). Plant membrane lipids. *Ann. Rev. Plant Physiol.*, **15**, 1–16.

Bernan, V., Filpula, D., Herber, W., Bibb, M. & Katz, E. (1985). The nucleotide sequence of the tyrosinase gene from *Streptomyces antibioticus* and characterization of the gene product. *Gene*, **37**, 101–10.

Bertrand, G. (1896). Sur une nouvelle oxydase, ou ferment soluble oxydant, d'origine vegetale. *C.R. Acad. Sci. Paris*, **122**, 1215.

Bertrand, G. (1908). Action of tyrosinase on some compounds similar to tyrosine. *C.R. Acad. Sci. Paris*, **145**, 1352–5.

Bourquelot, E. & Bertrand, G. (1895). *Compt. Rend. Soc. Biol.*, **47**, 579.

Bourquelot, E. & Bertrand, G. (1896). *J. Pharm. Chem.*, **3**, 177–80.

Boyer, R. F., Mascotti, D. P. & Schori, B. E. (1986). Ferroxidase activity of mushroom tyrosinase. *Phytochemistry*, **25**, 1281–3.

Brooks, D. W. & Dawson, C. R. (1966). Aspects of tyrosinase chemistry. In *The Biochemistry of Copper*, ed. J. Peisach, P. Aisen & W. E. Blumberg. Academic Press, New York, pp. 343–57.

Brown, J. L. (1985). Sulfite sensitivity. *J. Am. Med. Assoc.*, **254**, 825.

Cash, J. N., Sistrunk, W. A. & Stutte, C. A. (1976). Characteristics of Concord grape polyphenoloxidase involved in juice color loss. *J. Food Sci.*, **41**, 1398–402.

Chan, H. T. Jr & Yang, H. Y. (1971). Identification and characterization of some oxidizing enzymes of the McFarlin cranberry. *J. Food Sci.*, **35**, 169–73.

Cheynier, V. F., Trousdale, E. K., Singleton, V. L., Salgues, M. J. & Wylde, R. (1986). Characterization of 2-S-glutathionylcaftaric acid and its hydrolysis in relation to grape wines. *J. Agric. Food Chem.*, **34**, 217–21.

Cheynier, V. F., Osse, C. & Rigaud, J. (1988). Oxidation of grape juice phenolic compounds in model solutions. *J. Food Sci.*, **53**, 1729–32.

Chin, L. S. & Chen, H. E. (1988). Studies on the enzymatic browning capacity of grape. I. Characterization of polyphenol oxidase from Golden Muscat grape. *Food Sci. China*, **13**, 180–87.

Chu, H. L. & Shaw, J. F. (1989). Biochemical characterization of isozymes of bighead shrimp tyrosinase. *J. Chin. Agric. Soc.*, **27**, 360–72.

Clayton, R. A. (1959). Properties of tobacco polyphenol oxidase. *Arch. Biochem. Biophys.*, **81**, 404–17.

Cloughley, J. B. (1980). The effect of temperature on enzyme activity during the fermentation phase of black tea manufacture. *J. Sci. Food Agric.*, **31**, 920–23.

Conlon, M. & Murphy, R. F. (1976). The interaction of immobilized transition-metal ions with some gastrointestinal polypeptides. *Biochem. Soc. Trans.*, **4**, 860–61.

Coombs, J., Baldry, C., Bucke, C. & Long, S. P. (1974). *o*-Diphenyl:oxygen oxidoreductase from leaves of sugar cane. *Phytochemistry*, **13**, 2703–8.

Cornwell, C. J. & Wrolstad, R. E. (1981). Causes of browning in pear juice concentrate during storage. *J. Food Sci.*, **46**, 515–18.

Coseteng, M. Y. & Lee, C. Y. (1987). Changes in apple polyphenoloxidase and polyphenol concentrations in relation to degree of browning. *J. Food Sci.*, **52,** 985–9.

Czaninski, Y. & Catesson, A. M. (1972). Localisation ultrastructurale d'activites polyphenoloxydasiques dans les chloroplastes de *Nicotina glutinosa. J. Microscopie*, **15,** 409–14.

Czapski, J., Saniewski, M., Puchalski, J., Lange, E. & Nowacki, J. (1988). The effect of methyl jasmonate on activity and electrophoretic pattern of polyphenol oxidase isoenzymes in postclimacteric apples cv. Jonathan. *Fruit Sci. Rep.*, **15,** 103–10.

Dawson, C. R. & Magee, R. J. (1955). Plant tyrosinase (polyphenol oxidase). *Methods in Enzymology*, **2,** 817–27.

Della-Cioppa, G., Garger, S. J., Sverlow, G. G., Turpen, T. H. & Grill, L. K. (1990). Melanin production in *Escherichia coli* from a cloned tyrosinase gene. *Biotechnology*, **8,** 634–8.

Deverall, B. J. (1981). Phenolase and pectic enzyme activity in the chocolate spot disease of beans. *Nature* (*London*), **189,** 311.

Dewey, D. L., Butcher, F. W. & Galpine, A. R. (1977). Hydroxyanisole-induced regression of the Harding-Passey melanoma in mice. *J. Pathol.*, **122,** 117–28.

Dizik, N. S. & Knapp, F. W. (1970). Avocado polyphenoloxidase: purification, and fractionation on Sephadex thin layers. *J. Food Sci.*, **35,** 282–5.

Doherty, M. D., Cohen, G. M., Gant, T. W., Naish, S. & Riley, P. A. (1985). Metabolism of 1-naphthol by tyrosinase. *Biochem. Pharmacol.*, **34,** 3167–72.

Dressler, H. & Dawson, C. R. (1960). On the nature and mode of action of the copperprotein tyrosinase. II. Exchange experiments with radioactive copper and the functioning enzyme. *Biochim. Biophys. Acta*, **45,** 515–24.

Duckworth, H. W. & Coleman, J. E. (1970). Physicochemical and kinetic properties of mushroom tyrosinase. *J. Biol. Chem.*, **245,** 1613–25.

Dudley, E. D. & Hotchkiss, J. H. (1989). Cysteine as an inhibitor of polyphenol oxidase. *J. Food Biochem.*, **13,** 65–75.

Duke, S. O. & Vaughn, K. C. (1982). Lack of involvement of polyphenol oxidase in *ortho*-hydroxylation of phenolic compounds in mung bean seedlings. *Physiol. Plant.*, **54,** 381–5.

El-Bayoumi, M. A. & Frieden, E. (1957). A spectrophotometric method for the determination of the catecholase activity of tyrosinase and some of its applications. *J. Am. Chem. Soc.*, **79,** 4854–8.

Embs, R. J. & Markakis, P. (1965). The mechanism of sulfite inhibition of browning caused by polyphenoloxidase. *J. Food Sci.*, **30,** 753–8.

Eskin, N. A. M. (1990). *Biochemistry of Foods*. Academic Press, New York.

Estabrook, R. W. (1967). Mitochondrial respiratory control and the polarographic measurement of ADP:O ratios. *Methods in Enzymology*, **10,** 41–47.

Fan, A. M. & Book, S. A. (1987). Sulfite hypersensitivity: a review of current issues. *J. Appl. Nutr.*, **39,** 71–8.

Ferrer, O. J. (1988). Activation, characterization, inhibition and seasonal variation in the phenolase level of Florida spiny lobster (*Panulirus argus*). Dissert. Abst. Int., B 49, 963.

Ferrer, O. J., Koburger, J. A., Otwell, W. S., Gleeson, R. A., Simpson, B. K. & Marshall, M. R. (1989*a*). Phenoloxidase from the cuticle of Florida spiny lobster (*Panulirus argus*): mode of activation and characterization. *J. Food Sci.*, **54,** 63–7, 176.
Ferrer, O. J., Koburger, J. A., Simpson, B. K., Gleeson, R. A. & Marshall, M. R. (1989*b*). Phenoloxidase levels in Florida spiny lobster (*Panulirus argus*): relationship to season and molting stage. *Comp. Biochem. Physiol.*, **93B,** 595–9.
Flick, G. J. Jr, Ory, R. L. & St. Angelo, A. J. (1977). Comparison of nutrient composition and of enzyme activity in purple, green, and white eggplants. *J. Agric. Food Chem.*, **25,** 117–20.
Fling, M., Horowitz, N. H. & Heinemann, S. F. (1963). The isolation and properties of crystalline tyrosinase from *Neurospora*. *J. Biol. Chem.*, **238,** 2045–52.
Flurkey, W. H. (1985). *In vitro* biosynthesis of *Vicia faba* polyphenoloxidase. *Plant Physiol.*, **79,** 564–7.
Flurkey, W. H. (1986). Polyphenoloxidase in higher plants: immunological detection and analysis of in vitro translation products. *Plant Physiol.*, **81,** 614–18.
Flurkey, W. H. (1989). Polypeptide composition and amino-terminal sequence of broad bean polyphenoloxidase. *Plant Physiol.*, **91,** 481–3.
Flurkey, W. H. & Jen, J. J. (1978). Peroxidase and polyphenol oxidase activities in developing peaches. *J. Food Sci.*, **43,** 1828–31.
Flurkey, W. H. & Jen, J. J. (1980). Purification of peach polyphenol oxidase in the presence of added protease inhibitors. *J. Food Biochem.*, **4,** 29–41.
Friedman, M. & Molnar-Perl, I. (1990). Inhibition of browning by sulfur amino acids. 1. Heated amino acid-glucose systems. *J. Agric. Food Chem.*, **38,** 1642–7.
Fujita, S. & Tono, T. (1988). Purification and some properties of polyphenoloxidase in eggplant (*Solanum melangena*). *J. Sci. Food Agric.*, **46,** 115–23.
Galeazzi, M. A. M., Sgarbieri, V. C. & Constantinides, S. M. (1981). Isolation, purification and physicochemical characterization of polyphenoloxidases (PPO) from a Dwarft variety of banana (*Musa cavendishii,* L). *J. Food Sci.*, **46,** 150–55.
Golan-Goldhirsh, A. & Whitaker, J. R. (1984). Effect of ascorbic acid, sodium bisulfite, and thiol compounds on mushroom polyphenol oxidase. *J. Agric. Food Chem.*, **32,** 1003–9.
Golbeck, J. H. & Cammarata, K. V. (1981). Spinach thylakoid polyphenol oxidase. Isolation, activation and properties of the native chloroplast enzyme. *Plant Physiol.*, **67,** 977–84.
Goodenough, P. W. (1978). The *o*-diphenol-oxygen-oxidoreductase of *Agaricus bisporus*: activity and multiple forms during ageing. *Phytochemistry,* **17,** 633–6.
Goodenough, P. W., Kessell, S., Lea, A. G. H. & Loeffler, T. (1983). Mono- and diphenolase activity from fruit of *Malus pumila*. *Phytochemistry,* **22,** 359–63.
Graham, H. (1983). Tea. In *Encyclopedia of Chemical Technology,* Vol. 22. John Wiley and Sons, USA.

Gregg, D. C. & Nelson, J. M. (1940). Further studies on the enzyme, tyrosinase. *J. Am. Chem. Soc.*, **62,** 2500–505.

Gregory, R. P. F. & Bendall, D. S. (1966). The purification and some properties of the polyphenol oxidase from tea (*Camellia sinensis* L.). *Biochem. J.*, **101,** 569–81.

Gross, D. & Coombs, J. (1976). Enzymatic colour formation in beet and cane juices. I. *Int. Sugar J.*, **3,** 69–73.

Gutteridge, S. & Robb, D. A. (1973). Purification of mushroom tyrosinase by affinity chromatography. *Biochem. Soc. Trans.*, **1,** 519–20.

Gutteridge, S. & Robb, D. A. (1975). The catecholase activity of *Neurospora* tyrosinase. *Eur. J. Biochem.*, **54,** 107–116.

Ha, Y. D. & Lee, M. O. (1988). Some properties of polyphenol oxidase from potato (*Solanum tuberosum*) and inhibition of this enzyme system by sulfites. *J. Korean Soc. Food Nutr.*, **17,** 198–204.

Harel, E., Mayer, A. M. & Shain, Y. (1964). Catechol oxidases from apples, their properties, subcellular location and inhibition. *Physiol. Plant.*, **17,** 921–30.

Harel, E., Mayer, A. M. & Shain, Y. (1965). Purification and multiplicity of catechol oxidase from apple chloroplasts. *Phytochemistry*, **4,** 783–90.

Harel, E., Mayer, A. M. & Shain, Y. (1966). Catechol oxidases, endogenous substrates and browning in developing apples. *J. Sci. Food Agric.*, **17,** 389–92.

Harel, E., Mayer, A. M., Markus, Z. & Avigad, G. (1967). Inhibition of apple catechol oxidase by a galactose oxidase inhibitor from *Dactylium dendroides*. *Isr. J. Bot.*, **16,** 38–41.

Harel, E. & Mayer, A. M. (1968). Interconversion of sub-units of catechol oxidase from apple chloroplasts. *Phytochemistry*, **7,** 199–204.

Harel, E. & Mayer, A. M. (1970). The use of a fungal pectate lyase in the purification of laccase from peaches. *Phytochemistry*, **9,** 2447–52.

Harel, E. & Mayer, A. M. (1971). Partial purification and properties of catechol oxidases in grapes. *Phytochemistry*, **10,** 17–22.

Hasegawa, S. & Maier, V. P. (1980). Polyphenol oxidase of dates. *J. Agric. Food Chem.*, **28,** 891–3.

Hemdan, E. S. & Porath, J. (1985). Development of immobilized metal affinity chromatography. II. Interaction of amino acids with immobilized nickel iminodiacetate. *J. Chromatogr.*, **323,** 255–64.

Henry, E. W., De Poore, H. J. & O'Connor, M. N. (1982). Sorbitol disrupted spinach (*Spinacia oleracea* L.) chloroplasts: cytochemical localization of polyphenol oxidase in discontinuous sucrose density gradient fractions. *Micron*, **13,** 453–4.

Hilton, P. J. & Ellis, R. T. (1972). Estimation of the market value of Central African tea by theaflavin analysis. *J. Sci. Food Agric.*, **23,** 227–32.

Himmelwright, R. S., Eickman, N. C., LuBien, C. D., Lerch, K. & Solomon, E. I. (1980). Chemical and spectroscopic studies of the binuclear copper active site of *Neurospora* tyrosinase: comparison to hemocyanins. *J. Am. Chem. Soc.*, **102,** 7339–44.

Hsu, A. F., Kalan, E. B. & Bills, D. D. (1984). Partial purification and characterization of the soluble polyphenol oxidases from suspension cultures of *Papaver somniferum*. *Plant Sci. Lett.*, **34,** 315–22.

Hsu, A. F., Shieh, J. J., Bills, D. D. & White, K. (1988). Inhibition of mushroom polyphenoloxidase by ascorbic acid derivatives. *J. Food Sci.*, **53,** 765–7, 771.

Huber, M. & Lerch, K. (1986). Active site and protein structure of tyrosinase: comparison to hemocyanin. Invertebr. Oxygen Carriers Int. Conf., 1985, pp. 265–76.

Hutcheson, S. W., Buchanan, B. B. & Montalbini, P. (1980). Polyphenol oxidation by *Vicia faba* chloroplast membranes. Studies on the latent membrane-bound polyphenol oxidase and on the mechanism of photochemical polyphenol oxidation. *Plant Physiol.*, **66,** 1150–54.

Ingraham, L. L. (1957). Variation of the Michaelis constant in polyphenol oxidase catalyzed oxidations: substrate structure and concentration. *J. Am. Chem. Soc.*, **79,** 666–9.

Interesse, F. S., Ruggiero, P., D'Avella, G. & Lamparelli, F. (1980). Partial purification and some properties of wheat (*Triticum aestivum*) *o*-diphenolase. *J. Sci. Food Agric.*, **31,** 459–66.

Interesse, F. S., Ruggiero, P., Lamparelli, F. & D'Avella, G. (1981). Isoenzymes of wheat *o*-diphenolase revealed by column isoelectric focusing. *Z. Lebensm. Unters. Forsch.*, **172,** 100–103.

Interesse, F. S., Ruggiero, P., D'Avella, G. & Lamparelli, F. (1983). Characterization of wheat *o*-diphenolase isoenzyme. *Phytochemistry*, **22,** 1885–9.

Ishiguro, S., Yamamoto, H., Yanai, N. & Takeuchi, T. (1986). Quantitative analysis of mouse tyrosinase by enzyme-linked immunosorbent assay. *J. Biochem.*, **99,** 1081–5.

Ito, F. (1986). Reexamination of the structure of eumelanin. *Biochim. Biophys. Acta*, **883,** 155–61.

IUB–IUPAC Commission on Biochemical Nomenclature (1968). A one-letter notation for amino acid sequences. Tentative rules. *Biochemistry*, **7,** 2703–5.

Ivanov, T. P. (1966). Activity of polyphenol oxidase during ripening of Cherven muscat, dimyat, riesling and aligote grapes. *Lozar Vinar.*, **15,** 24–8.

James, W. O. (1953). The terminal oxidases of plant respiration. *Biol. Rev. Cambridge Philos. Soc.*, **28,** 245–60.

Janovitz-Klapp, A., Richard, F. & Nicolas, J. (1989). Polyphenoloxidase from apple, partial purification and some properties. *Phytochemistry*, **28,** 2903–7.

Janovitz-Klapp, A. H., Richard, F. C., Goupy, P. M. & Nicolas, J. J. (1990). Kinetic studies on apple polyphenol oxidase. *J. Agric. Food Chem.*, **38,** 1437–41.

Jayaraman, K. S., Ramanuja, M. N. & Vijayaraghavan, P. K. (1987). Studies on the purification of banana polyphenoloxidase. *Food Chem.*, **24,** 203–17.

Jen, J. J. & Flurkey, W. H. (1979). Hydrophobic chromatography of peach fruit polyphenol oxidase. *HortScience*, **14,** 516–18.

Jimbow, K., O'Hara, D. S. & Fitzpatrick, T. B. (1975). Identification and

purification of tyrosinase from melanosomes of mouse melanoma by tyrosine-dopa affinity chromatography. *J. Invest. Dermatol.*, **64**, 278.
Jimbow, K., Jimbow, M. & Chiba, M. (1981). Characterization of structural properties for morphological differentiation of melanosomes: I. Purification of tyrosinase by tyrosine affinity chromatography and its characterization in B16 and Harding Passey melanomas. *J. Invest. Dermatol.*, **77**, 213–18.
Jolley, R. L. Jr, Robb, D. A. & Mason, H. S. (1969*a*). Multiple forms of mushroom tyrosinase. Association-dissociation phenomena. *J. Biol. Chem.*, **244**, 1593–9.
Jolley, R. L. Jr, Nelson, R. M. & Robb, D. A. (1969b). The multiple forms of mushroom tyrosinase. Structural studies on the isoenzymes. *J. Biol. Chem.*, **244**, 3251–7.
Jolley, R. L. Jr, Evans, L. H. & Mason, H. S. (1972). Reversible oxygenation of tyrosinase. *Biochem. Biophys. Res. Commun.*, **46**, 878–84.
Jolley, R. L. Jr, Evans, L. H., Makino, N. & Mason, H. S. (1974). Oxytyrosinase. *J. Biol. Chem.*, **249**, 335–45.
Joshi, P. R. & Shiralkar, N. D. (1977). Polyphenolases of a local variety of mango. *J. Food Sci. Technol. India*, **14**, 77–9.
Kahn, V. (1975). Polyphenol oxidase activity and browning of three avocado varieties. *J. Sci. Food Agric.*, **26**, 1319–24.
Kahn, V. (1977). Latency properties of polyphenol oxidase in two avocado cultivars differing in their rate of browning. *J. Sci. Food Agric.*, **28**, 233–9.
Kahn, V. (1985*a*). Effect of proteins, protein hydrolyzates and amino acids on *o*-dihydroxyphenolase activity of polyphenol oxidase of mushroom, avocado, and banana. *J. Food Sci.*, **50**, 111–15.
Kahn, V. (1985*b*). Tropolone—a compound that can aid in differentiating between tyrosinase and peroxidase. *Phytochemistry*, **24**, 915–20.
Kahn, V. & Andrawis, A. (1985). Inhibition of mushroom tyrosinase by tropolone. *Phytochemistry*, **24**, 905–08.
Kahn, V. & Andrawis, A. (1986). Multiple effect of hydroxylamine on mushroom tyrosinase. *Phytochemistry*, **25**, 333–7.
Kahn, V. & Pomerantz, S. H. (1980). Monophenolase activity of avocado polyphenol oxidase. *Phytochemistry*, **19**, 379–85.
Kato, C., Uritani, I., Saijo, R. & Takeo, T. (1976). Cellular localization of particulate-bound polyphenol oxidase in tea leaves. *Plant Cell Physiol.*, **17**, 1045–52.
Kenten, R. H. (1957). Latent phenolase in extracts of broad-bean (*Vicia faba* L.) leaves. 1. Activation by acid and alkali. *Biochemical. J.*, **67**, 300–307.
Kenten, R. H. (1958) Latent phenolase in extracts of broad-bean (*Vicia faba* L.) leaves. 2. Activation by anionic wetting agents. *Biochemical. J.*, **68**, 244–51.
Kertesz, D. & Zito, R. (1962). Phenolase. In *Oxygenases*, ed. O. Hayaishi. Academic Press, New York, p. 307.
Kertesz, D. & Zito, R. (1965). Mushroom polyphenol oxidase. 1. Purification and general properties. *Biochim. Biophys. Acta*, **96**, 477–62.
Kertesz, D. (1966). The copper of polyphenoloxidase. In *The Biochemistry of Copper*, ed. J. Peisach, P. Aisen & W. E. Blumberg. Academic Press, New York, pp. 359–69.

Kertesz, D., Rotilio, G., Brunori, M., Zito, R. & Antonini, E. (1972). Kinetics of reconstitution of polyphenoloxidase from apoenzyme and copper. *Biochem. Biophys. Res. Commun.*, **49**, 1208–15.
Kidron, M., Harel, E. & Mayer, A. M. (1977). Copper content and amino acid composition of catechol oxidase from Clairette grapes. *Phytochemistry*, **16**, 1050–51.
Kikuchi, H. & Watanabe, M. (1981). Significance of use of amino acids and histamine for the elution of nonhistone proteins in copper-chelate chromatography. *Anal. Biochem.*, **115**, 109–12.
Kim, H. (1988). Isolation and characterization of polyphenol oxidase from cocoa beans. Dissert. Abst. Int., B 48, 2834.
King, R. S. & Flurkey, W. H. (1987). Effects of limited proteolysis on broad bean polyphenoloxidase. *J. Sci. Food Agric.*, **41**, 231–40.
Koller, A. & Neukom, H. (1969). Untersuchungen über den abbaumechanismus einer gereinigten polygalakturonase aus *Aspergillus niger*. *Eur. J. Biochem.*, **7**, 485–9.
Kubowitz, (1938). Spaltung und resynthese der polyphenoloxydase und des hamocyanins. *Biochem. Z.*, **299**, 32–57.
Kupper, U., Niedermann, D. M., Travaglini, G. & Lerch, K. (1989). Isolation and characterization of the tyrosinase gene from *Neurospora crassa*. *J. Biol. Chem.*, **264**, 17250–58.
Kuznetsova, L. S., Potavina, V. S. & Ovsyannikova, N. F. (1987). Activity of polyphenoloxidase in fresh and fermented cocoa beans. *Khlebopekarnayi i Konditerkaya Promyshlennost*, **3**, 34–5.
Lamikanra, O. (1988). Browning capacity of muscadine grapes. *Proc. Florida St. Hort. Soc.*, **101**, 224–8.
Langdon, T. T. (1987). Preventing of browning in fresh prepared potatoes without the use of sulfiting agents. *Food Technol.*, **41**, 64–7.
Lanker, T., Flurkey, W. H. & Hughes, J. P. (1988). Cross-reactivity of polyclonal and monoclonal antibodies to polyphenoloxidase in higher plants. *Phytochemistry*, **27**, 3731–4.
Lanzarini, G., Pifferi, P. G. & Zamorani, A. (1972). Specificity of an *o*-diphenoloxidase from *Prunus avium* fruits. *Phytochemistry*, **11**, 89–94.
Lee, V.-H. W., Chen, B. F., Wu, S.-Y., Leu, W.-M., Lin, J.-J., Chen, C. C. & Lo, S. J. (1988). A trans-acting gene b required for the phenotypic expression of a tyrosinase gene in *Streptomyces*. *Gene*, **65**, 71–81.
Lerch, K. (1976). *Neurospora* tyrosinase: molecular weight, copper content and spectral properties. *FEBS Lett.*, **69**, 157–60.
Lerch, K. (1978). Amino acid sequence of tyrosinase from *Neurospora crassa*. *Proc. Natl. Acad. Sci. U.S.A.*, **75**, 3635–9.
Lerch, K. (1981). Copper monooxygenases: tyrosinase and dopamine β-monooxygenase. In *Metal Ions in Biological Systems. Copper Proteins*, ed. H. Sigel. Marcel Dekker, New York, Basel, pp. 143–86.
Lerch, K., Hubner, M., Schneider, H.-J., Drexel, R. & Linzen, B. (1986). Different origins of metal binding sites in binuclear copper proteins, tyrosinase and hemocyanin. *J. Inorg. Biochem.*, **26**, 213–17.
Lerman, L. S. (1953). A biochemically specific method for enzyme isolation. *Proc. Natl. Acad. Sci. U.S.A.*, **39**, 232–6.

Lerner, A. B. (1953). Mammalian tyrosinase: effect of ions on enzyme action. *Arch. Biochem. Biophys.*, **36**, 473–81.

Lerner, H. R. & Mayer, A. M. (1976). Reaction mechanism of grape catechol oxidase—a kinetic study. *Phytochemistry*, **15**, 57–60.

Lerner, H. R., Harel, E., Lehman, E. & Mayer, A. M. (1971). Phenylhydrazine, a specific irreversible inhibitor of catechol oxidase. *Phytochemistry*, **10**, 2637–40.

Lerner, H. R., Mayer, A. M. & Harel, E. (1972). Evidence for conformational changes in grape catechol oxidase. *Phytochemistry*, **11**, 2415–21.

Lieberei, R., Biehl, B. & Voigt, J. (1981). Serological studies on phenolase from spinach leaves. *Phytochemistry*, **20**, 2109–16.

Lonnerdal, B. & Keen, C. L. (1982). Review. Metal chelate affinity chromatography of proteins. *J. Appl. Biochem.*, **4**, 203–8.

Loomis, W. D. & Battaille, J. (1966). Plant phenolic compounds and the isolation of plant enzymes. *Phytochemistry*, **5**, 423–38.

Loomis, W. D. (1968). Removal of phenolic compounds during the isolation of plant enzymes. *Methods in Enzymology*, **13**, 555–63.

Loomis, W. D. (1974). Overcoming problems of phenolics and quinones in the isolation of plant enzymes and organelles. *Methods in Enzymology*, **31**, 528–44.

Lowum, S. E. & Parkin, K. L. (1989). Characterization of dye-sensitized photooxidation of mushroom tyrosinase. *J. Food Biochem.*, **13**, 391–401.

McCord, J. D. & Kilara, A. (1983). Control of enzymatic browning in processed mushrooms (*Agaricus bisporus*). *J. Food Sci.*, **48**, 1479–83.

McIntyre, R. J. & Vaughan, P. E. T. (1975). Kinetic studies on the hydroxylation of *p*-coumaric acid to caffeic acid by spinach-beet phenolase. *Biochem. J.*, **149**, 447–61.

McManus, J. P., Davis, K. G., Beart, J. E., Gaffney, S. H., Lilley, T. H. & Haslam, E. (1985). Polyphenol interactions. Part 1. Introduction; some observations on the reversible complexation of polyphenols with proteins and polysaccharides. *J. Chem. Soc. Perkin Trans. II.*, **9**, 1429–38.

Macrae, A. R. & Duggleby, R. G. (1968). Substrates and inhibitors of potato tuber phenolase. *Phytochemistry*, **7**, 855–61.

Madhosingh, C. (1975). Mushroom inhibitors of dopa oxidation. *Can. J. Microbiol.*, **21**, 2108–11.

Makino, N., McMahill, P., Mason, H. S. & Moss, T. H. (1974). The oxidation state of copper in resting tyrosinase. *J. Biol. Chem.*, **249**, 6062–6.

Marsh, D. R. & Galliard, T. (1986). Measurement of polyphenol oxidase activity in wheat-milling fractions. *J. Cereal Sci.*, **4**, 241–8.

Marshall, M. R., Otwell, W. S. & Walker, B. (1984). Preliminary study on the isolation and activation of polyphenoloxidase from deep sea crab (*Geryon* sp.). *Proc. Ninth Ann. Trop. Subtrop. Conf. Am.*, **9**, 118–20.

Martinez, J. H., Solano, F., Penafiel, R., Galindo, J. D., Iborra, J. L. & Lozano, J. A. (1986). Comparative study of tyrosinases from different sources: relationship between halide inhibition and the enzyme active site. *Comp. Biochem. Physiol.*, **83B**, 633–6.

Martinez-Cayuela, M., De Medina, L. S., Faus, M. J. & Gil, A. (1988*a*). Cherimoya (*Annona cherimola* Mill,) polyphenoloxidase; monophenolase and dihydroxyphenolase activities. *J. Food Sci.*, **53**, 1191–4.

Martinez-Cayuela, M., Faus, M. J. & Gil, A. (1988*b*). Effects of some reductants on the activity of cherimoya polyphenol oxidase. *Phytochemistry*, **27,** 1589–92.

Martinez-Cayuela, M., Rodriguez-Vico, F., Faus, M. J. & Gil, A. (1989). Partial purification and intracellular localization of cherimoya (*Annona cherimolia* Mill.) polyphenoloxidase. *J. Plant Physiol.*, **133,** 660–63.

Mason, H. S. (1948). The chemistry of melanin. III. Mechanism of the oxidation of dihydroxyphenylalanine by tyrosinase. *J. Biol. Chem.*, **172,** 83–99.

Mason, H. S., Fowlks, W. L. & Peterson, E. (1955). Oxygen transfer and electron transport by the phenolase complex. *J. Am. Chem. Soc.*, **77,** 2914–15.

Matheis, G. & Belitz, H. D. (1977). Multiple forms of soluble monophenol, dihydroxyphenylalanine: oxygen-oxidoreductase (EC 1.14.18.1) from potato tubers (*Solanum tuberosum*). II. Partial characterization of the enzyme forms with different molecular weights. *Z. Lebensm. Unters. Forsch.*, **163,** 279–82.

Matheis, G. (1987*a*). Polyphenoloxidase and enzymatic browning of potatoes (*Solanum tuberosum*). I. Properties of potato polyphenol oxidase. *Chem. Mikrob. Technol. Lebensm.*, **11,** 5–12.

Matheis, G. (1987*b*). Polyphenoloxidase and enzymatic browning of potatoes (*Solanum tuberosum*). II. Enzymatic browning and potato constituents. *Chem. Mikrob. Technol. Lebensm.*, **11,** 33–41.

Matheis, G. & Whitaker, J. R. (1984). Modification of proteins by polyphenol oxidase and peroxidase and their products. *J. Food Biochem.*, **8,** 137–62.

Mathew, A. G. & Parpia, H. A. (1971). Food browning as a polyphenol reaction. *Adv. Food Res.*, **19,** 75–145.

Mayer, A. M. & Friend, J. (1960). Properties and solubility of phenolase in isolated chloroplasts. *Nature*, **185,** 464–5.

Mayer, A. M. (1966). Catechol oxidase: enzymatic liberation from sugar beet chloroplast. *Phytochemistry*, **5,** 1297–301.

Mayer, A. M. (1987). Polyphenol oxidases in plants—recent progress. *Phytochemistry*, **26,** 11–20.

Mayer, A. M. & Harel, E. (1979). Review: Polyphenol oxidases in plants. *Phytochemistry*, **18,** 193–215.

Mayer, A. M., Harel, E. & Shain, Y. (1964) 2,3-naphthalenediol, a specific competitive inhibitor of phenolase. *Phytochemistry*, **3,** 447–51.

Mayer, A. M., Harel, E. & Ben-Shaul, R. (1966). Assay of catechol oxidase—a critical comparison of methods. *Phytochemistry*, **5,** 783–9.

Mazzocco, F. & Pifferi, P. G. (1976). An improvement of the spectrophotometric method for the determination of tyrosinase catecholase activity by Besthorn's hydrazone. *Anal. Biochem.*, **72,** 643–7.

Menon, I. A. & Haberman, H. F. (1975). Use of affinity chromatography for purification of tyrosinase. *Acta Dermatovener* (*Stockholm*), **55,** 343–4.

Meyer, H. U. & Biehl, B. (1980). Relation between photosynthetic and phenolase activities in spinach chloroplasts. *Phytochemistry*, **21,** 9–12.

Miller, A. R., Kelley, T. J. & Mujer, C. V. (1990). Anodic peroxidase isoenzymes and polyphenol oxidase activity from cucumber fruit: tissue and substrate specificity. *Phytochemistry*, **29,** 705–9.

Molnar-Perl, I. & Friedman, M. (1990*a*). Inhibition of browning by sulfur amino acids. 2. Fruit juices and protein-containing foods. *J. Agric. Food Chem.*, **38**, 1648–51.

Molnar-Perl, I. & Friedman, M. (1990*b*). Inhibition of browning by sulfur amino acids. 3. Apples and potatoes. *J. Agric. Food Chem.*, **38**, 1652–6.

Montgomery, M. W. (1983). Cysteine as an inhibitor of browning in pear juice concentrate. *J. Food Sci.*, **48**, 951–2.

Montgomery, M. W. & Sgarbieri, V. C. (1975). Isoenzymes of banana polyphenol oxidase. *Phytochemistry*, **14**, 1245–9.

Moore, B. M. & Flurkey, W. H. (1989). Tyrosinase activities and isoenzymes in three strains of mushrooms. *J. Food Sci.*, **54**, 1377–8.

Moore, B. M. & Flurkey, W. H. (1990). Sodium dodecyl sulfate activation of a plant polyphenoloxidase. Effect of sodium dodecyl sulfate on enzymatic and physical characteristics of purified broad bean polyphenoloxidase. *J. Biol. Chem.*, **265**, 4982–8.

Moore, B. M., Kang, B. & Flurkey, W. H. (1988). Histochemical and immunochemical localization of tyrosinase in whole tissue sections of mushrooms. *Phytochemistry*, **27**, 3735–7.

Morgan, B. D. G., O'Neill, T., Dewey, D. L., Galpine, A. R. & Riley, P. A. (1981). Treatment of malignant melanoma by intravascular 4-hydroxyanisole. *Clin. Oncol.*, **7**, 227–34.

Morgan, B. D. G. (1984). Recent results of a clinical pilot study of intra-asterial 4HA chemotherapy in malignant melanoma. In *Hydroxyanisole: Recent Advances in Anti-melanoma Therapy*, ed. P. A. Riley. IRL Press, Oxford, pp. 233–41.

Muneta, P. & Wang, H. (1977). Influence of pH and bisulfite on the enzymatic blackening reaction in potatoes. *Am. Potato J.*, **54**, 73–81.

Nelson, J. M. & Dawson, C. R. (1944). Tyrosinase. *Advances in Enzymology*, **4**, 99–152.

Nelson, R. M. & Mason, H. S. (1970). Tyrosinase (mushroom). *Methods in Enzymology*, **17A**, 626–32.

Oda, Y., Kato, H., Isoda, Y., Takahashi, N., Yamamoto, T., Takada, Y. & Kudo, S. (1989). Purification and properties of polyphenoloxidase from spinach leaves. *Agric. Biolog. Chem.*, **53**, 2053–61.

Okamoto, A., Imagawa, H., Arai, Y. & Ozawa, T. (1988). Partial purification and some properties of polyphenoloxidases from sago palm. *Agric. Biol. Chem.*, **52**, 2215–22.

O'Neill, S. P., Graves, D. J. & Ferguson, J. J. Jr (1973). Affinity chromatography of mushroom tyrosinase. *J. Macromol. Sci.-Chem.*, **A7**, 1159–66.

Oszmianski, J. & Lee, C. Y. (1990). Inhibition of polyphenol oxidase activity and browning by honey. *Agric. Food Chem.*, **38**, 1892–5.

Overeem, J. C. (1976). Pre-existing antimicrobial substances in plants. In *Biochemical Aspects of Plant–Parasite Relationships*, ed. J. Friend & D. R. Threlfall. Academic Press, London, pp. 195–206.

Owuor, P. O. & Othieno, C. O. (1987). Studies on the use of shade in tea plantations in Kenya: Effects on chemical composition and quality of made tea. *J. Sci. Food Agric.*, **46**, 63–70.

Owusu-Ansah, Y. J. (1989). Polyphenoloxidase in wild rice. *J. Agric. Food Chem.*, **37,** 901–4.

Padron, M. P., Lozano, J. A. & Gonzales, A. G. (1975). Properties of *o*-diphenol:O_2 oxidoreductase from *Musa cavendishii*. *Phytochemistry*, **14,** 1959–63.

Palmer, J. K. (1963). Banana polyphenoloxidase. Preparation and properties. *Plant Physiol.*, **38,** 508–13.

Park, E. Y. & Luh, B. S. (1985). Polyphenol oxidase of kiwi-fruit. *J. Food Sci.*, **50,** 678–84.

Patil, S. S. & Zucker, M. (1965). Potato phenolases; purification and properties. *J. Biol. Chem.*, **240,** 3938–43.

Peñafiel, R., Galindo, J. D., Solano, F., Pedreño, E., Iborra, J. L. & Lozano, J. A. (1984). Kinetic study of the interaction between frog epidermis tyrosinase and chloride. *Biochem. Biophys. Acta*, **788,** 327–32.

Pfiffner, E. & Lerch, K. (1981). Histidine at the active site of *Neurospora* tyrosinase. *Biochemistry*, **20,** 6029–35.

Pierpoint, W. S. (1966). The enzymatic oxidation of chlorogenic acid and some reactions of the quinone produced. *Biochem. J.*, **98,** 567–80.

Pierpoint, W. S., Ireland, R. J. & Carpenter, J. M. (1977). Modification of proteins during the oxidation of leaf phenols: reaction of potato virus X with chlorogenoquinone. *Phytochemistry*, **16,** 29–34.

Pifferi, P. G. & Baldassari, L. (1973). A spectrophotometric method for the determination of the catecholase activity of tyrosinase by Basthorn's hydrazone. *Anal. Biochem.*, **52,** 325–35.

Pomerantz, S. H. (1966). The tyrosine hydroxylase activity of mammalian tyrosinase. *J. Biol. Chem.*, **241,** 161–8.

Pomerantz, S. H. & Li, J. P. C. (1970). Tyrosinases (hamster melanoma). *Methods in Enzymology*, **17A,** 620–26.

Ponting, J. D. (1960). The control of enzymatic browning of fruits. In *Food Enzymes*, ed. H. W. Schult. AVI Publishing Co., New York, pp. 109–24.

Porath, J. & Olin, B. (1983). Immobilized metal ion affinity adsorption and immobilized metal ion affinity chromatography of biomaterials. Serum protein affinities for gel-immobilized iron and nickel ions. *Biochemistry*, **22,** 1621–30.

Prabha, T. N. & Patwardhan, M. V. (1986). Polyphenol oxidase (PPO) and peroxidase (POD) enzyme activities and their isoenzyme patterns in ripening fruits. *Acta Aliment.*, **15,** 199–207.

Racusen, D. (1970). The catechol oxidase of bean leaves (*Phaseolus vulgaris*). *Can. J. Bot.*, **48,** 1029–32.

Randhawa, K. S., Khurana, D. S. & Bajaj, K. L. (1986). Effect of CIPC (Isopropyl-N-(3-chlorophenyl)carbamate) on total N content and polyphenol oxidase activity in relation to processing of potatoes (*Solanum tuberosum*) stored under different conditions. *Qualitas Plant.-Plant Foods Human Nutr.*, **36,** 207–12.

Redinbaugh, M. G. & Campbell, W. H. (1985). Quaternary structure and composition of squash NADH:nitrate reductase. *J. Biol. Chem.*, **260,** 3380–85.

Riley, P. A. (1970). Mechanism of pigment-cell toxicity produced by hydroxyanisole. *J. Pathol.*, **101**, 163–9.

Riley, P. A., Sawyer, B. & Wolff, M. A. (1975). The melanocytotoxic action of 4-hydroxyanisole. *J. Invest. Derm.*, **64**, 86–9.

Rivas, N. J. & Whitaker, J. R. (1973). Purification and some properties of two polyphenol oxidases from Bartlett pears. *Plant Physiol.*, **52**, 501–7.

Robb, D. A. (1984). Tyrosinase. In *Copper Proteins and Copper Enzymes*, ed. R. Lontie. CRC Press, Boca Raton, Florida, pp. 207–41.

Robb, D. A., Mapson, L. W. & Swain, T. (1964). Activation of the latent tyrosinase of broad bean. *Nature*, **4918**, 503–4.

Robb, D. A., Mapson, L. W. & Swain, T. (1965). On the heterogeneity of the tyrosinase of broad bean (*Vicia faba* L.). *Phytochemistry*, **4**, 731–40.

Roberts, E. A. H. (1952). The chemistry of tea fermentation. *J. Sci. Food Agric.*, **3**, 193–8.

Roberts, E. A. H. (1959). The interaction of flavanol orthoquinones with cysteine and glutathione. *Chem. Ind.*, August, 995.

Roudsari, M. H., Signoret, A. & Crouzet, J. (1981). Eggplant polyphenol oxidase:purification, characterization and properties. *Food Chem.*, **7**, 227–35.

Rubin, B. A. & Artskhoskaya, E. V. (1960). Biokhimiya i Fiziologiya Immuniteta Rastenni (Biochem. Physiol. Plant Immunity). *Izd. Akad. Nauk U.S.S.R., Chem. Abst.*, **55**, 8557d.

Ruiz-Hernandez, M. (1983). Blastomycetic studies on must fractions extracted from grapes at different pressures. *Food Sci. Technol. Abst.*, 1901972.

Sachde, A. G., Al-Bakir, A. Y. & Abdul-Raheem, J. A. K. (1989*a*). Polyphenol oxidase from Barhee and Zahdi dates. I. Purification. *J. Food Biochem.*, **12**, 227–40.

Sachde, A. G., Al-Bakir, A. Y. & Abdul-Raheem, J. A. K. (1989*b*). Polyphenol oxidase from Barhee and Zahdi dates. II. Characterization. *J. Food Biochem.*, **12**, 241–51.

Sánchez-Ferrer, A., Roque, B., Cabanes, J. & García-Carmona, F. (1988). Characterization of catecholase and cresolase activities of Monastrell grape polyphenol oxidase. *Phytochemistry*, **27**, 319–22.

Sánchez-Ferrer, A., Bru, R. & García-Carmona, F. (1989). Novel procedure for extraction of a latent grape polyphenoloxidase using temperature-induced phase separation in Triton X-114. *Plant Physiol.*, **91**, 1481–7.

Sánchez-Ferrer, A., Bru, R. & García-Carmona, F. (1990). Partial purification of a thylakoid-bound enzyme using temperature-induced phase partitioning. *Anal. Biochem.*, **184**, 279–82.

Sanderson, G. W. (1964*a*). Extraction of soluble catechol oxidase from tea shoot tips. *Biochim. Biophys. Acta*, **92**, 622–4.

Sanderson, G. W. (1964*b*). Changes in the level of polyphenoloxidase activity in tea flush on storage after plucking. *J. Sci. Food Agric.*, **15**, 634–8.

Sanderson, G. W. (1965). The action of polyphenolic compounds on enzymes. *Biochem. J.*, **95**, 24–5.

Sanderson, G. W. (1972). The chemistry of tea manufacturing. In *Structural and Functional Aspects of Phytochemistry, Rec. Adv. Phytochem.*, **5**, ed. V. C. Runeckles. Academic Press, London, 247–316.

Sapers, G. M., Douglas, F. W. Jr, Bilyk, A., Hsu, A. F., Dower, H. W., Garzarella, L. & Kozempel, M. (1989*a*). Enzymatic browning in Atlantic potatoes and related cultivars. *J. Food Sci.*, **54,** 362–5.

Sapers, G. M., Hicks, K. B., Phillips, J. G., Garzarella, L., Pondish, D. L., Matulaitis, R. M., McCormack, T. J., Sondey, S. M., Seib, P. A. and Ei-Atawy, Y. S. (1989*b*). Control of enzymatic browning in apple with ascorbic acid derivatives, polyphenol oxidase inhibitors, and complexing agents. *J. Food Sci.*, **54,** 997–1002, 1012.

Sapis, J. C., Macheix, J. J. & Cordonnier, R. E. (1983). The browning capacity of grapes. 1. Changes in polyphenol oxidase activities during development and maturation of the fruit. *J. Agric. Food Chem.*, **31,** 342–5.

Sapis, J. C., Romeyer, F., Macheix, J. J. & Cordonnier, R. (1984). Study of polyphenoloxidases and hydroxycinnamic derivatives in the grape in relation to browning of the fruit. *Bul. L'O.I.V.*, **57,** 502–14.

Satjawatcharaphong, C., Rymal, K. S., Dozier, Jr., W. A. & Smith, R. C. (1983). Polyphenol oxidase system in Red Delicious apples. *J. Food Sci.*, **48,** 1879–80.

Sato, M. & Hasegawa, M. (1976). The latency of spinach chloroplast phenolase. *Phytochemistry*, **15,** 61–5.

Sauchez de Medina, L., Plata, M. C., Martinez-Cayuela, M., Faus, M. S. & Gil, A. (1987). Changes in polyphenoloxidase, peroxidase, catalase and acid phosphatase activities for cherimoya fruit (*Annona cherimolia*) during ripening in controlled temperature and relative humidity. *Rev. Agr. Tecnol. Aliment.*, **26,** 529–38.

Savagaon, K. A. & Sreenivasan, A. (1978). Activation mechanism of pre-phenolase in lobster and shrimp. *Fish. Technol.*, **15,** 49–55.

Schoebein, C. F. (1856). *Phil. Mag.*, **11,** 137–9.

Schoot Uiterkamp, A. J. M. & Mason, H. S. (1973). Magnetic dipole–dipole coupled Cu(II) pairs in nitric oxide-treated tyrosinase: a structural relationship between the active sites of tyrosinase and hemocyanin. *Proc. Natl. Acad. Sci. USA*, **70,** 993–6.

Shannon, C. T. & Pratt, D. E. (1967). Apple polyphenols oxidase activity in relation to various phenolic compounds. *J. Food Sci.*, **32,** 479–83.

Sharma, R. C. & Ali, R. (1980). Isolation and characterization of catechol oxidase from *Solanum melongena*. *Phytochemistry*, **19,** 1597–600.

Sharon-Raber, O. & Khan, V. (1983). Avocado mesocarp; browning potential, carotenoid content, polyphenol oxidase, catalase and peroxidase activities: comparison between six avocado cultivars. *J. Food Sci.*, **48,** 1874–5.

Shaw, J. F., Chu, H. L. & Pan, B. S. (1989). Purification of isoenzymes of bigheaded shrimp tyrosinase. *J. Chin. Agric. Soc.*, **27,** 350–59.

Shimoda, T., Yonekura, M. & Funatsu, M. (1975). Affinity chromatography of phenoloxidase in the larvae of housefly. *Agr. Biol. Chem.*, **39,** 2423–4.

Siegenthaler, P. A. & Vaucher-Bonjour, P. (1971). Vieillessement de l'appareil photosynthetique. III. Variations et charactéristiques de l'activite o-diphénoloxydase (polyphenoloxydase) au cours du viellissement in vitro de chloroplastes isoles d'épinard. *Planta*, **100,** 106–23.

Simpson, B. K., Marshall, M. R. & Otwell, W. S. (1987). Phenoloxidase from shrimp (*Penaeus setiferus*): purification and some properties. *J. Agric. Food Chem.*, **35,** 918–21.

Simpson, B. K., Marshall, M. R. & Otwell, W. S. (1988). Phenoloxidases from pink and white shrimp: kinetic and other properties. *J. Food Biochem.*, **12,** 205–17.

Smith, D. M. & Montgomery, M. W. (1985). Improved methods for the extraction of polyphenol oxidase from d'Anjou pears. *Phytochemistry*, **24,** 901–4.

Söderhäll, K. & Häll, T. (1984). Lipopolysaccharide—induced activation prophenol oxidase activating system in crayfish haemocyte lysate. *Biochem. Biophys. Acta*, **797,** 99–104.

Söderhäll, K. & Unestam, T. (1979). Activation of serum prophenoloxidase in anthropod immunity. The specific cell wall glucan activation and activation by purified fungal glycoproteins of crayfish phenoloxidase. *Can. J. Microbiol.*, **25,** 406–16.

Söderhäll, K., Carlberg, I. & Eriksson, T. (1985). Isolation and partial purification of prophenoloxidase from *Daucus carota* L. cell cultures. *Plant Physiol.*, **78,** 730–33.

Stafford, H. A. & Dressler, S. (1972). 4-hydroxycinnamic acid hydroxylase and polyphenolase activities in *Sorghum vulgare*. *Plant Physiol.*, **49,** 590–95.

Stelzig, D. A., Akhtar, S. & Ribeiro, S. (1972). Catechol oxidase of Red Delicious apple peel. *Phytochemistry*, **11,** 535–9.

Stokes, D. M., Anderson, J. W. & Rowan, K. S. (1968). The isolation of mitochondria from potato—tuber tissue using sodium metabisulfite for preventing damage by phenolic compounds during extraction. *Phytochemistry*, **7,** 1509–12.

Strothkamp, K. G. & Mason, H. S. (1974). Pseudoperoxidase activity of mushroom tyrosinase. *Biochem. Biophys. Res. Commun.*, **61,** 827–32.

Strothkamp, K. G., Jolley, R. L. & Mason, H. S. (1976). Quaternary structure of mushroom tyrosinase. *Biochem. Biophys. Res. Commun.*, **70,** 519–24.

Swain, T., Mapson, L. W. & Robb, D. A. (1966). Activation of *Vicia faba* (L.) tyrosinase as effected by denaturing agents. *Phytochemistry*, **5,** 469–82.

Taira, S., Sugiura, A. & Tomana, T. (1987). Relationship between natural flesh darkening and polyphenol oxidase activity in Japanese persimmon (*Diospyros kaki* Thunb.) fruits. *J. Japan. Soc. Food Sci. Technol.*, **34,** 612–15.

Taneja, S. R. & Sachar, R. C. (1974*a*). Separate monophenolase and *o*-diphenolase enzymes in *Triticum aestivum*. *Phytochemistry*, **13,** 1367–71.

Taneja, S. R. & Sachar, R. C. (1974*b*). Modulation of *o*-diphenolase and monophenolase enzymes during wheat grain development. *Cereal Chem.*, **51,** 457–65.

Taoukis, P. S., Labuza, T. P., Lillemo, J. H. & Lin, S. W. (1990). Inhibition of shrimp melanosis (black spot) by ficin. *Lebensm.-Wiss. u. -Technol.*, **23,** 52–4.

Temesvari, J. (1986). Study of polyphenoloxidase activity in wheat varieties, meals and pasta. *Food Sci. Technol. Abst.*, 1960422.

Tengku, A. B. T. A., Augustin, M. A. & Ghazali, H. M. (1986). Polyphenoloxidase from starfruit (*Averrhoa carambola* L.). *Pertanika,* **9,** 219–24.

Thanaraj, S. N. S. & Seshadri, R. (1990). Influence of polyphenol oxidase activity and polyphenol content of tea shoot on quality of black tea. *J. Sci. Food Agric.,* **51,** 57–69.

Thomas, P. & Janave, M. T. (1973). Polyphenoloxidase activity and browning of mango fruits induced by gamma irradiation. *J. Food Sci.,* **38,** 1149–52.

Toiguchi, S., Hayashi, K., Adachi, Y., Motoki, M. & Haraguchi, K. (1989). Purification and characterization of soybean oxidase. *J. Japan. Soc. Food Sci. Technol.,* **36,** 597–602.

Tolbert, N. E. (1973). Activation of polyphenol oxidase of chloroplasts. *Plant Physiol.,* **51,** 234–44.

Trémolières, M. & Bieth, J. G. (1984). Isolation and characterization of the polyphenoloxidase from senescent leaves of black poplar. *Phytochemistry,* **23,** 501–5.

Udayasekhara Rao, P. & Deosthale, Y. G. (1987). Polyphenoloxidase activity in germinated legume seeds. *J. Food Sci.,* **52,** 1549–51.

Valero, E., Varón, R. & García-Carmona, F. (1988). Characterization of polyphenol oxidase from Airen grapes. *J. Food Sci.,* **53,** 1482–5.

Vámos-Vigyázó, L. (1981). Polyphenol oxidase and peroxidase in fruits and vegetables. *CRC Critic. Rev. Food Sci. Nutr.,* **15,** 49–127.

Van Driessche, E., Beeckmans, S., Dejaegere, R. & Kanarek, L. (1984). Thiourea: the antioxidant of choice for the purification of proteins from phenol-rich plant tissues. *Anal. Biochem.,* **141,** 184–8.

Van Lelyveld, L. J. & de Rooster, K. (1986). Browning potential of tea clones seedlings. *J. Hort. Sci.,* **61,** 545–7.

Van Loon, L. C. (1971). Tobacco polyphenoloxidases: a specific staining method indicating non-identity with peroxidases. *Phytochemistry,* **10,** 503–7.

Vanneste, W. & Mason, H. S. (1966). The binding of carbon monoxide by hemocyanin. In The Biochemistry of Copper, ed. J. Peisach, P. Aisen & W. E. Blumberg. Academic Press, New York, London, pp. 465–73.

Vanneste, W. H. & Zuberbühler, A. (1974). Copper-containing oxygenases. In *Molecular Mechanisms of Oxygen Activation,* ed. O. Hayaishi. Academic Press, New York, London, pp. 371–404.

Vaughan, P. F. T. & Butt, V. S. (1970). The action of *o*-dihydric phenols in the hydroxylation of *p*-coumaric acid by a phenolase from leaves of spinach beet (*Beta vulgaris* L.). *Biochem. J.,* **119,** 89–94.

Vaughan, P. F. T. & McIntyre, R. J. (1975). The action of hydrogen peroxide on the hydroxylation of *p*-coumaric acid by spinach-beet phenolase. *Biochem. J.,* **1511,** 759–62.

Vaughan, P. F. T., Eason, R., Paton, J. Y. & Ritchie, G. A. (1975). Molecular weight and amino acid composition of purified spinach beet phenolase. *Phytochemistry,* **14,** 2383–6.

Vaughn, K. C. & Duke, S. O. (1981*a*). Tentoxin-induced loss of plastidic polyphenol oxidase. *Physiol. Plant.,* **53,** 421–8.

Vaughn, K. C. & Duke, S. O. (1981*b*). Tissue localization of polyphenol oxidase in *Sorghum. Protoplasma,* **108,** 319–27.

Vaughn, K. C. & Duke, S. O. (1984*a*). Function of polyphenol oxidase in higher plants. *Minireview. Physiol. Plant.*, **60,** 106–12.

Vaughn, K. C. & Duke, S. O. (1984*b*). Tentoxin stops the processing of polyphenol oxidase into an active enzyme. *Physiol. Plant.*, **60,** 257–62.

Voigt, J. & Noske, R. (1966). Zur bestimmung der polyphenoloxidaseaktivitat. II. Orientierende versuche zur anwendbarkeit der methode mit Besthorns reagens in apfeln. *Z. Lebensm. Unters. Forsch.*, **130**, 9–14.

Walker, J. R. L. (1962). Studies on the enzymic browning of apple fruit. *N. Z. J. Sci.*, **5,** 316–24.

Walker, J. R. L. (1964). Studies on the enzymatic browning of apples. II. Properties of apple polyphenoloxidase. *Aust. J. Biol. Sci.*, **17,** 360–71.

Walker, J. R. L. (1975). Enzymatic browning in foods: A review. *Enzyme Technol. Dig.*, **4,** 89–100.

Walker, J. R. L. & Wilson, E. L. (1975). Studies on the enzymatic browning of apple. Inhibition of apple *o*-diphenol oxidase by phenolic acids. *J. Sci. Food Agric.*, **26,** 1825–31.

Walter, W. M. Jr & Purcell, A. E. (1980). Effect of substrate levels and polyphenol oxidase activity on darkening in sweet potato cultivars. *J. Agric. Food Chem.*, **28,** 941–4.

Wesche-Ebeling, P. & Montgomery, M. W. (1990). Strawberry polyphenoloxidase: its role in anthocyanin degradation. *J. Food Sci.*, **3,** 731–4.

Whitaker, J. R. (1972). Polyphenol oxidase. In *Principles of Enzymology for the Food Sciences*, ed. O. R. Fennema. Marcel Dekker, New York, pp. 571–82.

Whitaker, J. R. (1985). Mechanisms of oxidoreductases important in food component modification. In: *Chemical Changes in Food during Processing*, ed. T. Richardson & J. W. Finley. AVI Publishing Co., Westport, CT, pp. 121–76.

Wichers, H. J., Peetsma, G. J., Malingré, Th. M. & Huizing, H. J. (1984). Purification and properties of a phenol oxidase derived from suspension cultures of *Mucuna pruriens*. *Planta*, **162,** 334–41.

Wickremasinghe, R. L. (1978). *Tea Adv. Food Res.*, **24,** 229–86.

Wigfield, D. C. & Goltz, D. M. (1990*a*). Kinetics of reconstitution of apotyrosinase by copper. *Biochem. Cell Biol.*, **68,** 476–9.

Wigfield, D. C. & Goltz, D. M. (1990*b*). The pH dependence of the reconstitution of apotyrosinase: the question of Cu(I) versus Cu(II). *Biochem. Cell Biol.*, **68,** 648–50.

Wissemann, K. W. & Lee, C. Y. (1980). Purification of grape polyphenoloxidase with hydrophobic chromatography. *J. Chromatogr.*, **192,** 232–5.

Wongt, T. C., Luh, B. S. & Whitaker, J. R. (1971). Isolation and characterization of polyphenol oxidase isoenzymes of Clingstone peach. *Plant Physiol.*, **48,** 19–23.

Yamaguchi, M., Henderson, M. H., Hwang, P. M. & Campbell, J. D. (1969). Effect of oxygen concentration on *o*-diphenol oxidase activity. *Anal. Biochem.*, **32,** 178–81.

Yamaguchi, M., Hwang, P. M. & Campbell, J. D. (1970). Latent *o*-diphenol oxidase in mushrooms (*Agaricus bisporus*). *Can. J. Biochem.*, **48,** 198–202.

Yamaguchi, Y., Yamashita, I. & Aoki, S. (1988). Studies on preservation of shiitake mushrooms (*Lentinus edodes*) at the point of production. III. Polyphenoloxidase activity during storage in modified atmospheres and on subsequent exposure to air. *J. Japan. Soc. Cold Preserv. Food.*, **14,** 59–62.

Yamanishi, T. (1978). The aroma of various teas. In *Flavor of Foods and Beverages,* ed. G. Charalambous & G. E. Inglett. Academic Press, New York, pp. 305–28.

Yoshida, H. (1886). Chemistry of lacquer (Urushi). Part I. *J. Chem. Soc.*, **43,** 472–86.

Zawistowski, U., Sangster, K., Zawistowski, J., Langstaff, J. & Friesen, A. D. (1988). Immobilized metal affinity chromatography of wheat α-amylases. *Cereal Chem.*, **65**, 413–16.

Zawistowski, J., Blank, G. & Murray, E. D. (1986). Polyphenol oxidase activity in Jerusalem artichokes (*Helianthus tuberosus* L.). *Can. Inst. Food Sci. Technol. J.*, **19,** 210–14.

Zawistowski, J. (1987). Polyphenol oxidase of Jerusalem artichoke (*Helianthus tuberosus* L.) tubers: isolation, purification and properties. Dissert. Abst. Int., B 48, 3562.

Zawistowski, J., Weselake, R. J., Blank, G. & Murray, E. D. (1987*a*). Fractionation of Jerusalem artichoke phenolase by immobilized copper affinity chromatography. *Phytochemistry,* **26,** 2905–7.

Zawistowski, J., Blank, G. & Murray, E. D. (1987*b*). Inhibition of enzymatic browning in extracts of Jerusalem artichoke (*Helianthus tuberosus* L.). *Can. Inst. Food Sci. Technol. J.*, **20,** 162–5.

Zawistowski, J., Biliaderis, C. G. & Murray, E. D. (1988*a*). Purification and characterization of Jerusalem artichoke (*Helianthus tuberosus* L.) polyphenol oxidase. *J. Food Biochem.*, **12,** 1–22.

Zawistowski, J., Biliaderis, C. G. & Murray, E. D. (1988*b*). Isolation and some properties of an acidic fraction of polyphenol oxidase from Jerusalem artichoke (*Helianthus tuberosus* L.). *J. Food Biochem.*, **12,** 23–35.

Zemel, G. P., Sims, C. A., Marshall, M. R. & Balaban, M. (1990). Low pH inactivation of polyphenoloxidase in apple juice. *J. Food Sci.*, **55,** 562–3.

Zenin, C. T. & Park, Y. K. (1978). Isoenzymes of polyphenol oxidase from high L-DOPA containing velvet bean. *J. Food Sci.*, **43,** 646–7.

7

Carbohydrate Oxidases

S. A. Barker

School of Chemistry, The University of Birmingham, Edgbaston, UK

INTRODUCTION

In the UK, the Boots Company and later Sturge & Co. pioneered the industrial production of the first carbohydrate oxidase (glucose oxidase) mainly because both companies were actively researching on fungi like *Aspergillus niger* which produces glucose oxidase when grown on glucose. The recommendations of the Nomenclature Committee of the International Union of Biochemistry (Webb, 1984) put carbohydrate oxidoreductases in Class 1. They used a systematic name (see Table 1) based on a donor:acceptor oxidoreductase with a recommended usage of the terms 'dehydrogenase' and 'reductase' as an alternative. The term 'oxidase' was only to be employed where O_2 was the acceptor. The second figure in the code relates to the group in the hydrogen donor which undergoes oxidation. The third figure indicates, with a few exceptions, the type of acceptor involved. In Table 1, for ease of reference, enzyme substrates of similar type are grouped together and details of their products and acceptor given. It should be noted that with several oxidases, the enzymes themselves are flavoproteins (E.C. 1.1.3.4, 1.1.3.8, 1.1.3.24, 1.1.5.1, 1.1.99.10 and 11.99.13). These flavin coenzymes are usually bound very strongly to their proteins and do not dissociate, indeed they act as intermediates in the electron transfer chain. Enzymes that can use molecular oxygen as an electron acceptor contain either flavins or metal ions (or both). Among those in Table 1, where the enzyme is a copper glycoprotein, are E.C. 1.1.3.5 and E.C. 1.1.3.9.

Oxidised carbohydrates are sometimes recognisable (see Fig. 1) by the presence of carboxylic acids such as in aldonic acids where the aldose has been oxidised at its C1 aldehyde group. Since the latter normally exists as a cyclic semiacetal in a ring structure, the products with enzymes are the aldono lactones, i.e. D-glucose gives δ-D-

Table 1. Carbohydrate Oxidising Enzymes

Class 1.1.1.	*Polyhydric alcohol dehydrogenase*	*Ketulose reductase*	
No.	Substrate	Product	Acceptor
8	*R*-glycerol 3-phosphate	Glycerone phosphate	NAD^+
9	Xylitol	D-Xylulose	NAD^+
9	Erythritol	L-Erythrulose	NAD^+
10	Xylitol	L-Xylulose	$NADP^+$
11	D-Arabinitol	D-Xylulose	NAD^+
12	L-Arabinitol	L-Xylulose	NAD^+
13	L-Arabinitol	L-Ribulose	NAD^+
14	L-Iditol	L-Sorbose	NAD^+
14	D-Glucitol	D-Fructose	NAD^+
15	D-Iditol	D-Sorbose	NAD^+
15	Xylitol	L-Xylulose	NAD^+
15	L-Glucitol	L-Fructose	NAD^+
16	Galactitol	D-Tagatose	NAD^+
17	D-Mannitol 1-phosphate	D-Fructose 6-phosphate	NAD^+
56	Ribitol	D-Ribulose	NAD^+
67	D-Mannitol	D-Fructose	NAD^+
137	D-Ribitol 5-phosphate	D-Ribulose 5-phosphate	$NADP^+$
138	D-Mannitol	D-Fructose	$NADP^+$
140	D-Glucitol 6-phosphate	D-Fructose 6-phosphate	NAD^+
156	Glycerol	Glycerone	$NADP^+$
162	Erythritol	D-Erythrulose	$NADP^+$
Class 1.1.1	*Polyhydric alcohol*	*Aldehyde reductase*	*Acceptor*
No.			
21	Alditol	Aldose	$NADP^+$
114	D-Apiitol	D-Apiose	NAD^+
Class 1.1.1	*Cyclitol dehydrogenase*	*Ketone product*	*Acceptor*
18	Myo-inositol	2,4,6/3,5-Penta hydroxy cyclohexa-none	NAD^+
143	O^5-methyl-myo-inositol	5D-O^5-methyl-2,3,5/4,6-penta hydroxy hexanone	NAD^+
142	5D-O^5-methyl-chiro-inositol	5D-O^5-methyl-2,3,5/4,6-penta hydroxy cyclohexanone	$NADP^+$
Class 1.1.1	*Aldonate dehydrogenase*	*Uronate*	*Acceptor*
19	L-Gulonate	D-Glucuronate	$NADP^+$
20	L-Gulono-1,4-lactone	D-Glucurono-6,3-lactone	$NADP^+$
57	D-Mannonate	D-Fructuronate	NAD^+
58	D-Altronate	D-Tagaturonate	NAD^+
131	D-Mannonate	D-Mannuronate	$NADP^+$

continued

Table 1—*contd.*

Class 1.1.1	*Sugar nucleotide dehydrogenase*	*Hexuronate nucleoside*	*Acceptor*
22	UDP glucose	UDP glucuronate	NAD^+
132	GDP mannose	GDP mannuronate	NAD^+
136	UDP-*N*-acetyl-*D*-glucosamine	UDP-*N*-acetyl-2 amino-D-glucuronate	NAD^+
186	d TDP galactose	d TDP galacturonate	NAD^+

Class 1.1.1	*Aldonate dehydrogenase*	*Keto-product*	*Acceptor*
43	6-Phospho D-gluconate	6-phospho-2 dehydro D-gluconate	$NADP^+$
44	6-Phospho D-gluconate (decarboxylating)	D-Ribulose 5-phosphate $+CO_2$	$NADP^+$
45	L-Gulonate	3-Dehydro L-gulonate	NAD^+
69	D-Gluconate	5-Dehydro D-gluconate	$NADP^+$
125	2-Deoxy D-gluconate	2-Deoxy 3-dehydro D-gluconate	NAD^+
126	2-Dehydro-3-deoxy D-gluconate	(4S, 5S)-4,5-Dihydroxy-2,6-dioxohexanoate	$NADP^+$
127	2-Dehydro-3-deoxy-D-gluconate	(4S)-4,6-Dihydroxy-2,5-dioxohexanoate	$NADP^+$
128	L-Idonate	5-Dehydro-D-Gluconate	$NADP^+$
129	L-Threonate	3-Dehydro-L-threonate	NAD^+
130	3-Dehydro L-gulonate	(4R, 5S)-4,5,6-tri hydroxy-2,3-dioxo hexanoate	$NADP^+$

Class 1.1.1	*Aldose dehydrogenase*	*Aldono lactone*	*Acceptor*
46	L-Arabinose	L-Arabino-1,4-lactone	NAD^+
47	β-D-Glucose	D-Glucono-1,5-lactone	$NADP^+$
47	D-Xylose	D-Xylono-1,5-lactone	$NADP^+$
113	L-Xylose	L-Xylono-1,4-lactone	$NADP^+$
115	D-Ribose	D-Ribonate	$NADP^+$
116	D-Arabinose	D-Arabino-1,4-lactone	NAD^+
117	D-Arabinose	D-Arabino-1,4-lactone	$NADP^+$
118	D-Glucose	D-Glucono-1,5-lactone	NAD^+
119	D-Glucose	D-Glucono-1,4-lactone	$NADP^+$
120	D-Galactose	D-Galactonolactone	$NADP^+$
121	D-Aldose	D-Aldonolactone	NAD^+
122	D-Threo Aldose	D-Threo aldono-1,5-lactone	NAD^+
173	L-Rhamnofuranose	L-Rhamno-1,4-lactone	NAD^+
175	D-Xylose	D-Xylonolactone	NAD^+
179	D-Xylose	D-Xylono-1,5-lactone	$NADP^+$

Class 1.1.1	Substrate dehydrogenase	Product reductase	Acceptor
49	D-Glucose 6-phosphate	D-glucono-1,5-lactone 6-phosphate	$NADP^+$
133	d TDP-6-deoxy-L-mannose	d TDP-4-dehydro-6 deoxy-L-mannose	$NADP^+$
134	GDP-6-deoxy L-talose	GDP-4-dehydro-6-deoxy L-talose	$NADP^+$
135	GDP-6-deoxy D-talose	GTP-4-dehydro-6-deoxy D-talose	$NADP^+$
158	UDP-*N*-acetyl D-muramate	UDP-*N*-acetyl 3-O-(1-carboxy-vinyl)-D-glucosamine	$NADP^+$
187	GDP-6-deoxy-D-mannose	GDP-4-dehydro-6-deoxy D-mannose	$NADP^+$
203	D-galacturonate	D-Galactarate	NAD^+

Class 1.1.2	Polyhydric alcohol dehydrogenase	Ketulose	Acceptor
2	D-Mannitol	D-Fructose	Ferricyto-chrome

Class 1.1.3	Aldose oxidase	Aldonolactone	Acceptor
4	β-D-Glucose	D-Glucono-1,5-lactone	Oxygen
5	β-D-Glucose	D-Glucono-1,5-lactone	Oxygen
5	Also D-galactose, D-mannose, maltose, lactose, cellobiose	D-Galactonolactone, D-mannonolactone, maltobionolactone, lactobionolactone cellobionolactone	Oxygen

Class 1.1.3	Aldose oxidase	Product	Acceptor
10	D-Glucose (pyranose)	2-Dehydro D-Glucose	O_2
	Also D-xylose	(Oxidation at the	O_2
	L-sorbose	2-position with	O_2
	D-glucono-1,5-lactone	all)	O_2
9	D-Galactose	D-Galacto-hexodialdose	O_2

Class 1.1.3	Substrate oxidase	Product	Acceptor
8	L-Gulono-1,4-lactone	L-Ascorbate	O_2
11	L-Sorbose	5-Dehydro D-fructose	O_2
24	L-Galactono -1,4-lactone	L-ascorbate	O_2

continued

Table 1—*contd.*

Class 1.1.5	*Disaccharide dehydrogenase*	*Product*	*Acceptor*
1	Cellobiose	Cellobiono-1,5-lactone	Quinone
Class 1.1.99	*Substrate dehydrogenase*	*Product*	*Acceptor*
3	D-Gluconate	2-dehydro-D-gluconate	Acceptor
4	2-Dehydro-D-gluconate	2,5-Didehydro-D-gluconate	Acceptor
10	D-Glucose	D-Glucono-1,5-lactone	Flavo protein
11	D-Fructose	5-Dehydro D-fructose	2,6-Dichloro indo-phenol
12	L-Sorbose	5-Dehydro D-fructose	2,6-Dichloro indo-phenol
13	Sucrose	3-Dehydro α-D-glucosyl-β-D-fructofuranoside	Flavo protein
17	D-Glucose	D-glucono-1,5-lactone	Special protein
18	Cellobiose	Cellobiono-1,5-lactone	2,6-Dichloro indo-phenol
Class 1.10.2	*Substrate reductase*	*Product*	*Acceptor*
1	L-Ascorbate-cytochrome-b_5	Monodehydro-ascorbate	Ferri cyto chrome-b_5
Class 1.10.3	*Substrate oxidase*	*Product*	*Acceptor*
3	L-Ascorbate	2 dehydroascorbate	O_2 (a multi-copper protein)
Class 1.11.1	*Substrate peroxidase*	*Product*	*Acceptor*
11	L-Ascorbate	Dehydroascorbate	H_2O_2

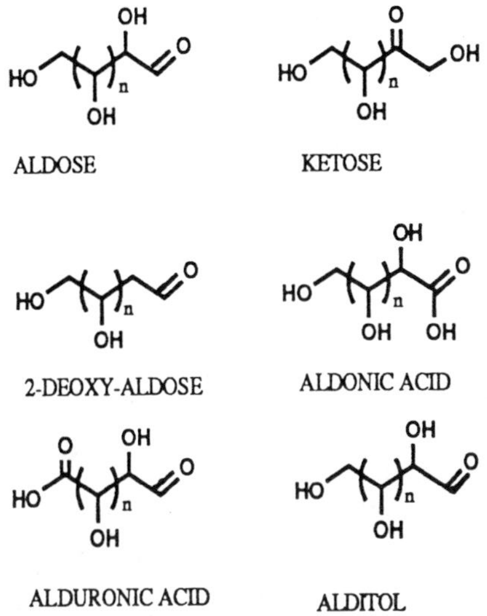

Fig. 1. Monosaccharide derivatives.

gluconolactone which later in aqueous solution exists in equilibrium with D-gluconic acid. Lactones are internal esters and with carbohydrates are δ- or six membered and γ- or five membered. Whereas aldonic acids are not components of polysaccharides, the product derived by oxidation of a hexose at C6, a hexuronic acid (e.g. glucuronic acid or galacturonic acid), is a common constituent. Not all uronic acids found in polysaccharides are directly derived from sugar nucleotides as substrates (e.g. UDP glucose → UDP glucuronate) and then incorporated into polysaccharides. One uronic acid may be converted to another within the polysaccharide. Whether in a cyclitol like myo-inositol (Fig. 2) or a polyhydric alcohol like D-arabinitol, simple oxidation of one secondary alcohol group can occur to give a keto adduct; in these cases 2,4,6/3,5 pentahydroxy cyclohexanone and D-xylulose, respectively. Oxidation can also occur at positions other than C1 or C6 in aldoses, e.g. D-glucose → D-glucosone, (2 keto-D-glucose). Much less well known are enzymes that oxidise disaccharides. Thus the enzyme that also attacks cellobiose and lactose is hardly recognisable as the D-glucoside 3-dehydrogenase which converts sucrose to 3-keto α D-glucosyl-β-D-fructo-furanoside. A rare exception among enzymes that oxidise carbohydrates is D-galactose oxidase

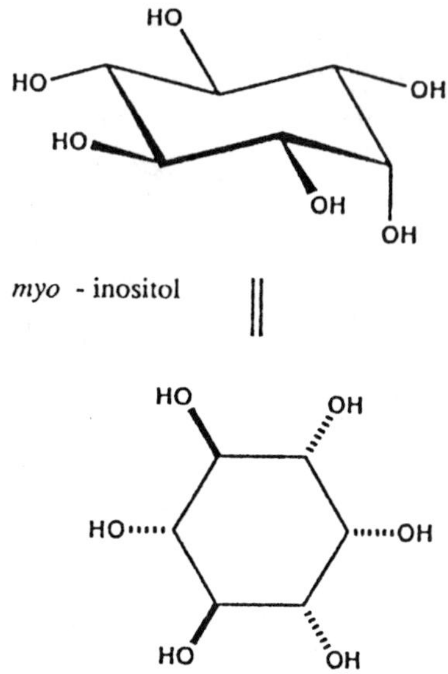

Fig. 2. Conformation of myo-inositol.

which oxidises the C6 primary alcohol group of D-galactose to an aldehyde group. This enzyme has the remarkable capacity to act on disaccharides like lactose, trisaccharides, polysaccharides, glycoproteins, glycolipids and polyhydric alcohols where there is an accessible D-galactose group or the stereochemical environment is similar.

We therefore see that there is no apparent limit to the ability of nature to devise oxidising enzymes which either are classed as oxidases (so generally require oxygen) or dehydrogenases (so require an acceptor like NAD). The exciting thing about carbohydrate oxidases is that we are only just at the beginning of defining their specificity whether it be very restricted like D-glucose oxidase or broad like D-galactose oxidase. Thus glucose oxidase attacks β-D-glucopyranose but not α-D-glucopyranose (Fig. 3). Oxidases convert a host of substances that are naturally occurring to give as a by-product hydrogen peroxide, which itself gives rise to a semi-sterile environment not encountered with dehydrogenases. Peroxides, through their connotations with radicals, are not desirable in all parts of a living cell. Fungi, however, have a widespread ability to oxidise glucose + O_2 to gluconolactone and hydrogen peroxide; by contrast many bacteria do

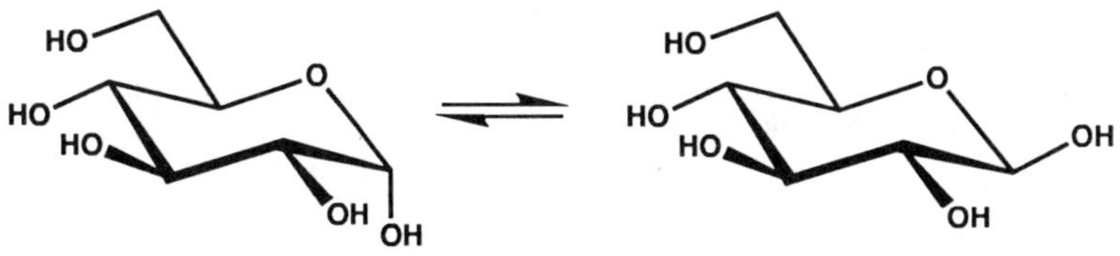

α - D - Glucopyranose β - D - Glucopyranose

Fig. 3. Equilibrium of D-glucopyranose.

not have a similar capacity. Hence here we see the possibility that the combination of acids and peroxide are used by fungi to protect their domain. We begin to see that some aspects of food preservation (Table 2) are to do with acid production arising either directly or indirectly from carbohydrates more often with much smaller molecules like lactic acid or acetic acid. The smallest carbohydrate is accepted as D-glyceraldehyde since it has all the features of an aldose; an aldehyde group, a secondary alcohol chiral centre and a primary alcohol group. The sequential addition of further secondary alcohol chiral centres gives tetroses (erythrose, threose), pentoses (xylose, arabinose, ribose, lyxose), hexoses and so on.

CARBOHYDRATE OXIDOREDUCTASES

In most cases the demands of metabolic pathways require a reversible reaction which is easily achieved with carbohydrate dehydrogenases that use NAD^+ and interconvert it to NADH or reductases that use $NADP^+$ and interconvert it to NADPH (see Table 1). A typical example of the latter is glucuronate reductase which converts L-gulonic acid (already an oxidised carbohydrate) to D-glucuronate. A similar reductase exists to interconvert the corresponding L-gulono-1,4-lactone to D-glucurono-6,3-lactone. To emphasise this reversibility the enzyme is often called an oxidoreductase like xylitol:NAD^+ 2-oxidoreductase which produces D-xylulose but also can act on erythritol as an L-erythrulose reductase. Such an enzyme offers a facile route from the *meso* structure of erythritol to what could be a valuable organic synthon. Equally, glycerol 3-phosphate dehydrogenase can be used to abolish a chiral centre in converting its substrate R-glycerol 3-

Table 2. Application of Oxidising Enzymes in Food

1. Fish meat treatment with glucose oxidase to prevent discoloration during pickling (Takebe, 1988).
2. Fish preservation in refrigeration using glucose oxidase (Koczot *et al.*, 1985).
3. Food preservative containing glucose oxidase, e.g. cooked noodles with glucose oxidase, glucose, ascorbic acid and lysozyme (Inamine & Watanabe, 1987).
4. Use of glucose oxidase in glucose-free fructose manufacture from sucrose (Goma & Seiller, 1987).
5. Fish rancidity prevention by glucose oxidase (Lin, 1986).
6. Oxygen removal from foods and beverages with compositions containing substrate, catalase, superoxide dismutase and glucose oxidase. Used with beer (Prieels *et al.*, 1986*a*).
7. Use of L-galactonate and L-galactonolactone dehydrogenase in the enzymic preparation of ascorbic acid with intrasequential cofactor regeneration. L-Ascorbic acid prepared by: (1) hydrolysis of alginic acid into D-mannuronic and L-guluronic acids; (2) enzymic reduction of D-mannuronic acid into D-mannonic acid; (3) enzymic oxidation of D-mannonic to D-fructuronic acid; (4) isomerisation of D-fructuronic acid to D-glucuronic acid and (5) conversion of D-glucuronic acid into L-ascorbic acid (Kulbe & Knopki, 1986).
8. Gluconate and mannitol manufactured with combined mannitol dehydrogenase and glucose (NAD) dehydrogenase in an enzyme membrane reactor (Kulbe *et al.*, 1986).
9. Use of glucose oxidase in low-alcohol wine manufacture by treatment of grape juice with the enzyme in the presence of oxidase which is then fermented (Villettaz, 1986).
10. Use of β-fructofuranosidase and glucose oxidase to reduce the sugar content of orange juice and other fruit juices (Donninger, 1986).
11. Use of glucose oxidase as an antibacterial food and feed additive. Typically lactoferrin, lactoperoxidase, sodium thiocyanate and glucose oxidase added to powdered milk based calf feed (Prieels *et al.*, 1986*b*).
12. Use of glucose oxidase as a bread quality improver. Addition of the enzyme, cystine and catalase advocated, Riken Vitamin Co., 1982, (Prieels *et al.*, 1986*b*).

phosphate with NAD^+ to glycerone phosphate + NADH. Equally in this section all the examples given so far act without changing the overall pH balance. Energy is conserved in the interchange of lactones, interchange of acids and the use of the coenzymes NAD^+ and $NADP^+$ as carriers of hydrogen. Such enzymes are listed under the classification 1.1.1 with a special number reserved for a specific enzyme. Within this category no less than 66 different carbohydrate enzymes have been recognised. A much smaller group 1.1.2 are able to use ferricytochrome as the acceptor like mannitol dehydrogenase 1.1.2.2 that produces fructose, with little evidence of reversibility. Equally final is the conversion of β-D-glucose with oxygen to D-glucono-1,5-lactone by glucose oxidase in the group of seven oxidases that use oxygen as an acceptor (1.1.3). The disaccharide cellobiose is the only recognised member of 1.1.5 where the enzyme cellobiose dehydrogenase uses quinone as an acceptor in oxidising it to cellobiono-1,5-lactone and a phenol. Another eight enzymes can use other acceptors (1.1.99) including 2,6-dichloroindophenol. An unusual enzyme is L-galactonolactone dehydrogenase that produces L-ascorbate with ferricytochrome C (Group 1.3.2). L-Ascorbate oxidase can itself oxidise ascorbate with oxygen to dehydroascorbate and is classed as 1.10.3.3. L-Ascorbate peroxidase produces the same product with hydrogen peroxide and is classed as 1.11.1.11. In food, besides acting as Vitamin C, ascorbate is nowadays widely added to mineral waters as antioxidant since PET (polyethyleneterephthalate) bottles tend to admit oxygen. Its antioxidant capacity is also important in food, hence enzymes like the last two mentioned that produce dehydroascorbate destroy that important property.

ENZYMES IN FOOD

Carbohydrates form a major proportion of the food we ingest. The enzymes present in food (see Table 3) arise from those naturally present in the food constituents and also from microbial cells sometimes added to ferment mainly the carbohydrate constituents e.g. yoghurts, beer and other alcoholic drinks. The microbial cells employed, such as *Lactobacillus* and yeasts, often contain quite different enzymes from those arising from mammalian and plant tissues. Microbial enzymes are becoming increasingly important as a third source of enzymes, namely those from microbial contamination as with *Salmonella* because of inadequate hygiene or as a deliberate action like hanging pheasants for a period of days before cooking them. Finally the fourth source is the deliberate addition of enzymes in food processing. In the USA alone the market for food enzymes is forecast to be worth $286m by 1993, some 33% larger than the 1988 level. The carbohydrases are the largest product group and one

Table 3. Active Oxidising Enzymes in Foods

Food	Enzyme	Reference
Dough	Ascorbate oxidase	Grant, 1987
Fruit (*Citrus medica*)	Ascorbate oxidase	Gautam, 1987
Apple (*Malus pumilla domestica*)	Sorbitol oxidase	Yamaki, 1987
Lettuce (*Lactuca*)	Ascorbate oxidase	Kesseli & Michelmore, 1986
Hops	Ascorbate oxidase	Reitmann, 1987
Lettuce	Glucose 6-phosphate dehydrogenase	Kesseli & Michelmore, 1986
Banana	Galactose dehydrogenase	Jarret & Litz, 1986
Banana	L-Iditol dehydrogenase	Jarret & Litz, 1986
Banana	Phosphogluconate dehydrogenase	Jarret & Litz, 1986
Grapes	Ascorbate oxidase	Nabiev & Velieva, 1986
Apples	L-Iditol dehydrogenase	Berueter, 1985

carbohydrate oxidase, namely glucose oxidase, is used extensively. Its initial use in the last World War was to treat dried egg exported to Britain. The dried egg first received here went brown because the glucose was reacting with the amino acids and proteins in the egg. This was very effectively prevented by selectively converting the glucose to gluconic acid with glucose oxidase. Later *A. niger* glucose oxidase was used to manufacture gluconate and δ-gluconolactone. They became permitted food additives. Gluconate is an excellent chelator of calcium ions. Glucono δ-lactone in water will produce gluconic acid so can be used to initiate setting processes with alginate and high methoxyl pectin. It also has the ability to produce CO_2 with sodium bicarbonate, once mooted as a new approach to bread making. Gluconate is hardly metabolised by humans so does not add to our calories. Hence if sucrose is inverted to glucose and fructose then treated with glucose oxidase we have a gluconate/fructose mixture with all the sweetness originally in the sucrose but only half of its calories. In the western world control of our calorific intake, coupled with a reduction in the cholesterol level of foods, will generate major markets for food enzymes. Quite a number of the acids arising from the oxidation of monosaccharides are not metabolised by humans, particularly if incorporated into food polysaccharides like alginate which contains D-mannuronic acid and L-guluronic acid. Alginates are initially produced as polymannuronic acids and another enzyme inverts the structure at C5 to give rise to the L-guluronic acids. In mammalian tissues we have the same structural relationship with D-glucuronic acid and L-iduronic acid in the polysaccharides of the ground substance. Many toxic and other substances are excreted by the body as D-glucuronides. With the current developments in exploiting Britain's hemicellulose reserves from wheat straw, normally burnt, it is logical to expect that in the coming years D-xylonic acid, the five-carbon analogue of D-gluconic acid, will be introduced into our food. Currently the most investigated oxidoreductases according to the *American Chemical Abstracts* are as follows.

Ascorbate Oxidase

E.C. 1.10.3.3 is one of the few unique enzymes found in plants, vegetables and cereals and produced by certain fungi like *Aspergillus* and *Penicillium* and bacteria like *Bacillus cereus*. It is not present in mammalian tissues. The enzyme isolated from *Aerobacter aerogenes* oxidises D-isoascorbate in addition to L-ascorbate. It has a K_m (M) of $2{\cdot}4 \times 10^{-3}$ with the latter at pH 6·5 and 37°C.

Cellobiose Oxidase

Cellobiose oxidase, with its connotations of cellulose, present also in plants but not mammals, is generally found as an adjunct to cellulase which gives cellobiose during cellulose breakdown. It is produced by *Sporotrichum pulverulentum*. Under anaerobic conditions rapid reduction of the associated flavin is followed by slower reduction of cytochrome *b*. Three two-electron donors (cellobiose molecules) reduce two-three-electron acceptors (the flavin-cytochrome *b* of cellobiose oxidase).

Galactose Oxidase

E.C. 1.1.3.9 again does not belong to the mammalian world, probably because its ability to produce C6-aldehydes from a wide variety of galactose containing metabolites could be a toxicological disaster to man. However it has been advocated in dentifrices along with lactoperoxidase. It can be found in *Fusarium gramineareum, Gibberella fujikuroi, Dactylium dendroides* and whether in soluble or immobilised form has proved immensely useful in biotechnological analyses. The *Penicillium circinatus* enzyme has a K_m (M) of $2{\cdot}4 \times 10^{-1}$ with D-galactose and its relative velocities are with D-galactose (1·0), 2-deoxy D-galactose (0·32), melibiose (0·80), raffinose (1·80), stachyose (6·10) and guaran (1·80).

Glucose Oxidase

E.C. 1.1.3.4 again belongs to the world of the fungi like *Aspergillus niger, Penicillium amagasakiense* and *Penicillium stipitatum.* Other organisms that produce it include *Actinobacillus actinomycetem comitans* and *Phanerochaete chrysoporium.* The *Penicillium notatum* enzyme oxidises β-D-glucose (1·00), 2-deoxy D-glucose (0·25), 6-deoxy 6-fluoro D-glucose (0·03) and others even less when some 80 sugars were tested.

Glucose 2-Oxidase

Glucose 2-oxidase has been derived from fungi also like *Aspergillus oryzae, Polyporus obtusus* and organisms like basidomycetes, *Coriolus versicolor* and *Phancrochaete chrysosporium.* The enzyme can be used to manufacture glucosone from glucose.

L-Gulonolactone Oxidase

E.C. 1.1.3.8 is an enzyme found in mammalian tissue and as an oxidase very rare in this respect. It is widely distributed also in fish, crabs and chickens including the egg yolk sac membrane. The kidney and liver are the best established sources. With rat liver enzyme its K_m (M) was $1{\cdot}9 \times 10^{-2}$ at pH 7·2 and 37°C. It can utilise L-galactonolactone (0·87), D-mannonolactone (0·68), D-altronolactone (0·47) and D-talonolactone (0·50).

Pyranose Oxidase

E.C. 1.1.3.10 is another microbial enzyme found in *Coriolus versicolor,* fungus, *Irpex lacteus* and *Polyporous obtusus*. This enzyme also oxidises D-xylose, L-sorbose and D-glucono-1,5-lactone. All have the same configuration at C2, C3 and C4.

D-Arabinitol Dehydrogenase

E.C. 1.1.1.11 is found in *Escherichia coli* and Mycobacteria and hence can be used in arabinitol determination in blood serum. The *Aerobacter aerogenes* enzyme only reduces D-xylulose of the ketopentoses. D-fructose is reduced at 2·5% the rate of D-xylulose with this enzyme at K_m (M) 2×10^{-3} at pH 6.

L-Arabinose Dehydrogenase

E.C. 1.1.1.46 is another microbial enzyme isolated from *Azospirillum brasiliense* but located in bean roots because of its presence in *Rhizobium*. The *Ps. saccharophilia* enzyme has K_m (M) $9{\cdot}2 \times 10^{-4}$ at pH 8·6 and 30° with L-arabinose and $3{\cdot}0 \times 10^{-3}$ under the same conditions with D-galactose (85% relative velocity).

Cellobiose Dehydrogenase

E.C. 1.1.5.1 like the oxidase belongs to the microbial world of *Chaetomium cellulolyticum, Fomes annosus, Dichomitus squalens, Sporotrichum pulverulentum, Sporotrichum thermophile, Sclerotium rolfsii,* and *Phanerochaete chrysosporium*. The enzyme uses a quinone as acceptor.

Fructose 5-Dehydrogenase

E.C. 1.1.99.11 has been purified from *Gluconobacter industrius*. The *Gluconobacter cerinus* enzyme has K_m (M) of $1{\cdot}0 \times 10^{-2}$ with D-fructose (1·00) but could be replaced by D-glucose (2·00), D-mannose (0·22), D-galactose (0·54), D-xylose (0·70) and L-arabinose (0·43).

L-Fructose Dehydrogenase

This has been isolated from *Pulluria pullulans*. However, the dehydrogenase using $NADP^+$ as an acceptor is present in the horseshoe crab.

L-Galactonate Dehydrogenase

This has been manufactured from *Bacillus*.

Galactitol Dehydrogenase

E.C. 1.1.1.16 is again from the world of *Klebsiella pneumoniae* *Fusarium* and *Pseudomonas cepacia*. The related L-galactitol 1-phosphate dehydrogenase has been also purified from *Klebsiella pneumoniae*. Galactitol dehydrogenase can act on dulcitol, L-iditol, L-arabinitol and very weakly on xylitol but not polyhydric alcohols where the L-threo configuration next to CH_2OH is absent.

Galactose Dehydrogenase

E.C. 1.1.1.48 is in *Acholeplasma, Pseudomonas cepacia* and *fluorescens* and *Rhizobium meliloti* but has confidently been used to assay galactose in human blood, and employed in analysis of soybean–meat blend. *Acholeplasma* is also the source of galactose 6-phosphate dehydrogenase which is found in goat liver. The corresponding glucitol 6-phosphate dehydrogenase is obtained from *Klebsiella pneumoniae.* The *Pseudomonas saccharophilia* enzyme (1.1.1.48) with NAD^+ does not oxidise D-glucose, L-glucose, D-mannose, D-ribose, D-xylose, D-galactose 6-phosphate, D-galacturonate or D-galactosamine.

Gluconate 2-Dehydrogenase

E.C. 1.1.99.3 is a microbial enzyme present in *Aspergillus niger, Gluconobacter oxydans, Pseudomonas aeruginosa, Klebsiella pneumonia* and the *Rhizobium* from cow pea. The *Pseudomonas aeruginosa* is highly specific for D-gluconate being inactive with D-glucose, glucuronate, 2-oxogluconate, saccharate, ribonate, arabonate, fructose and mannose.

Gluconate 5-Dehydrogenase

E.C. 1.1.1.69 again is microbial from *Gluconobacter suboxydans* and *Penicillium notatum*. The former had 28% of the activity with D-mannonate and 12% with L-idonate compared with that of D-gluconate. L-galactonate, D-ribonate and the arabonates not attacked.

Glucose Dehydrogenase

E.C. 1.1.1.118 is located widely in the world of microbes, particularly bacteria like *Acetobacter aceti, Acinetobacter calcoaceticus, Bacillus megaterium, Corynebacterium, Escherichia coli, Gluconobacter suboxydans*. Nevertheless it is produced by *Aspergillus flavus*. Glucose dehydrogenase is also located in liver endoplasmic reticulum and has been reported in cotton seeds. The variety (E.C. 1.1.1.47) using $NADP^+$ as acceptor is present in *Azospirullum lipoferum, Bacillus subtilis, Gluconobacter,* and *Thermoplasma acidophilum*. The liver enzyme with NAD^+ as cofactor has a K_m (M) that varies in different species, e.g. mammals, aves, reptilia and osteichthyes. Rat liver with glucose has K_m (M) $3–7 \times 10^{-1}$.

Glucose 6-Phosphate Dehydrogenase

E.C. 1.1.1.49 is our first ubiquitous enzyme of the mammalian world (adipose tissue, adrenal cortex, adrenal gland, aorta, muscle, artery, in blood, skin, blood platelets, serum, bone callus cells, brain, liver, epidermis, erythrocyte, eye lens, heart/atrium and ventricle), hepatocytes, fibroblasts, kidney, leucocytes, macrophage, mammary gland, mast cells, mitochondria, nerves, ovary, prostate gland, seminal vesicles, stomach chief cells, stomach fundus cells, thyroid gland, tooth root sheath, urine, uterus. Besides that, of course, it rightly is present in all other species where it has been sought including yeasts, fungi, bacteria, vegetables and crops of all kinds, worms, frogs, fish, seeds, flies, beetles, sea urchins, eggs. In fact it is difficult to suggest anything except a virus that does not contain it as a central enzyme of metabolism. Human erythrocyte enzyme has K_m (M) $3{\cdot}9 \times 10^{-5}$ with glucose 6-phosphate at pH 8 and K_m (M) $2{\cdot}1 \times 10^{-5}$ with NADP.

Glyceraldehyde Phosphate Dehydrogenase (E.C. 1.2.1.12)

This smallest sugar dehydrogenase is very widely distributed in the kidney, muscle, adipocyte, serum, cartilage, brain, muscle, heart, cornea, erythrocytes, eye lens, liver, thymus, thyroid glands and uterus in mammalian species. This widespread and essential enzyme for cell metabolism likewise is found in most other living species such as chicken, beans, algae, bacteria, moulds, yeast, asparagus, fish, crabs, cod, salmon, nematodes, trout, rice cells, snakes, wheat, just to illustrate its ubiquitous nature. The enzyme acts very slowly on D-glyceraldehyde.

The glyceraldehyde phosphate dehydrogenase having $NADP^+$ as coenzyme is a key enzyme in photosynthesis and so is associated with the chloroplast. Hence we find it in corn leaves, celery leaves, spinach leaves, the chloroplast of the cucumber, and sugar beet. As expected it is in the chloroplast of the photosynthetic algae such as *Scenedesmus* and *Chlorella.* However, yeast, bacteria and fish contain it.

Glycerol Dehydrogenase (EC 1.1.1.6)

An enzyme of the microbial world like *Bacillus, Candida, Escherichia coli, Glucono-bacter, Klebsiella pneumoniae, Cellulomonas flavigena, Pichia guilliermondii, Staphylococcus aureus, Aspergillus niger* and *Shizosaccharomyces pombe.* This enzyme also acts on 1,2-propanediol.

Glycerol Phosphate Dehydrogenase (EC 1.1.99.5)

Much more widely distributed in muscle, adipose tissue mitochondria, mammary gland, adipocyte, adrenal gland, muscle, aorta, arteriole and artery of spinal cord, blood platelets, kidney, liver, brain cortex and medulla, cytosol, cornea, colon mitochondria, epidermis, eye cornea epithelium, eye lens epithelium, fibroblast, granulocytes of bone marrow, heart, hepatocytes, leucocytes, lung, myelin, neurons, nerves, pancreatic islet skin and sperm mitochondria. Hence we also find it equally widespread outside this mammalian world in fish, frogs, ducks, flies, worms, yeasts, and the microbial world.

Inositol 2-Dehydrogenase (E.C. 1.1.1.18)

This is an enzyme reported in *Bacillus pumulus, Bacillus subtilis* and *Fusarium*. The *Aerobacter aerogenes* enzyme oxidises *meso*-inositol and scyllitol in the forward direction and *meso*-inosose-2 and DL-*epi*-inosose in the reverse direction.

Inositol 1-Phosphate Dehydrogenase

An important plant enzyme well characterised in mung beans.

L-Iditol Dehydrogenase (E.C. 1.1.1.14)

A widely distributed mammalian enzyme in blood serum, liver, brain, eye lens, microsomes, sperm, testis, semen, but equally important in plants and fruits like apples as well as cucumbers. Found in chickens, fish, flies, toads and microbes. The enzymes from a variety of sources exhibit low specificity able to use D-sorbitol, D-xylitol, D-ribitol, D-mannitol as well as L-iditol.

Less Well-Known Dehydrogenases

The dehydrogenase of 2-keto 3-deoxy fuconate has been characterised from liver and of 2-keto 3-deoxy gluconate from *Erwina chrysanthemi*. The latter is also the source of a keto deoxy uronate dehydrogenase. The dehydrogenase of ketogluconate itself has been characterised from *Gluconobacter melanogenus*. Lactose dehydrogenase was located in the blood serum of the common marmoset. The blood, liver and muscle of goldfish contained mannitol dehydrogenase also found in the microbial world of *Aspergillus*. Mannitol 1-phosphate dehydrogenase, as might be expected because of the importance of mannitol in algae, has been reported in Laminarea, Fucus and brown algae. Elsewhere it appears confined to the microbial world, mainly particular marine bacteria like *Pseudomonas* and *Vibrio* that take up mannitol and *Escherichia coli*. An isolated reference reports it in bishop pine. Ribitol dehydrogenase is mainly microbial and so is ribitol-5-phosphate dehydrogenase.

Phosphogluconate Dehydrogenase (Decarboxylating), (E.C. 1.1.1.44)

A widely distributed enzyme of mammals, plants, and microbes. Found in adipocytes, follicles, artery, blood, muscle, serum, liver, brain, hair, epididymis, erythrocyte, leukemic cells, foetal and maternal placenta, hepatocyte, heart, kidney, sperm, testis and uterus in the mammals. Found in the chloroplast, cytoplasm, wheat grains of plants but also in mushrooms, pink salmon as well as widely in bacteria like *Pseudomonas, Xanthomonas* and *Escherichia coli*. All enzymes are highly specific for 6-phosphogluconate. The enzymes utilise NADP as a cofactor except the *Leuconstoc mesenteroides* enzyme that uses NAD 25 times better than NADP.

D Sorbitol 6-Phosphate Dehydrogenase (EC 1.1.1.140)

This is an important enzyme in ripening fruit like apple and loquat. Also found in *Streptococcus*. The *Aerobacter aerogenes* enzyme is highly specific and oxidises D-dulcitol-6-phosphate at only 3% of the rate with D-sorbitol 6-phosphate. With the latter the K_m (M) is $6{\cdot}25 \times 10^{-3}$ at pH 9.

Xylitol 4-Dehydrogenase (EC 1.1.1.10)

Occurrence so far confined to organisms like *Erwinia uredovora, Klebsiella pneumoniae*. Of a number of other sugars tested only L-ribulose had slight activity with the enzyme from guinea pig liver where L-xylulose had K_m (M) $2{\cdot}9 \times 10^{-4}$ at pH 7.

Xylose Dehydrogenase [E.C. 1.1.1.175 (D)]

Both those for D-xylose and L-xylose were found in *Pachysolen tannophilus*. Yeasts contain the dehydrogenase for D-xylose which can be used in ethanol manufacture. The gene for D-xylose dehydrogenase (NADP) is found more widely and has been reported in mouse chromosome, pancreas and muscle of crab as well as yeast. In *Pseudomonas* both a cytoplasmic NAD specific xylose dehydrogenase and a membrane linked PQQ-linked dehydrogenase has been found. The above occurrences are located in the collected index of Vols 96–105 (1982–6) of the *American Chemical Abstracts*.

ISOLATION OF DEHYDROGENASES

Any isolation of an enzyme will depend on whether it is produced extracellularly or is held on or under the surface of the cells. Nowadays mild surfactants like octyl β-D-glucoside are available to ease the release of the enzyme from a cell surface. In the past the same role was played by deoxycholate. Otherwise the cells are smashed to release the intracellular enzymes. Affinity columns containing materials related to the substrate of the enzyme are the most specific means of purification but often prior to this the following options are considered. These include separation by size on molecular sieve columns such as the Sephadex or Sepharose range, and the use of more robust packings such as porous glass or porous silica. Thereafter derivatised polysaccharide derivatives like DEAE cellulose can be applied. The purifications now described illustrate and enlarge on these approaches which can be backed up by electrophoresis and other techniques.

E.C. 1.1.1.49

Glucose 6-phosphate dehydrogenase has been purified to homogeneity from chicken and rat brain (Farnararo *et al.*, 1978). This was achieved by affinity chromatography with Sepharose-bound 2′,5′ADP. The G6PDH from chick brains differed from the rat enzyme in affinity for this column packing, in pH optimum, in heat stability and was also differently affected by Mg^{2+} ions. Both enzymes were stable to urea treatment. Ropson & Powers (1989) have isolated H6PDH from the teleost *Fundulus heteroclitus* and three of the major allelic isoenzymes have been purified and compared for native molecular weight, subunit molecular weight, isoelectric point, thermal stability and steady state kinetic properties at pH 8 and 25°C. The predominant allelic isoenzyme in northern populations (H6PDH-AcAc) was found to be more sensitive to heat denaturation than were the predominant homozygous allelic isoenzymes isolated from southern populations along the east coast of North America (H6PDH-AaAa and H6PDH-AbAb). The H6PDH-AcAc had both a significantly greater K_m for glucose 6-phosphate than did either of the southern phenotypes and a significantly greater K_m for $NADP^+$ and K_i of NAD^+ than did one of the southern phenotypes (H6PDH-AaAa).

E.C. 1.1.99.17

Glucose dehydrogenase quinoprotein has been purified some 2430-fold from *Acinetobacter calcoaceticus* to electrophoretic homogeneity (Geyer & Görisch, 1986). The enzyme has a molecular weight of 110 000 and comprised a dimer of identical subunits each containing one molecule of pyrroloquinoline quinone. Its K_m with glucose was 3·3 mM. When the enzyme oxidises glucose it transfers the electrons to a cytochrome-*b*. The most effective step in the purification of the enzyme was chromatography on CM-Sepharose CL-6B. Prior treatment liberated the enzyme from the cells by lysozyme treatment and the crude extract was passed down DEAE-Sephacel on which the enzyme did not absorb. Final steps involved the use of chromatography on hydroxyapatite and phenyl-sepharose CL-4B.

E.C. 1.1.3.8

L-Gulonolactone oxidase has been purified from chicken kidney to homogeneity by a five-step method with a recovery of 30% (Kiuchi *et al.*, 1982). The molecular weight of the enzyme was 400 000 but when dissociated was 50 000. The absorption maxima of the enzyme were at 454, 352 and 278 nm, which changed to that of the reduced form when the substrate L-gulono-γ-lactone was added under anaerobic conditions. The reduced form of the enzyme returned to the oxidised form when aerated. Kiuchi *et al.* (1982) showed that the enzyme contained a covalently bound flavin moiety, FAD, by digesting the enzyme with trypsin/chymotrypsin and subsequently liberating 5′-AMP with nucleotide pyrophosphatase. The enzyme attacked a number of hexonic acid lactones whose hydroxyl group on C2 had the same configuration as that of L-gulono γ-lactone. The enzyme is believed to generate 2-oxo-L-gulono γ-lactone before this is transformed to L-ascorbic acid. The K_m was 7 μM at low concentrations of the substrate. Vertebrates higher than fish possess this enzyme. The major steps in the purification of the enzyme were a 20-fold enrichment on DEAE-Sephadex A-50 and the homogeneous enzyme was obtained by chromatography on hydroxylapatite. Thus the L-gulonolactone oxidase was purified 240-fold from chicken kidney microsomes with a final yield of 29%.

STRUCTURE AND PROPERTIES OF DEHYDROGENASES

E.C. 1.1.1.14

The sorbitol dehydrogenases from livers of cow, man, rat and sheep have molecular weights of about 140 000 (Walsall *et al.*, 1978). This L-iditol:NAD oxidoreductase has subunits of molecular weight 40 000 in beef, rat and sheep liver and in each of these species they each possess an isoelectric point of 7·3. All four have identical mobilities upon disc-gel electrophoresis and display a single band of enzyme activity. This activity is activated by increasing ionic strengths up to 0·5 M salt concentration and also by ampholines. Microassays for sorbitol dehydrogenase have been developed to permit its precise location in tissues, e.g. present in renal proximal tubeles of rats but not in the inner medullary collecting ducts (Sands *et al.*, 1989). Balla *et al.* (1989) have shown that L-iditol dehydrogenase acts on [3H] iditol to yield [3H] sorbose. Examination of sorbitol dehydrogenase occurring elsewhere has shown three phenotypes in human seminal plasma as revealed by electrophoresis (one slow band, one fast and both bands). In sperm, Perea *et al.* (1989) found a phenotype of one band faster than the one of seminal plasma. Further study showed that the isoenzymes were due to conformational isomerism or to molecular aggregates and that thiol groups participate in the mobility of the seminal plasma fast band.

Using the Abbott Bichromatic Analyser, Pakuts *et al.* (1988) have applied an automated single reagent microassay for the determination of sorbitol dehydrogenase to follow its level in plasma. Thus experimentally induced hepatotoxicity by acetaminophen produced a dose-dependent increase in blood sorbitol dehydrogenase activity. The kinetic rate was linear up to 250 U/litre and had a CV of 2·65%. The assay required only 25 μl of sample.

E.C. 1.1.1.49

Shantz *et al.* (1989) reported that dehydroepiandrosterone and certain structural analogues are potent uncompetitive inhibitors of mammalian glucose 6-phosphate dehydrogenases. Cartana *et al.* (1989) have shown that the effect of nickel on G-6-PDH activity is consistent with a mixed type inhibition pattern with a competitive character. Pre-incubation of yeast G6PDH with Al(III) produced an inactive enzyme containing 1 mol of Al(III)/mol of enzyme subunit (Cho & Joshi, 1989). Citrate, $NADP^+$, EDTA and NaF protected the enzyme against the Al(III) inactivation. Modification of histidine and lysine residues of the enzyme with diethylpyrocarbonate and acetylsalicylic acid respectively, inactivated the enzyme. Circular dichroism studies showed that the binding of Al(III) to the enzyme induced a decrease in alpha-helix and beta-sheet and an increase in random coil. It was concluded that the inactivation was due to the conformational change induced by Al(III) binding.

MECHANISM OF DEHYDROGENASES

The range of aldose dehydrogenases acting on sugars is remarkable in that they often only act on one anomer and not the other. Thus Berkowitz & Benner (1987) have pointed out that D-glucose dehydrogenase prefers the β-anomer as does D-glucose 6-phosphate dehydrogenase, D-galactose dehydrogenase, L-fucose dehydrogenase and D-abequose dehydrogenase. D-Arabinose dehydrogenase uses the α-configuration. All the sugars are in the aldopyranose form and require nicotinamide cofactors to produce aldono-1,5-lactones. All these enzymes conform to a mechanism with a stereochemically selective substrate, whereby the enzyme transfers an axial hydrogen on C1 in that chair conformation of the sugar of lowest energy. The most efficient catalysis of the expulsion of this hydride requires that the transition state has two oxygen lone pairs oriented antiperiplanar to the departing hydride. This rationalisation fitted in with the data found by Berkowitz & Benner (1987) for pig fucose dehydrogenase. It acts on an L sugar with a $^1C_4$ ground state conformation in contrast with the D-hexoses which have a distinct $^4C_1$ ground state conformation.

The sorbitol dehydrogenase (E.C. 1.1.1.14) from human liver is tetrameric and carries one zinc per subunit. It has a residue identity with Class 1 alcohol dehydrogenases of 25%, but three of the four cysteine ligands of the non-catalytic zinc of the ADHs is absent in SDH (Maret, 1989). The polyol substrates of SDH have the S configuration at C2 and the R configuration at C4. The corresponding C2 and C4 epimers do not serve as substrates. For C3, either the R or S configuration is compatible with reaction, but the 3(R) substrates are better substrates. A hydroxyl at C3 is not essential; 3-deoxyxylitol is a good substrate. The configuration at C5 has very little effect on substrate behaviour. However, since D-erythritol scarcely serves as a substrate, the occupation of the 5-position by a CHOH moiety is clearly important. The 1-hydroxy group is essential since 1-deoxysorbitol is not a substrate. It is believed that the oxygen atom of C1 of the sorbitol (or analogue) is hydrogen bonded to Lys-319 and Glu-174 and this stabilises the active conformation at O_2 (Eklund *et al.*, 1985).

Galactose oxidase (EC 1.1.3.9) is a copper-containing enzyme and the only known mononuclear Cu(II) protein that catalyses a two-electron redox reaction involving molecular oxygen without the involvement of another metal ion or organic cofactor. Driscoll & Kosman (1987) concluded that a single solvent-derived proton was involved in a galactose oxidase rate determining step. Their data indicated that a Cu(II)-bound-H_2O was required in the protonated aquo state for catalysis. They postulated that proton transfer from the Cu-bound H_2O to O_2 occurred concurrent with, or following, a one-electron transfer from substrate alcohol.

An early example showing the constraints imposed in substrate conformation was that of Barker *et al.* (1959) using a mannitol dehydrogenase from a ropy-cider organism able to interconvert mannitol and fructose. Only one of the chair forms of fructose is suitable (Fig. 4), while polyols written in the zig-zag form require the 2- and the 5-hydroxyl groups and the 3-hydrogen to be directed below the plane of the paper.

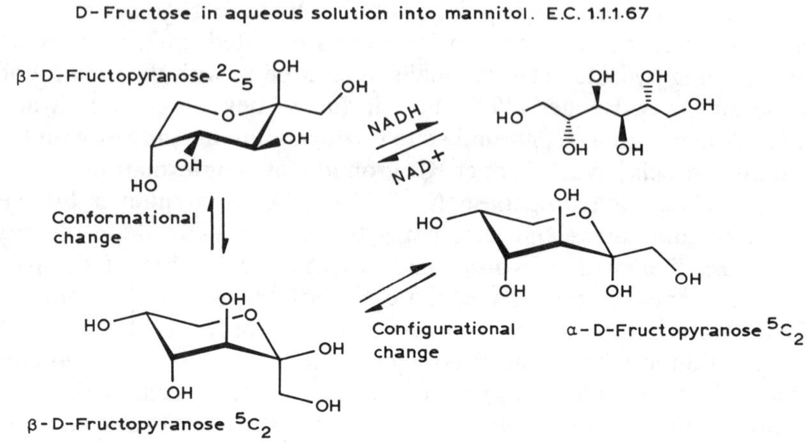

Fig. 4. Enzymic conversion of D-fructose.

ALTERNATIVE ACCEPTORS

With many enzymes, the electron acceptor cannot be replaced. Further in the reverse reaction, often neither can the electron donor. Examples of these in Group 1.1.1. are L-xylulose reductase: NAD could not replace NADP (oxidised or reduced). D-Arabitol dehydrogenase; reduced NADP cannot replace reduced NAD. L-Iditol dehydrogenase; NAD cannot be replaced by NADP. Galactitol dehydrogenase; NADP cannot replace NAD. Mannitol 1-phosphate dehydrogenase; reduced NADP cannot replace reduced NAD. Inositol dehydrogenase; NADP cannot replace NAD. L-Arabinose dehydrogenase; NADP cannot replace NAD. D-Glucuronolactone dehydrogenase; NADP could not replace NAD. Aldose dehydrogenase; NADP could not replace NAD. 5-Keto D-fructose reductase; NAD could not replace NADP. GDP-Mannose dehydrogenase; NADP could not replace NAD.

By contrast there are examples of replacement albeit in varying degrees. Phosphogluconate dehydrogenase (decarboxylating); NAD used 25 times more effectively than NADP. Glucose 6-phosphate dehydrogenase; NAD and NADP can be used equally well. Mannitol dehydrogenase; reduced NADP has 50% of the activity of reduced NAD and NADP 2·5% that of NAD. 5-Ketogluconate reductase; *Klebsiella* or *Escherichia coli* enzymes promote reactions twice as fast with reduced NAD as with reduced NADP. Galactose 6-phosphate dehydrogenase; NAD could be replaced by NADP. Mannitol dehydrogenase; NAD or reduced NAD had 10–15% the activity of NADP (or reduced NADP). 2-Keto 3-deoxy D-gluconate dehydrogenase; reduced NAD had 7% the activity of reduced NADP. Ribitol 5-phosphate dehydrogenase; NAD and NADP both active.

In Group 1.1.2 replacement is much more flexible. Mannitol dehydrogenase: ferricytochrome can be replaced by ferricyanide, 2,6-dichlorophenolindophenol 2,6-dichlorophenolindophenol, and endogenous cytochromes.

Oxygen can sometimes be replaced in Group 1.1.3 either to a limited extent such as with glucose oxidase and dichlorophenolindophenol or more freely as with L-sorbose oxidase and the same dye.

Glucose dehydrogenase of group 1.1.99 is very flexible, able to employ dichlorophenolindophenol or phenazine methosulfate with equal facility. Another member of this group D-glucoside 3-dehydrogenase was able to use cytochrome-*c*, 2,6-dichlorophenolindophenol, 5-methylphenazinium methyl sulphate and ferricyanide.

ROLE OF DEHYDROGENASES

E.C. 1.1.1.14

Jans *et al.* (1989) used rabbit renal papillary tissue to demonstrate the routes by which D-sorbitol was synthesised. Thus D-[6-13C] fructose were converted directly into D-sorbitol via the aldose reductase and sorbitol dehydrogenase pathway respectively. D-[1-13C] Ribose and [2-13C] glycerol labelled the D-glyceraldehyde pool that in turn caused a labeling of the D-sorbitol. These reactions could be investigated by ^{13}C NMR spectroscopy.

In erythrocytes, Nagasaka *et al.* (1988) showed sorbitol dehydrogenase played an important role inducing rapid fluctuation in erythrocyte sorbitol in response to changes in plasma glucose. With intact erythrocytes from healthy subjects and diabetic patients, erythrocyte sorbitol increased rapidly in response to increased ambient glucose concentration. In the same manner, the accumulated sorbitol easily decreased according to the rapid reduction of ambient glucose concentration.

E.C. 1.1.1.49

Jeffery *et al.* (1989) have located a reactive lysine residue in a segment Ile–Asp–His–Tyr–Leu–Gly–Lys*–Glu–Met–Val–Lys of the glucose 6-phosphate dehydrogenase from the yeast *Pichia ladinii.* This structure differs from that of other characterised glucose 6-phosphate dehydrogenases but outside yeasts the segment does not vary in known mammalian, insect and bacterial forms.

Thus in the *Saccharomyces cerevisiae* enzyme a lysine residue has been associated with the activity of the enzyme and identified as a reactive nucleophile possibly required for functional interactions involving glucose 6-phosphate. A similar lysine residue in a closely similar amino acid sequence has been located in human glucose 6-phosphate dehydrogenase. The rat, fruit fly and *Leuconostoc mesenteroides* enzymes have identical segments here equivalent to that of the human. To locate the lysine in the glucose 6-phosphate dehydrogenase of *Pichia jadinii,* the enzyme was labelled by acetylation with acetyl salicylate which causes loss of activity and selective labelling of a lysine residue (Jeffery *et al.*, 1989).

Hung *et al.* (1989) showed that juvenile white sturgeon fed either glucose or maltose diets retained the highest per cent energy related in part to significantly higher liver G6PDH. This contrasted with the poor ability of sturgeon to utilise either sucrose or lactose due to low intestinal sucrase and lactase.

For fruits as might be expected, during the initial development of apple fruits, sucrose was not detectable and the level of the other sugars sorbitol, glucose and fructose were low. Then in the 2 weeks before the onset of the June drop period, the sucrose accumulates with a sharp decrease in acid invertase. Sorbitol dehydrogenase is absent at this stage while the level of sorbitol was relatively high. Sorbitol 6-phosphate dehydrogenase did not change during fruit development. The results were interpreted on the basis that sucrose is the predominant carbon source for growth in the young fruit, whereas during the sugar storage phase sorbitol is the major substrate for fructose accumulation. Thus with the appearance of sorbitol dehydrogenase in sugar-accumulating fruits, the sorbitol concentration decreased and the storage of fructose continued until fruit development ceased (Berueter, 1985).

REFERENCES

Balla, T., Hunyady, L., Baukal, A. J. & Catt, K. J. (1989). Structures and metabolism of inositol tetrakisphosphates and inositol pentakisphosphate in bovine adrenal glomerulosa cells. *J. biol. Chem.*, **264**(93), 86–90.

Barker, S. A., Bourne, E. J., Salt, E. & Stacey, M. (1959). Ketose-Polyol interconversions by a ropy-cider organism. *J. Chem. Soc.*, **120,** 593–98.

Berkowitz, D. B. & Benner, S. A. (1987). Anomeric specificity of fucose dehydrogenase, a stereochemical imperative in aldopyranose dehydrogenases? *Biochemistry,* **26,** 2606–11.

Berueter, J. (1985). Sugar accumulation and changes in the activities of related enzymes during development of the apple fruit. *J. Plant. Physiol.*, **121,** 331–41.

Cartana, J., Arola, L. & Romeu, A. (1989). Characterisation of the inhibition effect induced by nickel ion on glucose-6-phosphate dehydrogenase and glutathione reductase. *Enzyme,* **41,** 1–5.

Cho, S. W. & Joshi, J. G. (1989). Inactivation of bakers yeast glucose 6-phosphate dehydrogenase by aluminium. *Biochemistry,* **18,** 3613–8.

Donninger, C. (1986). Reduction of the sucrose content of a fruit juice product. S. African ZA 85 04,565.

Driscoll, J. J. & Kosman, D. J. (1987). Solvent and solvent proton dependent steps in the galactose oxidase reaction. *Biochemistry,* **26,** 3429–36.

Eklund, H., Horjales, E., Jörnvall, H., Brändon, C. V. & Jeffery, J. (1985). Molecular aspects of functional differences between alcohol and sorbitol dehydrogenases. *Biochemistry,* **24,** 8005–12.

Farnararo, M., Favilli, F., Tonelli, M. F. & Bruni, P. (1978). Some properties of glucose 6-phosphate dehydrogenase from rat and chick brain; a comparative study. *Comp. Biochem. Physiol. B.*, **61,** 351–6.

Gautam, S. P. (1987). Studies on oxidising enzymes of citrus fruits infected with Thielaviopsis paradoxa. *Natl. Acad. Sci. Lett.* (*India*), **10,** 9–11.

Geyer, O. & Görisch, H. (1986). Crystalline quinoprotein glucose dehydrogenase from Acinetobacter calcoaceticus. *Biochemistry,* **25,** 6043–8.

Goma, G. & Seiller, I. (1987). Glucose-free fructose preparation. Fr. Demande FR 2 588,568. *Amer. Chem. Abs.*, **107,** 174749.

Grant, D. R. (1987). Ascorbate oxidase inhibition in dough by fluoride ion and its effect upon dough rheology. *Cereal Chem.*, **64,** 403–6.

Hung, S. S., Fynn-Aikens, F. K., Lutes, P. B. & Xu, R. P. (1989). Ability of juvenile white sturgeon to utilise different carbohydrate sources. *J. Nutr.*, **119,** 727–33.

Inamine, S. & Watanabe, S. (1987). Food preservative composition. J.P. 62 143672 [87 143,672]. *Amer. Chem. Abs.*, **107,** 153168.

Jans, A. W., Grunewald, R. W. & Kinne, R. K. (1989). Pathways for the synthesis of sorbitol from ^{13}C-labelled hexoses, pentose and glycerol in renal papillary tissue. *Magn. Reson. Med.*, **9,** 419–22.

Jarrett, R. L. & Litz, R. E. (1986). Isoenzymes as genetic markers in bananas and plantains. *Euphytica,* **35,** 539–49.

Jeffery, J., Wood, I., Macleod, A., Jeffery, R. & Jornvall, H. (1989). Glucose 6-phosphate dehydrogenase. Characterisation of a reactive lysine residue in the Pichia jadinii enzyme reveals a limited structural variation in a functionally significant segment. *Biochem. Biophys. Res. Comm.*, **160,** 1290–5.

Kessli, R. V. & Michelmore, R. W. (1986). Genetic variation and phylogenies detected from isoenzyme markers in species of Lactuca. *J. Hered.*, **77,** 324–31.

Kiuchi, K., Nishikimi, M. & Yagi, K. (1982). Purification and characterisation of L-gulonolactone oxidase from chicken kidney microsomes. *Biochemistry,* **21,** 5076–82.

Koczot, A. B., Szewczuk, M. A. & Madler, J. (1985). Protecting fish against colour changes during refrigeration. Pol PL 129,338. *Amer. Chem. Abs.*, **107,** 114549.

Kulbe, K. D. & Knopki, G. (1986). Intrasequential cofactor regeneration in enzymatic synthesis, particularly when producing Vitamin C. PCT Int. Appl. WO 86 04,353.

Kulbe, K. D., Schwab, U., Howaldt, M. & Kimmerle, K. (1986). Use of membrane processes in the simultaneous preparation of mannitol and gluconic acid from sucrose by conjugated NAD^+ related hydrogenase. *GBF Monogr. Ser.*, **9,** 189–200.

Lin, Y. H. (1986). Inhibition of lipid oxidation in fatty fish by glucose oxidase preservation. *Diss. Abstr. Int. B.*, **47,** 3184.

Maret, W. (1989). Cobalt (II) substituted Class III alcohol and sorbitol dehydrogenase from human liver. *Biochemistry,* **28,** 9944–9.

Nabiev, A. A. & Velieva, E. G. (1986). Influence of degree of ripeness on the quality of table grapes in storage. *Izv. Vyssh. Uchebn. Zaved. Pishch. Tekhnol.*, **3,** 58–60. *Amer. Chem. Abs.*, **105,** 96221.

Nagasaka, Y., Fujii, S. & Kaneko, T. (1988). Human erythrocyte sorbitol metabolism and the role of sorbitol dehydrogenase. *Diabetologia,* **31,** 766–70.

Pakuts, A. P., Whitehouse, L. W. & Paul, C. J. (1988). Plasma sorbitol dehydrogenase determination in experimental hepatotoxicity using the Abbott Bichromatic Analyser. *J. Clin. Chem. Clin. Biochem.*, **26,** 693–5.

Perea, F. J., Vaca, G., Alvarez, C., Cantu, J. M. & Ibarra, B. (1989). Electrophoretic pattern of sorbitol dehydrogenase (E.C. 1.1.1.14) in human seminal plasma and spermatozoa. *Ann. Genet.*, **32,** 33–5.

Prieels, J. P., Maschelein, C. & Heilporn, M. (1986*a*). Elimination of oxygen from food and beverages and an enzymic composition used for this effect. Eur. Pat. Appl. EP 207,039.

Prieels, J., Paul, H. P. & Perraudin, J. P. (1986*b*). Antibacterial food and feed additive. Ger. Offen. DE 3,525,902. Riken Vitamin Co. Ltd (1982). JP 82,86,235.

Reitmann, I. G. (1987). Oxidoreductase activity in hops during post harvest treatment, processing and storage. *Fermentn. Spirt. Prom-st,* (**2**), 33–7. *Amer. Chem. Abs.*, **106,** 194,704.

Ropson, I. J. & Powers, D. A. (1989). The allelic isozymes of hexose-6-phosphate dehydrogenase isolated from *Fundulus heteroclitus*: physical characteristics and kinetic properties. *Mol. Biol. Evol.*, **6,** 171–85.

Sands, J. M., Terada, Y., Bernard, L. M. & Knepper, M. A. (1989). Aldose reductase activities in microdissected rat renal tubele segments. *Am. J. Physiol.*, **256,** F563–9.

Shantz, L. M., Talalay, P. & Gordon, G. B. (1989). Mechanism of inhibition of growth of 3T3-L1 fibroblasts and their differentiation to adipocytes by dehydroepiandrosterone and related steroids; role of glucose-6-phosphate dehydrogenase. *Proc. Nat. Acad. Sci.* (*USA*), **86,** 3852–6.

Takebe, M. (1988). Discoloration prevention of fish meat. Jpn. Kokai Tokkyo Koho JP 63 44,866 [88 44,866]. *Amer. Chem. Abs.*, **109,** 36894c.

Villettaz, J. C. (1986). Low alcoholic wine and agent for its production. Europ. Pat. Appl. EP 194,043.

Walsall, E. P., Lyons, S. A. & Metzger, R. P. (1978). A comparison of selected properties of hepatic sorbitol dehydrogenases [L-iditol:NAD oxidoreductases] from four mammalian species. *Comp. Biochem. Physiol. B*, **59,** 213–8.

Webb, E. C. (1984). *Enzyme Nomenclature. International Union of Biochemistry.* Academic Press, London, pp. 20–141.

Yamaki, S. (1987). ATP-promoted sorbitol transport into vacuoles isolated from apple fruit. *Plant Cell. Physiol.*, **28,** 557–64.

Index